ALLE ZEIT WACH
1842

Günter Dietmar Roth (Ed.)

Compendium of Practical Astronomy

Volume 3: Stars and Stellar Systems

Translated and Revised by
Harry J. Augensen and Wulff D. Heintz

With 128 Figs. and 54 Tables

Springer-Verlag
Berlin Heidelberg New York
London Paris Tokyo
Hong Kong Barcelona Budapest

Dipl.-Kfm. Günter Dietmar Roth
Ulrichstrasse 43, Irschenhausen, D–82057 Icking/Isartal, Germany

Dr. Harry J. Augensen
Department of Physics and Astronomy, Widener University, Chester, PA 19013, USA

Professor Dr. Wulff-D. Heintz
Department of Physics and Astronomy, Swarthmore College, Swarthmore, PA 19081, USA

Completely Revised and Enlarged Translation of the 4th German Edition of the title "Roth (Ed.), Handbuch für Sternfreunde, Vols. 1 and 2".

ISBN 3-540-54886-6 Springer-Verlag Berlin Heidelberg New York
ISBN 0-387-54886-6 Springer-Verlag New York Berlin Heidelberg

ISBN 3-540-56273-7 Volumes 1, 2, and 3
ISBN 0-387-56273-7 Volumes 1, 2, and 3

Library of Congress Cataloging-in-Publication Data
Handbuch für Sternfreunde. English
Compendium of practical astronomy / Günter Roth, ed. :
translated by Harry J. Augensen and Wulff D. Heintz.
p. cm.
Rev. translation of: Handbuch für Sternfreunde (4th ed.).
Includes index.
Contents: v. 1. Instrumentation and reduction techniques -- v.
2. Earth and solar system -- v. 3. Stars and stellar systems.
ISBN 0-387-56273-7 (New York). -- ISBN 3-540-56273-7 (Berlin)
1. Astronomy--Handbooks, manuals, etc. I. Roth, Günter Dietmar.
QB64.H3313 1993 520--dc20 93-27023

Cover Design: Erich Kirchner, Heidelberg
Typesetting: Data conversion by Lewis & Leins Buchproduktion, Berlin
Production: PRODUserv Springer Produktions-Gesellschaft, Berlin
SPIN 10051891 55/3020 - 5 4 3 2 1 0 – Printed on acid-free paper

Contributing Authors to Volume 3

Drechsel, Horst, Dr.
Dr. Remeis Sternwarte, Sternwartstrasse 7, D–96049 Bamberg, Germany

Feitzinger, Johannes V., Prof. Dr.
Planetarium der Sternwarte Bochum, Castroper Strasse 67, D–44777 Bochum, Germany

Herczeg, Tibor J., Prof. Dr.
Department of Physics and Astronomy, University of Oklahoma, 440 W. Brooks, Norman, OK 73019, USA

Heintz, Wulff-Dieter, Prof. Dr.
Department of Physics and Astronomy, Swarthmore College, Swarthmore, PA 19081, USA

Neckel, Thorsten, Dr.
Max-Planck-Institut für Astronomie, Königstuhl 17, D–69117 Heidelberg, Germany

Preface to the Second English Edition

It has been a particular pleasure to produce this revised English edition of the German *Handbuch für Sternfreunde*, thus making it available to a wider readership. I should like to express my gratitude to the authors and translators, who have contributed invaluably to the process of revision and translation.

I am deeply indebted to Prof. W.D. Heintz and Prof. H.J. Augensen, who not only made the translation but also assisted in the critical reading and improvement of various chapters. Moreover, Prof. Augensen has written for Vol. 1 a chapter which says a great deal about "Astronomy Education and Instructional Aids." This contains full information with regard to the situation in the United Kingdom, Canada, and the United States. It is an extremely helpful survey for both teachers and students in colleges and high schools, and the staff of planetaria and public observatories.

Welcomed as a new author is R. Kresken, who wrote the chapter "Artificial Earth Satellites" especially for this English edition.

Last, but not least, I gratefully acknowledge the helpfulness of Springer-Verlag, Heidelberg, where Prof. W. Beiglböck always gave every possible consideration to the translators' and my suggestions.

Irschenhausen
Summer 1994

Günter D. Roth

Translators' Preface

It is a pleasure to present this work, which has been well received in German-speaking countries through four editions, to the English-speaking reader. We feel that this is a unique publication in that it contains valuable material that cannot easily—if at all—be found elsewhere. We are grateful to the authors for reading through the English version of the text, and for responding promptly (for the most part) to our queries. Several authors have supplied us, on their own initiative or at our suggestion, with revised and updated manuscripts and with supplementary English references. We have striven to achieve a translation of *Handbuch für Sternfreunde* which accurately presents the qualitative and quantitative scientific principles contained within each chapter while maintaining the flavor of the original German text. Where appropriate, we have inserted footnotes to clarify material which may have a different meaning and/or application in English-speaking countries from that in Germany.

When the first English edition of this work, *Astronomy: A Handbook* (translated by the late A. Beer), appeared in 1975, it contained 21 chapters. This new edition is over twice the length and contains 28 authored chapters in three volumes. At Springer's request, we have devised a new title, *Compendium of Practical Astronomy*, to more accurately reflect the broad spectrum of topics and the vast body of information contained within these pages. It should be noted that, while much of this information is directed toward the "amateur," it is equally applicable to the professional astronomer or physicist who teaches at a small college and is searching for suitable astronomical projects to give to his or her students.

The *Compendium of Practical Astronomy* is structured somewhat differently from its German counterpart. The former consists of three volumes, the latter two. Volume 1 is essentially the same as the corresponding volume in the *Handbuch für Sternfreunde*, but the chapters have been reordered in a sequence which, we feel, presents the topics in more homogeneous groups. In addition, "Astronomy Education and Instructional Aids" (H.J. Augensen) has been added as the last chapter. Volume 2 contains chapters covering the Earth and Solar System, including "The Terrestrial Atmosphere and Its Effects on Astronomical Observations" (F. Schmeidler), which appeared in Vol. 1 of the *Handbuch*. Also, "Artificial Earth Satellites" (R. Kresken) replaces the corresponding chapter by Petri in the German edition. Volume 3 is devoted to topics of stellar, galactic, and extragalactic astronomy. In the fourth German edition, these latter two volumes appear as one very thick second volume.

At the end of each volume, we have included a "Supplemental Reading List" for each of the chapters in that volume. We have prepared a new appendix, "Educational Resources in Astronomy" (Appendix A in Vol. 1), as a supplement to Chap. 12, and have also updated some tables, primarily those which constitute Appendix B, "Astronomical Data," in Vol. 3. Recognizing the fact that many readers will want to utilize computer techniques in the various reduction procedures presented in this *Compendium*, it is our hope that the references on computers and programmable calculators given in Sect. 12.4.4 of Vol. 1 will prove helpful.

The superb guidance provided by Prof. W. Beiglböck and the able assistance of his secretary Ms. S. Landgraf (Springer-Verlag, Heidelberg) has been invaluable in the completion of this project. We are also indebted to Mark Seymour at Springer for his thorough proofreading of the entire manuscript, and to Dr. Fred Orthlieb of Swarthmore College for his help in translating several technical terms. Finally, we gratefully acknowledge the unwavering assistance of Computing Services at Widener University, in particular Barry Poulson, James Connalen, John Neary, Kim Stalford, Lynn Pollack, and David Walls, who provided expert advice on the preparation of this entire manuscript in TEX.

One of us (WDH) still cherishes his acquaintance with A. Güttler, W. Jahn, R. Kühn, R. Müller, and K. Schütte—the now-deceased authors of the first edition, whose enthusiasm helped Günter Roth's project off to a good start over 30 years ago.

Swarthmore/Chester
June 1994

Wulff D. Heintz
Harry J. Augensen

Preface to the Fourth German Edition

The ability to employ objective techniques to appreciate the wide variety of cosmic phenomena quantitatively is not restricted to professional astronomers. It is the principal aim of the *Handbuch für Sternfreunde* to provide the astronomically interested public—amateur observers as well as teachers—with instruction and guidance in practical astronomical activities. This goal has remained unchanged since the first edition, which appeared in 1960.

What has changed is the technical content and organizational structure in various areas. Larger and more effective telescopes are currently within the reach of non-professionals. The professional accessories used in photography, photometry, and spectroscopy are now being operated by amateurs; schools and private observatories own electronic equipment. Thus equipped, the amateur can now engage in observational tasks ranging from photoelectric photometry of planets and variable stars to studies of high-resolution photographs of distant galaxies.

These developments are reflected in every chapter of the present work. The presentation of new tools, techniques, and tasks has required a significant expansion, so that the *Handbuch für Sternfreunde* now appears in two volumes.

Volume 1 provides the technical basis for astronomical observations and measurements with amateur equipment. These include fundamental methods for recording and processing light intensities in photography, photometry, and spectroscopy. The optical range is augmented by radio-astronomical observations, whose instrumental basics are described. Also included within this volume are instructions on how to organize astronomical observations and subsequently process the results by mathematical methods, a guide to the literature, and a brief history of astronomy.

The following chapters in Vol. 1 have been completely rewritten or newly added for the fourth edition: "Optical Telescopes and Instrumentation" (H. Nicklas), "Telescope Mountings, Drives, and Electrical Equipment" (H.G. Ziegler), "Astrophotography" (B. Koch, N. Sommer), "Principles of Photometry" (H.W. Duerbeck, M. Hoffmann), and "Historical Exploration of Modern Astronomy" (G.D. Roth).

Volume 2 presents the objects of astronomical study in detail, commenting on the execution and evaluation of observational tasks. The topics covered include, among others, the various Solar System bodies, the stars, the

Milky Way, and extragalactic objects. This volume contains an expanded section of tables, general literature references, and the cumulative subject index for both volumes. Also included as an appendix is the contribution "Instructional Aids in Astronomy" (A. Kunert).

The following chapters in Vol. 2 have been completely rewritten or newly added for the fourth edition: "The Sun" (R. Beck et al.), "Lunar Eclipses" (H. Haupt), "Noctilucent Clouds, Polar Aurorae, and the Zodiacal Light" (C. Leinert), "Stars" (T. Neckel), "Variable Stars and Novae" (H. Drechsel, T. Herczeg), "The Milky Way and the Objects Composing It" (T. Neckel), and "Extragalactic Objects" (J.V. Feitzinger).

I also wish to thank all the authors on this occasion for their successful collaboration. Welcomed as new contributors are: Dr. R. Beck and his coworkers V. Gericke, H. Hilbrecht, C.H. Jahn, E. Junker, K. Reinsch, and P. Völker (the "Sun" Working Group of the Vereinigung der Sternfreunde); Dr. H. Drechsel (Bamberg), Dr. H. Duerbeck (Münster), Prof. J.V. Feitzinger (Bochum), Prof. H. Haupt (Graz), Prof. T.J. Herczeg (Norman, OK), Dr. M. Hoffmann, B. Koch (Düsseldorf), Dr. C. Leinert (Heidelberg), Dr. T. Neckel (Bochum), Dr. H. Nicklas (Göttingen), and N. Sommer (Düsseldorf).

The preparation of the fourth edition was aided by the valuable advice given by Prof. F. Schmeidler (Munich) and Dr. H.J. Staude, managing editor of the magazine *Sterne und Weltraum*. Dr. W. Kruschel (Konstanz) supported the preparation of the tables in Vol. 2 by providing updated material. Photographs and tables were made available by C. Albrecht (Freiburg), H. Haug and coworkers at the Wilhelm-Foerster-Sternwarte (Berlin), and J. Meeus (Erps-Kwerps, Belgium).

As the representative of Springer-Verlag, Prof. W. Beiglböck has attended to this large project with care and provided numerous suggestions. The somewhat tedious task of preparing the manuscripts for printing was in the capable hands of Mrs. C. Pendl, to whom the editor and the authors are very grateful.

Irschenhausen
Summer 1989

Günter D. Roth

Contents of Volume 3

Contents of Volume 1

Contents of Volume 2

24 The Stars

T. Neckel

24.1 The Positions of the Stars

To identify a star, its spherical position can be given. Its coordinates in the equatorial system (see Chap. 2 in Vol. 1) are called *right ascension*, α, and *declination*, δ. Owing to precession and to the proper motions of stars, α and δ depend on time; thus in addition to these coordinates, their equinox and epoch must both be stated. Precession depends only upon the coordinates themselves, and can be approximately read from Appendix Table B.10. Proper motions, however, are individual features of stars, and have to be determined separately for each case. For faint stars, they are often (but not always) so small as to be for most purposes negligible.

Different star catalogues and sky atlases exist for a variety of applications [24.1,2]. As an initial guide around the sky, a rotatable star map or a planisphere is best, but these, however, contain only the bright stars. Maps containing all stars perceivable with the naked eye can be found in star atlases such as those of Clark [24.3] or the *B.A.A. Star Charts* [24.4]. If a star is seen in the sky where none was noted previously, then a glance at the chart will quickly answer whether a nova has been found, or whether there is a variable star, perhaps caught just at the maximum of light, near the position in question.

To determine the position of, for instance, a minor planet or a comet on a photograph, a number of reference stars with precise coordinates for the time of the photograph are needed (see Sect. 3.4 in Vol. 1). For this purpose, the proper motions of the reference stars must also be considered. The required data for a large number of stars can be found in the PPM and in the SAO catalogues.

24.2 Stellar Magnitudes and Colors

Since ancient times the brightnesses of stars have been expressed in *astronomical magnitudes*. The brightest stars were assigned magnitude "1," the faintest reached by the naked eye magnitude "6." The presently used mathematical definition of astronomical magnitudes was given in 1850 by Pogson and states that the magnitudes m_1 and m_2 of two stars differ by 1 magnitude when the logarithms (base 10) of their radiation fluxes I_1 and I_2 differ by 0.4. In general,

$$I_1/I_2 = 10^{0.4(m_2-m_1)} \tag{24.1a}$$

or

$$m_2 - m_1 = 2.5 \log(I_1/I_2). \tag{24.1b}$$

A first-magnitude star is therefore 100 times brighter than a sixth magnitude star.

Having defined the "step-width" of the magnitude system, a zero-point must then be chosen. In practice, this point is provided by a number of *standard stars* of known magnitudes, about equally distributed over the sky, so that several of them can be observed at any one time.

The system of magnitudes is quite practicable in many ways. For instance, atmospheric extinction diminishes the perceived intensity of starlight in such a way that there exists a linear relation between the *air mass* (essentially the secant of the zenith distance) and the apparent brightness of a star expressed in magnitudes. Therefore, if the brightness of a star is measured at various zenith distances and graphed against the air masses, then the measured points will scatter around a straight line, the slope of which is the extinction coefficient (i.e., the extinction in the zenith direction). The effect of interstellar extinction on brightness and color expressed in magnitudes is also linear, and thus easily calculated.

When a stellar magnitude is given, the spectral range in which it was measured must also be specified. Until about 35 years ago, the predominant usage was to state *visual* and *photographic magnitudes.* Visual magnitudes are those as perceived by the human eye or by a device with a spectral sensitivity distribution similar to that of the eye. "Photographic" magnitudes are those which were recorded by the early photographic plates, which were blue-sensitive (see also Chap. 8 in Vol. 1).

This international magnitude system m_v and m_{pg} has been superceded by the *UBV* system of Johnson and Morgan, where U, B, and V are apparent magnitudes measured in certain well-defined ranges of the spectrum, each about 100 nm wide. U is the magnitude of a star in the ultraviolet part of the spectrum, B that in the blue, and V in the yellow, the last of which is the spectral range of highest sensitivity of the human eye. So, the V-magnitude corresponds approximately to the former m_v. The zero-points for B and V are chosen so that for A0 stars $U = B = V$.

The magnitudes U, B, and V of stars can be measured photoelectrically with high precision (errors of only 1% to 2%). With a photomultiplier (e.g., EMI 6256) not sensitive to red light, the filter combination UG 1 (2 mm) for U, BG3 (1 mm) + GG385 (2 mm) for B, and GG495 (3 mm) for V gives approximate *UBV* magnitudes. A more precise relation between the magnitudes measured and the exact magnitudes in the *UBV* system is established by measuring a larger number ($n \geq 20$) of *UBV* standard stars. Using a photomultiplier which is also sensitive to red (e.g., RCA 31034), it must be kept in mind that the UG1 filter is transparent to red light. Its "red leak" can be suppressed by a second filter which is transparent in the ultraviolet and blue but completely opaque in red. This may be achieved with a copper-sulfate filter; such filters have over the past few years been manufactured in solid form.

UBV magnitudes can also be obtained photographically using, for instance, the following emulsion/filter combinations:

IIa-O + UG1 (2 mm) for U;
IIa-O + GG385 (2 mm) for B;
IIa-O + GG495 (3 mm) for V.

For the purpose of calibration, the star field which is photographed must contain a number of stars with known magnitudes in the UBV system. These are then usually measured photoelectrically in a separate process.

The difference of two magnitude values of a star in different color ranges is known as the *color index* (or color for short). The $U-B$ and $B-V$ indices formed from UBV magnitudes are the ones usually catalogued. The color indices of a star indicate the spectral distribution of its light: a red M-type star is faint in the blue but substantitally brighter in the visual (yellow). The numerical value of its B-magnitude is therefore larger than that of the V-magnitude, and thus the color index $B - V$ is positive. A blue O-star is brighter in the blue than in the visual, its B-magnitude thus has a smaller numerical value than in V, and consequently the color index is negative. An O-type star which has not been reddened by interstellar extinction has a color index $B-V = -0.3$, while, on the other hand, a late M-type star has $B-V = +1.65$. The Sun lies near the middle of this range, with $B-V = +0.65$.

In recent years, the UBV system has been extended by adding other wavelength ranges. The red magnitude R and the near-infrared I can still be measured with red-sensitive photomultipliers. The magnitudes J, H, K, L, and M follow for even longer wavelengths. The centers of each of these photometric wavelength ranges are compiled in Table 24.1. The following definition is valid: all magnitudes $R, I, \ldots, M$ are (approximately) equal, and hence all color indices are zero, for an A0 star.

The system of astronomical magnitudes has sometimes been criticized as lacking an astrophysical foundation. Yet, those physicists who are familiar with stellar magnitudes often use the system with great pleasure simply because it is so convenient to handle. For some purposes, however, it is necessary to convert astronomical magnitudes into physical units, a task which is readily performed using the numbers given in Table 24.1.

In addition to much other useful information, the *Catalogue of Bright Stars* [24.5] lists the photoelectrically measured V, $B-V$, $U-B$, and $R-I$ for the majority of the 9110 stars contained within it. A compilation of all UBV magnitudes measured to the present is the *UBV Photoelectric Photometry Catalogue* by Mermilliod [24.6]; see also Appendix Table B.28.

24.3 Trigonometric Determination of Distances to Nearby Stars

To obtain the first clues as to the nature of stars, it is of the utmost importance to have information on their distances. Fortunately, some stars are near enough that their distances can be obtained using trigonometric methods. These directly measured distances serve to calibrate other methods which reach further out. Trigonometric distances are therefore the foundation of all current knowledge of the structure of the universe.

Table 24.1. Effective wavelengths λ_{eff} and absolute radiation fluxes F_ν of a star with magnitude 0.0 in each of the photometric bands.

Band	λ_{eff} (μm)	F_ν ($\times 10^{-24}$ W m^{-2} Hz^{-1})
U	0.36	188
B	0.44	44.4
V	0.55	38.1
R	0.70	30.1
I	0.90	24.3
J	1.25	17.7
H	1.62	11.2
K	2.2	6.3
L	3.4	3.1
M	5.0	1.8
N	10.2	0.43

The largest available baseline for trigonometry in space is the semimajor axis of the Earth's orbit around the Sun. If, after a particular star has been observed during the course of a year from various points in the Earth's orbit, it is found that the direction in which it is seen is continually shifted by a small angle, then its distance can be deduced trigonometrically. If the star is located, say, at the pole of the ecliptic, then over the course of a year it describes a circle whose radius π_* is called the *parallax* of the star.[1] A star on the ecliptic would move back and forth in linear fashion between $\pm\pi_*$. Thus, the parallax π_* is the angle under which the radius of Earth's orbit (= 1 AU) appears as seen from the star. If this angle is exactly 1″, then the distance to the star is $r = 1$ AU $\times 360 \times 3600/2\pi = 206\,265$ AU. (The product 360×3600 is the number of arcseconds contained in 360°.) This distance is defined as 1 *parsec*, abbreviated as 1 pc. This name indicates that this is the distance to a star which exhibits, owing to the Earth's orbital motion around the Sun, a *par*allax of one arc*sec*ond. It can be converted into kilometers or light years using

$$1 \text{ pc} = 3.086 \times 10^{13} \text{ km} = 3.26 \text{ light years.}$$

Thus, the light ray needs 3.26 years to travel a distance of 1 parsec.

The first trigonometric measurement of a star was made in 1838 by F.W. Bessel in Königsberg, Germany. He found a parallax for 61 Cygni which is close to the modern value of $\pi_* = 0.''293$. The distance is thus $r = 1/\pi_* = 3$ pc. The parallaxes of the nearest stars α Centauri and Proxima Centauri are $0.''75$ and $0.''76$, respectively, and hence their distances are 1.33 pc.

At a distance $r = 1$ pc, the radiation flux from the Sun would be reduced by a factor $(1/206\,265)^2 = 2.35 \times 10^{11}$ from the value it has at the distance of the Earth.

1 The stellar parallax π_* has here been given the subscript "*" to distinguish it from the familiar mathematical constant $\pi = 3.141\,592\,654$.

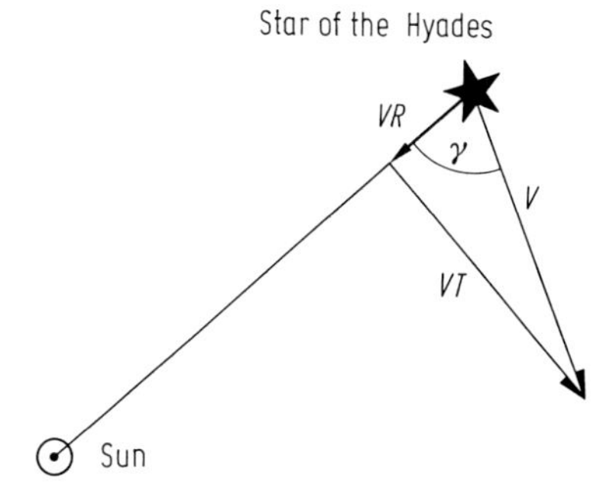

Fig. 24.1. By determining the velocity components V_R and V_T of the Hyades member stars, the distance to this moving cluster can be obtained.

According to Eq. (24.1b), this corresponds to a magnitude difference of 26.57. As the apparent brightness of the Sun is $V = -26.70$, it would appear from a distance of 1 pc as a star of apparent magnitude $V = -0.13$. These numbers immediately underscore the fact that the Sun would not be distinguished from the other stars if it were as far away as they are. In other words, most stars evidently are celestial bodies comparable with the Sun.

Even the nearest stars all have parallaxes under $1''$. Such small displacements over a year are measured relative to other stars (or quasars) which are so distant that they do not possess substantial parallactic shifts. The majority of faint stars satisfy this condition.

The accuracy of parallax measurements is about $\pm0\overset{''}{.}005$, so that parallaxes can be found to an accuracy of $\pm10\%$ out to a distance of 20 pc. The *method of moving cluster parallaxes* reaches a bit further. The best-known moving cluster is the Hyades, the loose grouping of stars around the star Aldebaran. The proper-motion vectors of all Hyades member stars intersect in a *convergence point* at $\alpha = 6^{\mathrm{h}}12^{\mathrm{m}}$, $\delta = +9°$. This means that the Hyades all move in that direction on nearly parallel paths. Thus, the direction in which the Hyades move is known, and this permits the computation of their distances by combining radial-velocity data with proper motions. The basic idea behind the method is as follows: knowing the direction of motion of a star from the radial component of its velocity, the *space velocity* can be computed, and subsequently also the tangential component of motion at right angles to the line of sight. The latter, however, is also shown in the proper motion of the star and depends on the distance, sincc the farther away the star is, the smaller the proper motion. Figure 24.1 displays quantitatively what has been said. If γ is the angle between the direction of motion of a Hyades star and a line toward the Sun, then the space velocity V follows from the radial velocity V_R:

$$V = V_R/\cos\gamma = V_R \sec\gamma. \tag{24.2}$$

The tangential velocity is

$$V_T = V \sin\gamma, \tag{24.3}$$

and thus it follows from the radial velocity that

$$V_T = V_R \tan\gamma. \tag{24.4}$$

Furthermore, V_T is obtained from the proper motion μ of the star measured in arcseconds per year. During a year the Earth moves a length 2π AU with an average speed of 30 km s^{-1}. Thus, the distance 1 AU is passed in one year with $v = 30$ km s$^{-1}/2\pi = 4.74$ km s^{-1}. A star at a distance of 1 pc and with a proper motion of 1$''$ yr^{-1} thus has a tangential velocity of 4.74 km s^{-1}. In general, for a parallax π_* in $''$ and a proper motion μ in $''$ yr^{-1}, the tangential speed V_T is

$$V_T = 4.74\mu/\pi_* \text{ km s}^{-1}. \tag{24.5}$$

Using the previous relations, the distance is obtained:

$$r = \frac{V_R \tan\gamma}{4.74\mu} \quad \text{(pc)}. \tag{24.6}$$

The distance to the Hyades thus comes out to be 45 pc with a relative error of at most 10%.

24.4 Absolute Magnitudes and Distance Moduli

The notion of *absolute magnitudes* was introduced in order to obtain a measure of stellar brightnesses which is independent of distance. The absolute magnitude is defined as that apparent magnitude which a star would have if it were placed at a distance of $r = 10$ pc. The relation between apparent magnitude m and absolute magnitude M is

$$m - M = 5 \log r - 5. \tag{24.7}$$

The difference $m - M$ is known as the *distance modulus*. This relation holds only if there is no interstellar extinction (see Chap. 27); otherwise, a value for the extinction $A(r)$ has to be added to the right-hand side.

Apparent magnitudes in the photometric *UBV* system are symbolized by the letters U, B, and V. The respective absolute magnitudes are M_U, M_B, and M_V.

From the Sun's distance, 1 AU =1/206 265 pc, its distance modulus is found to be −31.57. Thus from the apparent magnitudes of the Sun in the *UBV* system,

$$U = -25.85, \quad B = -26.03, \quad V = -26.70,$$

the corresponding absolute magnitudes are

$$M_U = +5.72, \quad M_B = +5.54, \quad M_V = +4.87,$$

so the Sun is really quite a modest star relative to most. O-type stars reach absolute magnitudes of around −5, which means that they are 10 000 times brighter than the Sun. The most luminous supergiants are 100 times brighter still. On the other hand, there are also much fainter stars. The faintest M-type dwarfs have absolute magnitudes $M_V \approx +15$. Thus, the brightest and faintest stars on the absolute scale differ in brightness by about 25 magnitude steps, which corresponds to 10 orders of magnitude in luminosity.

24.5 Stellar Spectral Types

A wide variety of information can be gleaned from the spectra of stars. All stars possess a continuum, but it is overlaid by a varying number of dark (Fraunhofer) lines. These lines are caused by absorptions in the stellar photospheres by neutral atoms and a variety of ions filtering certain wavelengths out of the continuous stellar radiation. In certain rare cases, emission lines, whose intensities can grossly exceed that of the neighboring continuum, also occur.

To establish order in the enormous variety of stellar spectra, similar spectra have been entered together in groups coded with capital letters. From this the common spectral classification of stars originated, normally in the sequence O B A F G K M, with a side branch R. A finer subdivision appends the numbers 0 to 9 to the code letters, and in some cases half subdivisions are used. In the portion of the sequence between O9 and B3, the types O9.5, B0.5, B1.5, and B2.5 are inserted between classes O9, B0, B1, B2, and B3. Stars near the beginning of the sequence (i.e., O- and B-type stars) are, for historical reasons, referred to as "early" types, the K- and M-stars as "late" types. The subdivision of O-stars began originally at O5, but hotter stars have since been discovered, and so the sequence now begins at O3; see Chap. 7 in Vol. 1.

Actually, this sequence of types from O to M expresses a *temperature* range. O-type stars are the hottest, with surface temperatures as high as 50 000 K, while M-types are the coolest, with temperatures of only 3000 K. The type O3 includes the hottest and most massive stars; only a very few such objects are known to exist. To classify a stellar spectrum, the intensities of certain characteristic spectral lines must be observed. Thus, the very hottest stars—the early O-types—are characterized by He II lines. He I lines become stronger in the later O-stars. The Balmer lines of hydrogen are very weak in O-type spectra, but increase in intensity in B-type spectra, reaching maximum intensity in type A0. In later types, the Balmer lines continually decrease again, while Ca II lines become stronger. In G- and K-type spectra, more and more lines of neutral elements appear, and, finally, the coolest stars of type M are characterized by molecular bands (e.g., of TiO).

The determination of the spectral type of a star does not depend on its distance nor upon the extinction between star and Earth. Contrary to apparent brightness and color index, the spectral type is a parameter that is determined exclusively by stellar properties.

24.6 The Hertzsprung–Russell Diagram

The absolute magnitudes of stars are defined so as to be independent of distance. If one plots the absolute magnitudes of stars against their spectral types, the resulting *Hertzsprung–Russell diagram* (or simply HR diagram) is determined only by the physical properties of the stars. Figure 24.2 shows such a diagram for those nearby stars whose M_V has been determined by trigonometric methods to a deviation not exceeding 0.7 mag. This diagram displays a fundamental finding: not all combinations of spectral type and luminosity occur equally frequently in nature, but all stars lie on a narrow band beginning at early-type stars of high absolute luminosity and extending

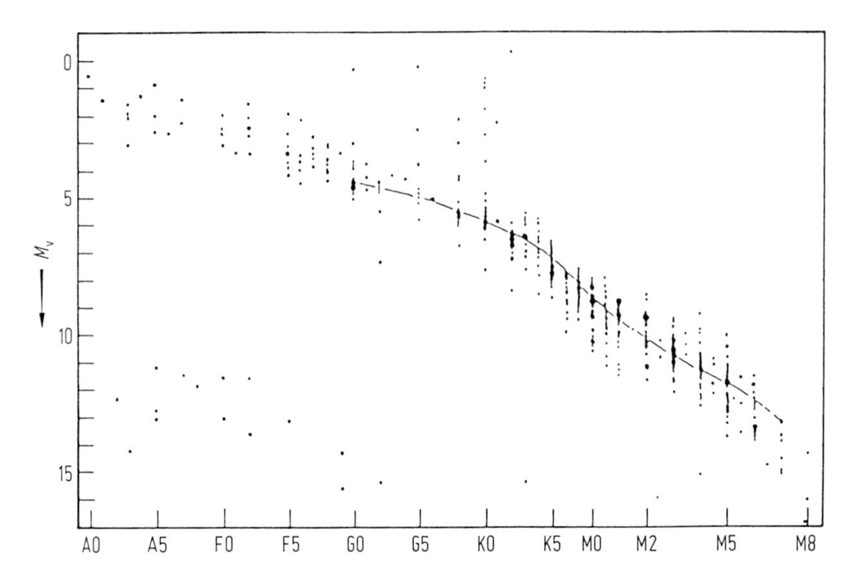

Fig. 24.2. The Hertzsprung–Russell diagram for stars near the Sun. Note that the majority of the stars fall along the main sequence. After W. Gliese.

to late-type stars of low luminosity. This band is called the *main sequence*, and those stars which fall along it are known as *main-sequence stars*. The apparent width of this band in Fig. 24.2 is due in large part to uncertainties in measuring M_V. This is seen when the graph is restricted to stars with the best-determined M_V, such as are presented in Fig. 24.3, where only the nearby stars ($R < 10$ pc) have been used and supplemented by stars from clusters with the best determined distances; see Chap. 7.

A small number of stars are found below the main sequence; these are the *white dwarfs*. Some *giants* are seen above the main sequence.

As is the case for most stars in the solar neighborhood, stars which are members of star clusters are also located along the main sequence. Very young clusters generally contain a few stars substantially brighter than the bright giants in Figs. 24.2 and 24.3. These are the *supergiants*, which are not represented among the nearby stars. By comparing HR diagrams of stars with trigonometrically known distances and of open clusters (see Sect. 27.5.2), the absolute magnitudes of supergiants have been deduced.

24.7 Luminosity Classes

Compared with the high concentration of main-sequence stars in the HR diagram, the areas of giants and supergiants are much more uniformly populated. How far above the main sequence a star is placed in the HR diagram is ascertained from the characteristic features in its spectrum. These include the intensity ratios of certain absorption lines as well as the widths of the Balmer lines, which are largest for main-sequence stars and smallest for supergiants. These luminosity criteria permit the placement of the star in a system of luminosity classes known as the *MK System* or

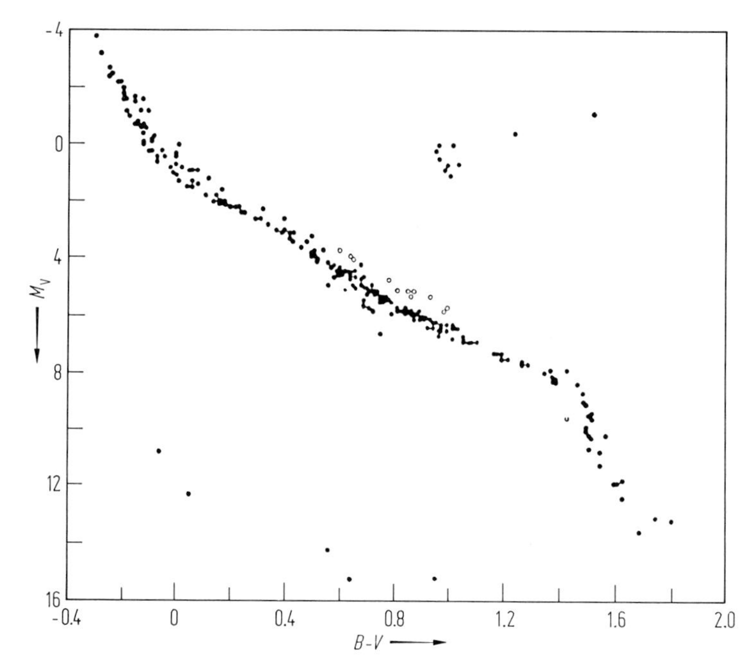

Fig. 24.3. The color–magnitude diagram for stars with the best-known distances, showing how narrow the true main sequence is.

Yerkes System. Members of class I are supergiants, and are subdivided into subclasses Ia, Iab, and Ib. Extremely bright supergiants are coded Ia+ or Ia-0. Then follow, in order of diminishing luminosity, the *bright giants* of class II, normal giants of class III, *subgiants* of class IV, and the main-sequence stars of class V. The latter are still sometimes called "dwarfs," which is most certainly misleading when applied to the bright and massive O-stars. The rare *subdwarfs*, which lie about 1 magnitude below the main sequence in the HR diagram, are usually coded as class VI. The dependence of average absolute magnitudes on spectral type and luminosity class have been determined from photometry of open clusters (see Chap. 27) and are compiled in Appendix Table B.27. Also, the average unreddened colors of the various MK types have been derived (see Appendix Table B.28); they are usually given in the *UBV* system and coded as $(B-V)_0$, $(U-B)_0$, and so on.

Since the unreddened colors $(B-V)_0$ vary monotonically with spectral type from $(B-V)_0 = -0.32$ for early O-types down to $(B-V)_0 = +1.65$ for late M-types, the traditional HR diagram with the spectrum as the abscissa can be replaced with one using $(B-V)_0$. This then is called a *color-luminosity diagram*. Members of a star cluster can be graphed with the apparent magnitude as ordinate. Since they all have the same distance modulus, the characteristics of the color–luminosity diagram are preserved.

If the resolution of a spectrum is sufficient to determine the MK type of a star, then its unreddened color and absolute magnitude can be determined. From these values, the amount of visual extinction suffered by the light as it traverses the path to Earth can be deduced (see Chap. 27). This information then makes it possible to compute the distance to the star. Compared with trigonometric distances, this method constitutes an enormous increase in the accessible range of objects. With this technique, the highest-luminosity stars can be reached out to as far as 10 kpc and more, assuming, of course, that interstellar extinction is not excessive.

24.8 Two-Color Diagrams

From the observed color indices $(B - V)$ and $(U - B)$, a *two-color diagram* can be constructed which may be useful for many applications; $(U-B)$ (counted as increasing positive downward) is graphed against $(B - V)$. Unreddened main-sequence stars lie along a well-defined, S-shaped curve (see Fig. 24.4). Giants and supergiants differ by as much as 0.20 magnitudes from main-sequence stars of the same type.

The influence of interstellar extinction displaces stars in the two-color diagram along a path seen by comparing Fig. 24.4 with the graph of Fig. 27.7 in Chap. 27. It is to be noted that reddening displaces the early-type stars from O to B3 into a region of the diagram which does not coincide with possible positions of later spectral types. Thus, from the positioning in the diagram and by "shifting back" this segment to the main sequence in the direction opposite to that of the reddening displacement, the spectral type and the *color excess* E_{B-V} (the horizontal component of the displacement) for such stars can be unambiguously found. Since by far the majority of stars belong to the main sequence, it can often safely be assumed that a star for which only photometric data are available is a main-sequence star. From its position in the two-color diagram together with the ordinary spectral type, some information on its luminosity can then be deduced, and thus the distance estimated. Should the star belong to one of the luminosity classes I to IV, then the distance derived from this assumption will be too small. Thus, the assumption that a certain star lies on the main sequence yields, when only *UBV* data are available, only a lower limit for its distance.

24.9 Bolometric Magnitudes

The total energy output of a star can be expressed as its *bolometric magnitude* M_{bol}. This quantity corresponds to the total radiation emitted over the entire wavelength range from 0 to ∞. The bolometric magnitude of the Sun is defined to equal its V magnitude. The difference between m_{bol} and V (or between M_{bol} and M_V) is called the *bolometric correction* BC, where

$$\text{BC} = M_{\text{bol}} - M_V. \tag{24.8}$$

For stars of about the solar type (between A0 and K0) the bolometric correction is small, but it increases for early-type as well as for late-type stars to about +3 mag (see Table 24.2). This indicates that very early-type and very late-type stars radiate most of their energy at wavelengths far removed from the V band. The luminosity of

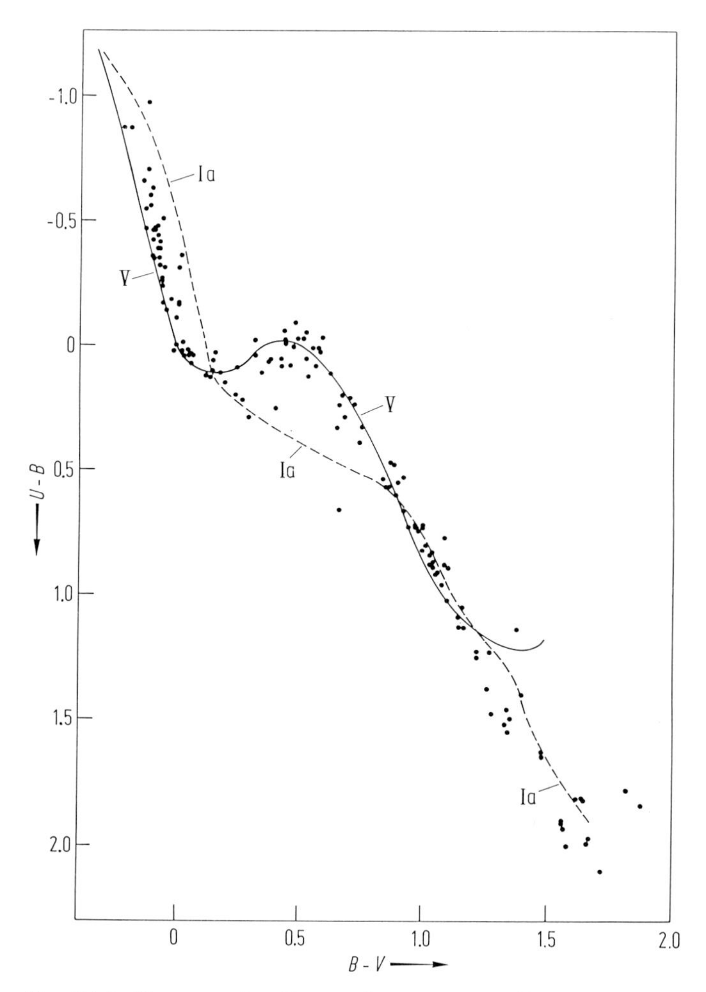

Fig. 24.4. The two-color diagram for some stars listed in the *Catalogue of Bright Stars* is little affected by interstellar reddening. The lines show the positions of unreddened stars of luminosity classes V and Ia.

Table 24.2. The bolometric correction $BC = M_{bol} - M_V$ for stars with various spectral types. After Schmidt-Kaler, Landolt-Börnstein: *Astronomy and Astrophysics*, Vol. 2b, Springer, Berlin Heidelberg New York 1982, p. 182.

Spectral Type	O5	B0	A0	F0	G0	K0	M0	M5
BC	−4.4	−3.2	−0.3	−0.1	−0.2	−0.3	−1.4	−2.7

a star expressed in units of the Sun's luminosity $L_{\odot}$ is

$$L = 10^{0.4(4.77 - M_{\text{bol}})}\, L_{\odot}, \tag{24.9}$$

where $L_{\odot} = 3.82 \times 10^{33}$ erg s^{-1}.

24.10 Stellar Diameters

Were the Sun 1 pc distant from Earth, its apparent diameter would be less than 0.″01. Such small angles cannot be directly measured even with the largest of Earth-based telescopes, as the atmosphere does not permit such a high resolution as this. Since virtually all stars are much farther from Earth than 1 pc, no stellar diameter can be seen direct, and no star can be distinguished from a mathematically exact point source.

Stellar diameters were first measured with a Michelson stellar interferometer at the 2.5-m telescope of the Mount Wilson Observatory. In this experiment, light rays of up to 15 m apart were brought to interference. This method, however, was limited to just a few (11) nearby giants and supergiants with comparatively large angular diameters (about 0.″01). Substantially smaller stellar diameters were determined using the intensity interferometer of Hanbury-Brown and Twiss, which boasted an error of only 0.″0001 and permitted measurements of the diameters of bright main-sequence stars in addition to giants. Other results for the stellar diameters were obtained for the components of eclipsing binaries as derived from their light curves.

The stellar diameters that have been measured are mostly those which are large compared with the Sun's diameter. The largest stars are late-type supergiants, whose radii reach to several hundred times larger than the solar radius, or several AU. Diameters of some star types are given in Table 24.3.

24.11 Stellar Temperatures

From the luminosity L and radius R of a star, the *radiation flux* passing through 1 cm^2 of surface can be computed: $F = L/4\pi R^2$. According to the Stefan–Boltzmann law, the *effective temperature* T_{eff} follows from

$$\frac{L}{4\pi R^2} = \sigma T_{\text{eff}}^4. \tag{24.10}$$

The constant has the value $\sigma = 5.67 \times 10^{-5}$ erg cm^{-2} s^{-1} K^{-4}. Effective temperatures of stars vary between 50 000 K for the earliest O-type stars and 3000 K for late M-types. Table 24.3 again provides examples.

The determination of the effective temperature requires data on L and R, which are available with sufficient accuracy for only a few bright stars. It is simpler to determine the *color temperature* of a star, which is that temperature T representing the stellar continuum approximately by a Planck curve over a certain wavelength interval:

$$B(\lambda, T) = \frac{2hc^2}{\lambda^5} \mathrm{e}^{-c_2/\lambda T}, \tag{24.11}$$

Table 24.3. Physical parameters of the stars. d = diameter, $T_{\rm eff}$ = effective temperature, $\mathfrak{M}$ = mass, ρ = density, g = surface gravity. $T_{\rm eff}$ is given in K; all other parameters are in solar units. After Schmidt-Kaler, Landolt-Börnstein: *Astronomy and Astrophysics*, Vol. 2b, Springer, Berlin Heidelberg New York 1982, pp. 31, 453.

Spectral type	$d/d_\odot$	$T_{\rm eff}$ (K)	$\mathfrak{M}/\mathfrak{M}_\odot$	$\rho/\rho_\odot$	$g/g_\odot$
O5 V	12	44 500	60	0.03	0.40
B0 V	7.4	30 000	17.5	0.04	0.32
A0 V	2.4	9520	2.9	0.20	0.50
F0 V	1.5	7200	1.6	0.50	0.79
G0 V	1.1	6030	1.05	0.79	0.89
K0 V	0.85	5250	0.79	1.26	1.12
M0 V	0.60	3850	0.51	2.24	1.41
M5 V	0.27	3240	0.21	10.0	3.16
B0 III	15	29 000	20	0.006	0.08
A0 III	5	10 100	4	0.32	—
G0 III	6	5850	1.0	0.004	0.03
K0 III	15	4750	1.1	0.0003	0.005
M0 III	40	3800	1.2	0.000 02	0.001
O5 I	30	40 300	70	0.0025	0.079
A0 I	60	9730	16	0.000 08	0.0050
F0 I	80	7700	12	0.000 03	0.0020
G0 I	120	5550	10	0.000 01	0.0008
K0 I	200	4420	13	0.000 002	0.0003
M0 I	500	3650	13	0.000 0001	0.000 05

where c is the speed of light, h is Planck's constant, and the constant c_2 has the value 1.438 88 cm K.

24.12 Stellar Masses

The most important single parameter of any star is its *mass*, which—together with its chemical composition—determines its future destiny. It is possible to obtain masses of stars, as for all other celestial bodies, by measuring their gravitational influence on the motions of other bodies. Specifically, the mass of a star can be determined directly only if it is a component in a binary star system. In the simplest case of mass determination, not only is the semimajor axis a of one component about the other known, but also the individual semimajor axes a_1 and a_2 of each component about the common center of gravity. This permits the determination of the orbits of both components about a common center, and the definition of that center provides the relation

$$\mathfrak{M}_1 a_1 = \mathfrak{M}_2 a_2, \qquad (24.12)$$

where $\mathfrak{M}_1$ and $\mathfrak{M}_2$ are the masses of the two stars. Also, by Kepler's Third Law,

$$\mathfrak{M}_1 + \mathfrak{M}_2 = a^3/P^2. \qquad (24.13)$$

The masses are in units of the solar mass, the semiaxes in AU, and the period P in years. Combining the two equations permits the determination of $\mathfrak{M}_1$ and $\mathfrak{M}_2$. An *eclipsing variable* is a close binary with the orbital plane oriented such that, as seen from the Earth, mutual eclipses occur. The period of the light curve permits the determination of stellar masses if the radial velocities of both components can be measured over the orbital phases.

The resulting masses range from 0.1 $\mathfrak{M}_\odot$ to 60 $\mathfrak{M}_\odot$. This interval is substantially smaller than that for luminosities. A relatively small mass difference between two stars is mapped into a much larger difference in their luminosities. This is shown by the *mass–luminosity relation*:

$$\begin{aligned} L &\sim \mathfrak{M}^4 \quad \text{for} \quad 0 < M_{\text{bol}} < +7, \\ L &\sim \mathfrak{M}^{1.5} \quad \text{for} \quad M_{\text{bol}} > +7. \end{aligned} \tag{24.14}$$

This relation holds only for main-sequence stars.

A knowledge of stellar masses permits the computation of mean densities and surface gravities of stars. The range of both parameters is very large: white dwarfs show mean densities of up to 10^6 g cm^{-3}, while M-type supergiants reach only to 10^{-7} g cm^{-3}. Correspondingly, extreme values of the gravitational acceleration are 10^8 cm s^{-2} and 2 cm s^{-2}. Values for masses, densities, and gravities of some stellar types are compiled in Table 24.3.

24.13 Energy Generation and Lifetime

Until 1939 it was not clear how the energy which is continuously radiated in large amounts by the Sun and stars was produced. Geological studies had shown the age of the Earth to be several 10^9 years, during which time the solar radiation could not have changed substantially. If it had, life could not have evolved continuously over long time spans. The early attempts to explain the solar luminosity by the infall of meteorites or by gravitational contraction yielded wholly inadequate time scales.

In 1939, Bethe and Weizsäcker realized that at very high temperatures and pressures, such as exist in stellar interiors, hydrogen is converted into helium, liberating large amounts of energy. When four hydrogen nuclei (each a single proton) merge to form one helium nucleus, which is about 0.7% lighter than the sum of the four hydrogen nuclei, the small mass loss Δm in the reaction is converted into energy

$$E = \Delta m c^2 .$$

The hydrogen fusion proceeds by the *proton–proton reaction* (abbreviated as simply the p–p process) or by the *CNO cycle*. In the p–p reaction, first two hydrogen nuclei combine to form a deuteron nucleus, which then captures a third hydrogen atom to form ^{3}He (the so-called light helium). Two such ^{3}He nuclei then form a ^{4}He nucleus, where again two protons are released. The CNO cycle begins with $^{12}_{6}$C capturing a proton. This forms a nitrogen nucleus $^{14}_{7}$N. After capturing two more protons and emitting two more antiprotons, or positrons, it generates $^{15}_{7}$N nuclei, which, after capture of a fourth proton, decay back to a $^{12}_{6}$C atom and an α particle, the latter being just another name for a ^{4}He nucleus. In this reaction, the number of carbon nuclei is ultimately unchanged, so that carbon serves only as a catalyst.

The higher the temperature, the faster the two reactions run. The efficiency of the p–p reaction increases with the fifth power of the temperature, that of the CNO cycle to the seventeenth power. Such strong dependencies of the reaction rates on temperature are unknown in cases other than nuclear fusion reactions.

Although the Sun radiates an enormous amount of energy into space (3.8×10^{33} erg s^{-1}), the energy generation per unit volume is quite small. It can be easily calculated that production of 1 kilowatt via nuclear reactions requires a cube of average solar matter 15 m on a side. It is often heard said that the much-sought-after nuclear fusion reactor, which could mimic the energy generation in the solar interior, could ultimately solve humanity's energy problems. This statement is misleading: in a fusion reactor, the energy would have to be generated a million times more vigorously than it is in the solar interior.

The surface temperatures of the Sun and stars (3000 K to 50 000 K) are far too low to permit nuclear fusion. The requisite temperature of around 10^7 K occurs only deep in the interiors of stars. A photon generated there is absorbed and re-emitted by surrounding matter after traveling only about 1 cm. While the general flow of photon radiation is, of course, directed radially outward, each photon travels in a zigzag path, so that it takes about a million years for the photon to traverse the distance from the solar center to its surface.

As long as a star has converted less than 10% of its hydrogen supply into helium, it is in a state of equilibrium. During this time it is found on the main sequence in the HR diagram. The main-sequence lifetime is easily estimated; this requires the luminosity and mass of the star, and the energy generation rate for the complete conversion of hydrogen into helium, which is 6.3×10^{18} erg g^{-1}. In the case of the Sun, the main-sequence lifetime is found to be about 10×10^9 years, so the Sun, with its present age of over 4×10^9 years, can radiate unchanged for several more 10^9 years.

From the evolution times computed for other main-sequence stars, it is found that the earliest types live for only a few 10^6 years, the latest star types, however, for around 100×10^9 years. The reason for this large difference is the higher central temperatures of the early types produced by their higher masses. Such high temperatures result in much more intense nuclear burning, which, as has been said, depends on a high power of the temperature.

Since the earliest-type stars can be only a few million years old, far less than the age of the Galaxy, the conclusion is that such young stars are continually forming. During this relatively short period of time, they cannot have traveled far away from their place of birth. An indication as to where new stars are born is provided by the observation of giant molecular clouds. When these clouds are ionized by the hottest young stars, emission nebulae or HII regions are produced. Reflection nebulae are parts of molecular clouds which are illuminated by later-type stars recently formed within them.

24.14 The Chemical Composition of the Stars

Detailed analyses of the spectra of bright stars and of the Sun permit astronomers to find the abundances of various chemical elements in stellar atmospheres. The results for the Sun as for most Population I stars and for galactic gaseous nebulae are very similar, as several examples in Table 24.4 show. Objects of Population II, such as high-velocity stars and globular clusters, show smaller amounts of heavy elements than in Population I objects. This difference is caused by the chemical evolution of the Galaxy. Population II objects originated soon after the Milky Way system was formed, and consisted originally mostly of hydrogen (90%) and helium (10%). Heavier elements were present only in traces, if at all. They were gradually built up by nuclear fusion inside stars.

The nuclear processes in stars first convert hydrogen into helium. After consuming the major part of hydrogen in the stellar interior, the helium is converted into carbon via the *triple-α process*. This reaction, however, requires higher temperatures (10^8 K) than the hydrogen fusion reactions. At still higher temperatures, the carbon also "burns" to form oxygen. Heavier elements—often collectively called *metals*—are forged by a multitude of nuclear reaction. During the course of time, they are partly expelled into the interstellar medium, essentially by two possible processes: some stars, such as red giants, continuously lose matter in the form of stellar winds, or by expelling their outer shells in the form of planetary nebulae. High-mass stars explode at the end of their evolution to become supernovae, and scatter the bulk of their matter, which has been enriched by heavy elements through nuclear processes, into their interstellar environment. Certain other elements which cannot form during the normal lifetime of a star are produced during the supernova itself. Owing to the gradual, ongoing enrichment of the interstellar medium with heavier elements, those stars which are just now being born already contain, at the beginning of their existence, a few percent of heavy elements.

24.15 The Evolution of Stars with Time

Aside from the more or less periodic variations in the brightnesses of some stars and an occasional supernova collapse, no changes in the physical parameters of stars can be observed within the brief span of a human life; a stellar lifespan is too long and its

Table 24.4. Logarithms of the frequency of certain chemical elements in the Sun as well as in α Cygni (both Population I objects) and in the Population II object HD 140283. The values are normalized to $\log N(\mathrm{H}) = 12.0$.

	^{1}H	^{6}C	^{11}Na	^{12}Mg	^{13}Al	^{14}Si	^{20}Ca	^{26}Fe	^{38}Sr
Sun	12.0	8.6	6.3	7.5	6.4	7.6	6.4	7.6	2.9
α Cygni	12.0	8.1	–	7.9	6.8	8.0	6.7	7.6	3.4
HD 140283	12.0	6.3	3.7	7.2	3.8	5.3	4.0	5.1	0.6

evolution too slow. It is known that some stars, including the Sun, are already quite old, while others have apparently formed in relatively recent times. Nevertheless, the manner in which a star evolves between its birth and its death can be stated with great reliability, this being true for three reasons. First, the physical laws governing the processes inside stars are now largely known. It can be stated which processes generate energy in the interiors, and, in so doing, transmute lighter elements into heavier ones. It is also known how this energy gets from the depths of a star to the outside—partly by radiation and partly by convection.

Second, thanks to decisive contributions by fast, modern computers, the time scale of stellar evolution is fairly well known. It is only by these computers that theorists are able to perform, within tolerable time requirements, the tedious integrations of the mathematical equations describing stellar structure.

Finally, it is possible to check the correctness of the results of such model calculations with observations, which is a significant advantage. The crucial criteria can be found, in particular, in star clusters. The HR diagrams of clusters show how a group of stars of different masses but with similar chemical compositions evolve within the same time span. The theoretical reproduction of HR diagrams of young open clusters as well as of old globular clusters considerably strengthens the confidence in stellar model calculations.

The first chapter in a stellar life describes its birth from a cloud of interstellar matter. The matter density is initially so low that the heat generated during the incipient contraction can be radiated from the center without obstruction. Only from that moment when the contracting part of the cloud has become massive and dense enough that the heat radiation emitted is largely reabsorbed within itself (self-absorption) is the cloud then referred to as a *protostar*. Henceforth, further contraction leads to a continuous increase in the temperature.

The subsequent evolution is characterized by the rapid formation of a small, dense core onto which the matter from the extended parts of the collapsing cloud rain down. At the beginning of this phase, the core possesses only a few percent of the final stellar mass.

Once the temperature of the protostar reaches the range 1000–2000 K, it will have evolved into an infrared or very red object with a luminosity of about 100 $L_\odot$ (assuming a final mass of between 1 and 2 solar masses). Since star formation always occurs deep inside a molecular cloud, this stage of the protostar is concealed from view.

Heat energy generated by further contraction first serves to dissociate H_2 molecules, and later to ionize hydrogen, and then also helium. During this stage, the temperature initially does not rise, although the protostar shrinks in size. Its luminosity thus drops continuously, and the protostar moves downward in the HR diagram at about constant color (i.e., constant spectral type and effective temperature). This part of the evolutionary track is termed the *Hayashi line*. The contraction slowly continues at a gradually rising temperature, moving the star to the left in the HR diagram; after some time, it approaches the main sequence. Shortly before reaching the main sequence, the first nuclear reactions are ignited in the interior. This marks the actual time of birth of the new star. The energy generation by nuclear fusion raises the pressure in the interior and halts any further contraction. Then follows the longest phase in the lifetime of any star, namely the time spent on the main sequence. During this period, the hydrogen is consumed in a core region which may contain over 10% of the mass of the star. Gravitation and gas pressure balance at every point in the star so that the star does not undergo changes with time. As mentioned in Sect. 24.13, the main-sequence lifetime of a star is determined by its mass, and reaches around 100×10^9 years for late-type stars.

When finally, in the core of a star, the hydrogen has been largely converted into helium, the fading energy generation and hence the diminished pressure in the core leads to its further contraction. As was the case in the contraction of the protostar, the temperature again is raised until now new nuclear processes occur, thereby resulting in increased energy generation, which halts the contraction; in this case it is the triple-α process which converts three helium nuclei into one carbon nucleus.

The temperature increase in the interior subsequently causes an expansion of the outer portions of the star. The stellar diameter can grow to as much as 50 times the original value, and it is accompanied by an enormous increase in luminosity. The outer layers of the star, which are actually the only directly observable part, then attain a temperature less than that of the stellar photosphere during the main-sequence phase, when the photosphere was much closer to the hot core. Both effects (the increase in total luminosity and the decrease in effective temperature) taken together cause the movement of the star toward the upper right in the HR diagram and into the region of *red giants*. Given a sufficiently high mass, the star can always, once the fusion process in the core has been depleted of fuel, initiate a new contraction phase and—when the temperature has increased sufficiently—a new fusion process. This scenario is repeated until a core of iron has formed in the interior. Thermal fusion cannot generate elements heavier than iron. Once the iron core begins to contract, this collapse cannot be stopped by incipient new reactions; instead it marks the first step toward a *supernova* event.

Stars with lesser mass do not end their lives in so spectacular a fashion. During or just after the red-giant phase they expel a portion of their outer envelope, which then surrounds the star as a gaseous shell called a *planetary nebula*. The mechanism which leads to this mass loss is not known in detail. The remnant of the star ultimately evolves into a *white dwarf*, in which nuclear reactions have ceased to occur. The radiation emitted by a white dwarf is supplied by contraction only. After several 10^9 years, even this supply of energy becomes exhausted, and only a dead *black dwarf* remains.

24.16 Subjects for Amateur Observations

Brighter stars are reddened only in rare cases; thus the measured colors largely correspond to the true ones and are direct indicators of the spectral types. Determining colors is visually rather difficult, but is handled conveniently by photography. The whole range of colors from blue to red can be documented on color film, in particular by not imaging the stars as points. To this end, either the camera may be slightly defocused, showing the stars as small disks, or the camera tracking may be turned off so that the stars are photographed as trails (cf. Sect. 6.4.1 in Vol. 1).

An idea of various astrophysical relations can be gleaned by those amateurs who can measure magnitudes on astrophotographic plates obtained with filter sets which match the RGU system of Becker, or reproduce the Johnson *UBV* system (cf. Sect. 8.3 in Vol. 1). It is then possible to construct color–magnitude diagrams of star clusters which serve to obtain various data: the age of the cluster, the amount of reddening, and also the distance. Similar studies in any star field will show how the young stellar population is concentrated toward the Milky Way, while the directions to the galactic poles are dominated by members of the older population.

References

24.1 Hirshfeld, A., Sinnott, R.W.: *Sky Catalogue 2000.0 Vol. 1: Stars of Visual Magnitude 8.05 and Brighter*, Cambridge University Press, Cambridge 1982.
24.2 Burnham, R.: *Burnham's Celestial Handbook: An Observer's Guide to the Universe Beyond the Solar System*, Dover Publ., New York 1978.
24.3 Clark, R.N.: *Visual Astronomy of the Deep Sky*, Cambridge University Press, Cambridge 1989.
24.4 Tirion, W.: *B.A.A. Star Charts*, British Astronomical Association, 1882.
24.5 Hoffleit, D.: *Catalogue of Bright Stars* (4th edn.), Yale University Observatory, New Haven 1982.
24.6 Mermilliod, J.C.: *A General Catalogue of UBV Photoelectric Photometry*, Lausanne Observatory, Geneva 1977.

25 Variable Stars

T.J. Herczeg and H. Drechsel

25.1 Introduction

The subject of *variable stars*, or simply *variables*, originally (about 200 to 250 years ago) referred to just a handful of stars noted for substantial changes in brightness. Present-day astrophysical studies have demonstrated, however, that brightness changes are only part of the variability picture. The definition of variability in a star has been generally extended to include *time variability of any parameter*, especially the luminosity, temperature, and radius. With the development of radio and satellite astronomy, the changes are meant to refer more generally to a time variation of radiation considered over the entire electromagnetic spectrum from radio to X-rays.

Since it is now known with certainty that such complex structures as stars can never be "absolutely constant," there emerges the question at what point a star is considered variable. It is often heard said that at sufficiently high resolution every star displays some variability, but this statement is rather a trivial one. The real difficulty lies in defining accurate limits for the variability. Take, for example, the best-known star, the Sun. Active regions in the surface layers of the Sun form and then evolve and decay in an 11-year cycle. During times of high activity, both the ultraviolet radiation and the particle emission of the Sun show substantial and mostly short-period changes, while the solar magnetic field and the shape of the corona change more slowly. Also, satellite observations from space have recently provided scientists with the answer to the much-debated question of whether or not the solar constant is really constant. It is not, but the changes seem to be at about the 0.1% level.

Hence, the Sun is technically a variable star, but *not* in the sense used in the following discussion. This is the case, first, because similar changes in other, much more distant stars would be detected either not at all or only with supreme effort and patience. Second, and more importantly, the mentioned variations in the uppermost layers of the Sun are apparently a normal phenomenon accompanying a stable, stationary phase of evolution (as are also strong stellar winds and the accompanying mass loss from massive stars). Variable stars thus will be defined as those stars whose observed changes are of a deeper nature and are dependent on the particular structure of the stellar interior, which, at least in certain stages of stellar evolution, is directly related to the characteristics of these objects. Stars that are variable in this sense form several distinct classes which will be discussed in detail in the following pages.

25.1.1 Initial Discoveries

The earliest-known variable stars were primarily "new stars," known in present terminology as novae or supernovae, which, by their sudden appearance and spectacular development in brightness, were most easily detected. Stars which changed their brightness in a more or less regular fashion had been recognized for a long time, but were considered curious—and even doubtful—cases of lesser importance.

From about the middle of the 18th century, the notion of variable stars gradually gained acceptance, and since the beginning of the 20th century, the "new stars" were considered an independent class. It is now known that the nova phenomenon is linked to a specific configuration of close binaries, and that novae are a subtype within the system of variability.

The history of variable stars in astronomy, at least in the western hemisphere, began in 1572 with the appearance of a very bright new star in the constellation Cassiopeia, now known as Tycho's Supernova. The indisputable change in the "translunar" realm shook the Aristotelian notion of the universe and triggered vehement philosophical discussions. In 1595, Fabricius discovered the "Miracle Star of Cetus," Mira Ceti. It was also originally thought to be a "new star," but its periodic nature was soon uncovered. Further discoveries followed, and by the early 18th century, four "new stars" and four variables were known to exist. (The variability of Algol was not considered established, and its nature was not clarified until the studies of Goodricke in 1782.) By 1800, seven more variables had been found.

Although this was a short list, with only 15 objects, it provided a good cross section of all major classes of variables. After 1800, the number of identified variable stars grew, but still at a slow rate. Thus, by the middle of the 19th century, when Argelander introduced the coding of capital letters, astronomers were convinced that it would be necessary to make use of only a few letters from R to Z in each constellation. Nowadays, the number of known variables is in the tens of thousands—the new edition of the *General Catalogue* contains about 28 500 certain cases of variability[2]—and so the old system of nomenclature obviously had to be extended.

25.2 Nomenclature and Classification

25.2.1 Nomenclature

Variable stars are designated according to a somewhat cumbersome convention: brighter stars, such as β Lyrae, γ Cassiopeiae, δ Cephei, or 44 Bootis, which already have a name or number, receive no further designation if found to be variable. Other variables are coded by Latin capital letters in the sequence of discovery in each constellation. First, the letters R, S, T, ..., Z are used in succession, then RR, RS, RT,

2 This number does not include the over 2500 variables in globular clusters, nor the approximately 3500 variables in the Magellanic Clouds, nor the several thousands of variables in other extragalactic systems.

..., YZ, ZZ, and finally from AA, AB, AC, ..., BB, BC, ... to QY, QZ, adding the constellation name in genitive form. Examples include T Coronae Borealis, RR Lyrae, or HZ Herculis. This coding scheme suffices for 334 variables in each constellation. For the remaining variable stars, the sequence number preceded by the letter "V," which indicates variability, is used, and thus, beginning with V 335, there is virtually no limit to the number of available codes.

This system is not followed for variables in extragalactic star systems, which usually carry only an *ad hoc* designation.

Newly discovered variables first receive a provisional coding. After the variability has been examined, the final coding is given by the Astronomical Council of the Academy of Sciences of the Soviet Union on behalf of the International Astronomical Union. The lists of names are published from time to time in the *Information Bulletin of Commission 27* (Budapest).

Compilation catalogues of variable stars (GCVS = *General Catalogue of Variable Stars*) have also been prepared by the Astronomical Council of the Soviet Academy. The first edition appeared in 1948. The fourth edition, in four volumes, is edited by N. Kholopov and coworkers; the first three volumes list the variables according to constellations, from Andromeda to Vulpecula, the fourth (N.N. Samus, editor) gives the reference tables. Among special catalogues, a particularly important one is the *Finding List for Observers of Interacting Binary Stars*, published by the Universities of Florida and Pennsylvania and compiled by F.B. Wood and coworkers. For further bibliographical material, the list of references at the end of this chapter should be consulted.

25.2.2 Classification

The growing number of discoveries gradually revealed the wide variety of variable star types, each presumably with a different mechanism of variability. The development of modern physics and astrophysics permitted a better understanding of the structure and evolution of stars and hence also the variability of certain star types. Although there were early attempts at classifying the variables into groups, it is only in recent decades that physically meaningful classification schemes have been worked out. The ordering principle usually depends on the physical nature of the variability (e.g., pulsations of a star or eruptions in the outer layers of the atmosphere), and also on the evolutionary status.

Two more recent classification schemes will be briefly mentioned: one is the classification presented by H.W. Duerbeck and W.C. Seitter in *Astronomy and Astrophysics*, Vol. 2b of the tabular compilation "Landolt-Börnstein," which introduces two primary classes, *pulsating* and *eruptive variables*. *Eclipsing variables* are excluded, as their variability is due merely to geometrical and not physical causes. Necessarily included in both main classes are some "foreign" variables: the *rotational variable stars*, which are placed into the pulsating variable class, and the R CrB stars, which are included among the class of eruptive variables. (The R CrB group is actually characterized by sudden, strong losses in brightness caused by optically thick dust clouds which form around the star.) The attractive feature of this scheme is that the dichotomy emphasizes the two primary mechanisms which are responsible for the variability. As the vast majority of stars fall into one or the other class, one may, when attempting to characterize the variability of stars "in a nutshell," refer to only these two basic groups.

The second new classification has been elaborated by N. Kholopov and coworkers, and is used in the latest edition of the GCVS. It contains six basic types of variability: eruptive, pulsating, rotating, cataclysmic, and, in addition, eclipsing and X-ray binaries. Young irregular variables of the Orion and T-Tauri types are subsumed into the class of eruptive variables. Most primary classes are subdivided into numerous subclasses. It is quite important for cataloguing purposes to have such a refined scheme.

It will be preferable, for the purpose of this chapter, to use a scheme of classification lying about midway between the two schemes just mentioned. As such, it can be considered a modernized version of the noteworthy classification system proposed in 1938 by C. Payne-Gaposchkin and S. Gaposchkin in their book *Variable Stars*. Rotational variables appear here as a distinct primary class (as they do in the GCVS) although the subclasses show a rather inhomogeneous selection of types. Also, the young and irregular variables of the Orion and T-Tauri types are assigned a new class since they represent, despite occasional eruptions, a wholly different evolutionary stage from the eruptive variables.

With very few exceptions, the known variables can be ordered naturally into the following classification:

I. *Eclipsing Variables* or photometric binaries (geometric, not necessarily physical variables).
   1. Detached systems (without interaction between components).
   2. Weakly interacting systems (tidally deformed or ellipsoidal).
   3. RS Canum Venaticorum stars.
   4. Semi-detached systems (with mass transfer and mass loss).
   5. Contact systems (early-type, high-mass contact systems and later-type W Ursae Majoris stars).

II. *Physical Variable Stars*

A. *Pulsational Variables*
   1. Short-period variables: β Cephei (or β CMa) stars, δ Scuti and RR Lyrae stars, or cluster variables.
   2. δ Cephei and W Virginis stars.
   3. Mira stars, or long-period variables, OH/IR variables.
   4. RV Tauri stars and semi-regular variables.
   5. irregular or red variables.
   6. R Coronae Borealis stars (dust-shell stars).

B. *Rotational Variable Stars*
   1. α_2 Canum Venaticorum stars, or magnetic variables.
   2. BY Draconis stars ("spotted" stars).
   3. Radio pulsars.

C. *Eruptive Variables*

1. Supernovae.
2. Cataclysmic variables (novae, recurrent novae, dwarf novae, nova-like variables).
3. X-ray binaries (high-mass systems like Cen X-3, and low-mass systems, including the systems found in the central bulge of the Galaxy).
4. Symbiotic stars.
5. Flare stars (UV Ceti stars).

D. *Irregular Variables*

1. Orion variables (nebular variables) and RW Aurigae stars (not associated with gaseous nebulae), pre-main-sequence objects.
2. T Tauri stars (later-type nebular variables with spectral peculiarities such as high lithium abundance), also pre-main-sequence objects.
3. Variable Be stars, not pre-main sequence objects (γ Cas), or peculiar eruptive objects, such as η Car.

It is hoped that this scheme conveys the idea of which star types and by which process variability occurs. Radio pulsars are usually not counted among variables. The inclusion of pulsars in the present classification seems reasonable on the grounds that one should consider not only variable brightness in the visible range but variability in general.

As is true for any classification, it will be found that there exist some objects which fit into two or even three different classes. Ellipsoidal stars, for instance, are here treated as photometric binaries, but they could also be considered rotational variables. Novae and dwarf novae are in all likelihood close binaries, and many of them exhibit eclipsing light variations. In such cases, a somewhat subjective judgement can be made as to what the primary characteristic feature of the observed class is.

Not all classes and subclasses of variable stars can be discussed here with equal thoroughness. In some cases, the definition and brief indications for the physics of variability must suffice. The following chapter concentrates upon three primary subjects: eclipsing variable stars, pulsating variables, and cataclysmic variables as a subclass of eruptive variables. Owing to the propitious appearance of Supernova 1987A in the Large Magellanic Cloud and also because of the significance of supernovae for cosmic evolution, supernovae will also be considered in somewhat more detail. A handful of peculiar objects, such as η Carinae, are mentioned along with the irregular variables.

25.3 Eclipsing Variables

Eclipsing-type variable stars are close binaries with the orbital plane oriented in space such that an observer on Earth sees mutual occultations of the two components. This gives rise to an apparent periodic variation in the integrated light of the system. During one revolution, the light curve typically shows a deeper *primary minimum* and a shallower *secondary minimum*. The primary minimum is caused by the eclipse of the hotter (and usually more luminous) component of the photometric binary by the fainter secondary star; one-half period later, the reverse configuration creates the secondary minimum. The ratio of depths of both minima is determined by the ratio of the surface brightnesses of the two components, which is determined essentially by the effective surface temperatures.

25.3.1 Structure of Close Binaries

The revolution of the two stellar components about the common center of gravity proceeds generally on elliptical orbits with the radius vector

$$r = \frac{a(1-e^2)}{1+e\ \cos\theta}, \tag{25.1}$$

where a is the semimajor axis of the relative orbit, e the orbital eccentricity, and θ the phase angle (true anomaly).

For non-circular orbits, the distance between the two stars varies between the maximum $a(1+e)$, called *apastron*, and the minimum $a(1-e)$, which is the *periastron*. The special case of a circular orbit ($e = 0$) is approximately fulfilled in many close binaries. The period P of revolution is related to the scale of the orbit (semimajor axis a) and the masses $\mathfrak{M}_{1,2}$ of the components via Kepler's third law:

$$P = 2\pi \left[\frac{a^3}{G(\mathfrak{M}_1 + \mathfrak{M}_2)}\right]^{1/2}, \tag{25.2}$$

where G is the universal gravitational constant.

A "close" binary is any system in which the distance between the components is of the same order as their radii. Owing to the mutual influences on structure and evolution of the stars, these systems are also called *interacting binaries*. The most important interacting process is the mass exchange between components, which is responsible for the peculiar properties of several different kinds of variable stars (e.g., X-ray binaries and cataclysmic variables).

The particular significance of eclipsing variables lies in the fact that photoelectric observations can be combined with simultaneously determined spectroscopic measurements of the radial velocities to obtain the masses of the two stars.

When the absolute orbits (semimajor axes a_1 and a_2) are observed through periodic Doppler shifts of spectral lines of both components, then the ratio of radial velocity amplitudes K_1/K_2 found from the observed radial velocity curves $V_{1,2}(t)$ gives the mass ratio:

$$\mathfrak{M}_2/\mathfrak{M}_1 = K_1/K_2. \tag{25.3}$$

With the orbital inclination i, which is known from the analysis of the light curve, and with $a = a_1 + a_2$, the two masses can be found.

The more common case in which the lines of only the more luminous component are identified in the spectrum yields only a relation between the two masses and the angle of inclination i, the so-called *mass function*:

$$F(\mathfrak{M}_1, \mathfrak{M}_2, i) = \frac{(\mathfrak{M}_2 \sin i)^3}{(\mathfrak{M}_1 + \mathfrak{M}_2)^2} = \frac{4\pi^2}{G}\frac{(a_1 \sin i)^3}{P^2}. \tag{25.4}$$

For relatively well-separated eclipsing binaries, i (defined as the angle between the line of sight and the normal to the orbital plane) is usually very close to 90° and in any event lies above a certain lower limit, so that a good value of $\mathfrak{M}_2$ can be obtained from a value of $\mathfrak{M}_1$ estimated from the spectral type.

Knowledge of stellar masses is of great importance, since the mass of a star of given chemical composition determines all of its other parameters such as luminosity, surface temperature, and radius, and the observed quantities can be compared with the theory of stellar structure and evolution.

Kepler's third law and the motion of the components on closed orbital ellipses holds strictly only for centrally condensed spherical stars (sufficiently well approximated with point masses) with gravitational potentials proportional to r^{-1}. Tidal interaction in close pairs, however, leads to a deformation of the stars and a subsequent modification of the potentials, which, in an eccentric orbit, causes a rotation of the apsides (the direction of the semimajor axis). Through the viscous damping of stellar oscillations, tidal forces also effect a rapid decay of the orbital eccentricity, which approaches a circular shape in the stable state of minimum total energy at constant total angular momentum. The damping of the vibrations caused by periodically variable tidal actions occurs on a short time scale compared with the evolutionary age of the system, so that most close systems are in circular orbits and in synchronous rotation. That is to say, the angular velocity of the star's rotation coincides exactly with its orbital motion.

The closer the pair, the more pronounced is the departure of the shape of the two stars from spherical symmetry. With decreasing distance between the mass centers, the influence of tidal interaction increases proportional to a^{-3}, and also the angular speed of the bound rotation increases, in accordance with Kepler's third law, in proportion to $a^{-3/2}$, which increases the centrifugal action in proportion to a^{-2} and hence the rotational flattening of the stars. The three-dimensional shape of the surface of each stellar component corresponds to a surface of constant pressure and constant density, and hence a surface of constant potential. It is therefore called an *equipotential surface*. The spatial shapes of equipotential surfaces in close binaries are described by the *Roche model*.[3] Three important assumptions permit a simple description of the shapes of the equipotential surfaces:

- The mass distributions in both stars can be approximated by point masses;
- The orbit is circular;
- The rotation of the stars is synchronous with the orbital motion.

In a corotating Cartesian coordinate system with its origin at $\mathfrak{M}_1$ ($\mathfrak{M}_1 \geqq \mathfrak{M}_2$), where the x-axis is along the line joining $\mathfrak{M}_1$ and $\mathfrak{M}_2$ and the z-axis is perpendicular to the orbital plane, the potential ψ composed of the graviational potentials of $\mathfrak{M}_1$ and $\mathfrak{M}_2$ and of the centrifugal action at a point $P(x, y, z)$ is given by

$$\psi = -G\left(\frac{\mathfrak{M}_1}{r_1} + \frac{\mathfrak{M}_2}{r_2}\right) - \frac{\omega^2}{2}\left[\left(x - \frac{\mathfrak{M}_2 a}{\mathfrak{M}_1 + \mathfrak{M}_2}\right)^2 + y^2\right], \tag{25.5}$$

where

$$r_1 = (x^2 + y^2 + z^2)^{1/2} \quad \text{and} \quad r_2 = \left[(a - x)^2 + y^2 + z^2\right]^{1/2}$$

3 After the mathematician E.A. Roche (1820–1883).

are the distances of P from $\mathfrak{M}_1$ and $\mathfrak{M}_2$ respectively. The angular velocity $\omega = 2\pi/P$ is equal to the Keplerian value $[G(\mathfrak{M}_1 + \mathfrak{M}_2)]^{1/2}a^{-3/2}$.

The *Roche equipotential surfaces* $\Omega =$ const are connected with the potential ψ via

$$\Omega = \frac{a\psi}{G\mathfrak{M}_1} + \frac{q^2}{2(1+q)}. \tag{25.6}$$

Normalizing the potential by putting $a = 1$, then Ω depends only on the mass ratio $q = \mathfrak{M}_2/\mathfrak{M}_1$, and not upon the individual masses. Often the Roche potential Ω is also expressed by

$$C = \frac{2}{1+q}\Omega, \tag{25.7}$$

and the surfaces $C =$ const are the *Jacobi-Hill surfaces* of the restricted three-body problem.

An intersection of equipotential surfaces ($\Omega =$ const) with the orbital plane ($z = 0$) is shown in Fig. 25.1. Near the mass center of each star the equipotential surfaces are nearly spherical as the gravitating influence of the other star on its shape there is negligible; the gravitational potential of the near mass center is much larger than other terms. With increasing volume or with decreasing distance between the stars, the potential of the other component causes a distortion toward it; the line joining the centers forms the symmetry axis of the approximately ellipsoidally deformed body. The effect becomes stronger the more the surface of the star approaches the so-called *inner critical Roche surface*, which is the largest closed equipotential surface around each component, and thus defines the maximum possible volume in a detached configuration.

In addition to the tidal action, the synchronous rotation of the stars also leads to a distortion owing to the centrifugal force; the stars are flattened on the rotation axis

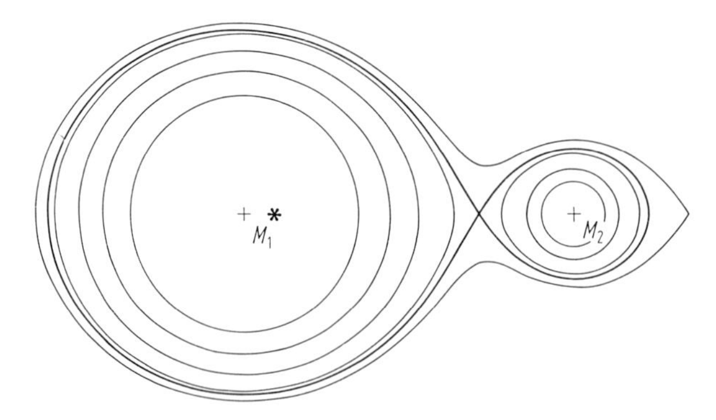

Fig. 25.1. Intersection of equipotential surfaces with the orbital plane of a binary system with mass ratio $q = \mathfrak{M}_2/\mathfrak{M}_1 = 0.1$. The mass centers of the two components are marked by plus (+) signs, the system center of mass with an asterisk (*). Shown are some equipotential surfaces representing detached systems, the figure-of-eight inner critical surface, and the outer critical surface.

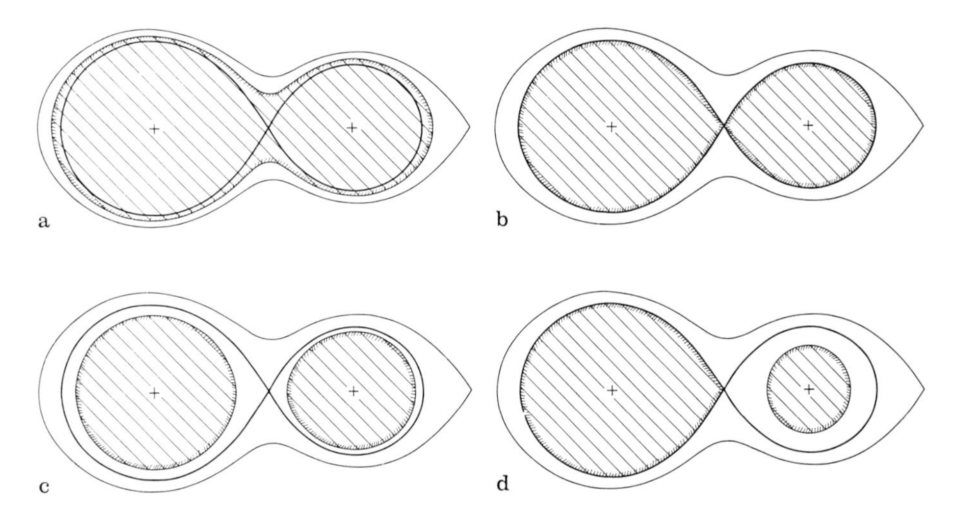

Fig. 25.2a–d. Different configurations of close binary systems, the equipotential structure graphed for the mass ratio $q = \mathfrak{M}_2/\mathfrak{M}_1 = 0.5$. **a** Over-contact system; **b** contact system; **c** detached system; **d** semi-detached system.

perpendicular to the orbital plane. The equipotential surfaces graphed in Fig. 25.1 include the centrifugal potential. The outermost surface graphed is the largest closed envelope of the system, and thus is called the *outer critical Roche surface*. With the aid of the equipotential surfaces, a classification of various configurations of closed binary stars can be effected (see Fig. 25.2).

I Both stars are smaller than their maximum volumes bounded by the inner critical surface; this is a *detached system* (Fig. 25.2 c);

II One of the stars fills its Roche limit and the system is *semi-detached* (Fig 25.2 d);

III Both stars just fill their maximum volumes, thus making a *contact system* (Fig. 25.2 b);

IV *Over-contact systems* (Fig. 25.2 a) have both components further expanded so that matter overflows the inner Roche limit and fills part of the space between the inner and outer Roche surface where the matter is still bound to the system but cannot be associated with either of the two components.

Over-contact systems can be characterized by a parameter known as the *degree of contact* f:

$$f = \frac{\Omega_\mathrm{i} - \Omega}{\Omega_\mathrm{i} - \Omega_\mathrm{o}} \qquad 0 \leqq f \leqq 1, \tag{25.8}$$

where Ω, Ω_i, and Ω_o are the values of the potentials for the actual stellar surface and for the inner and outer critical surface, respectively. When the stars expand beyond the outer Roche surface, material is lost from the system, and thus there is no stationary solution for the shapes of the surfaces.

25.3.2 Classification and Analysis of Light Curves

Light curves of eclipsing variables can in most cases be ordered phenomenologically into one of three classes. These are the light curves of

- Algol-type (EA),
- β Lyrae-type (EB), and
- W UMa-type (EW).

Figure 25.3 shows the light curves of the prototypes, not all of which are typical representatives of their classes. Indeed, Algol (β Persei) is actually a semi-detached system with a red subgiant as the contact component, whereas many members of the pho-

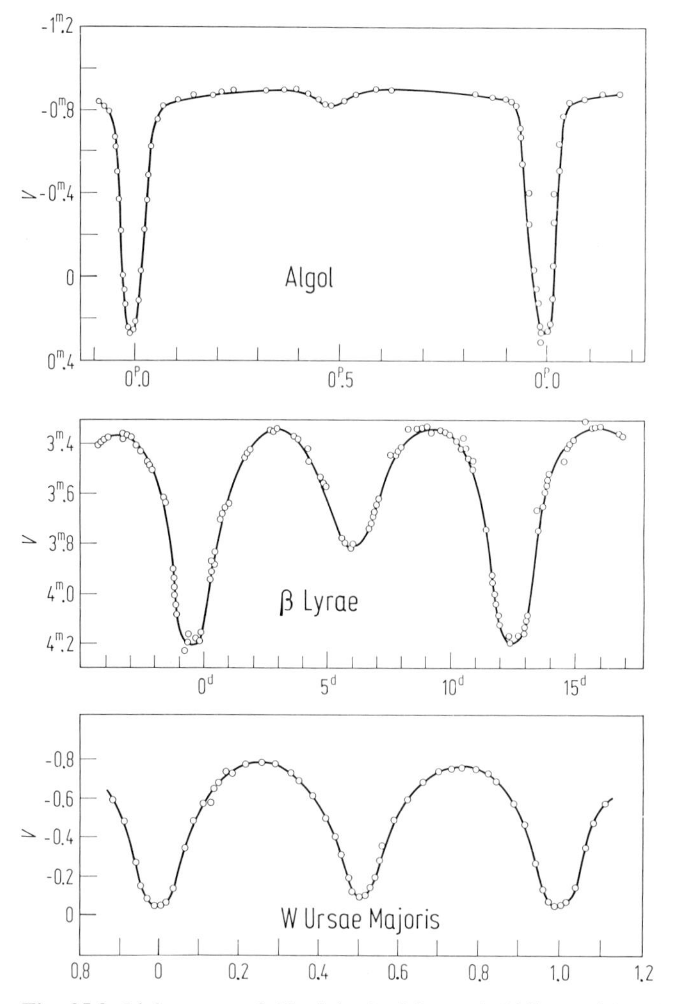

Fig. 25.3. Light curve of Algol (*top*), β Lyrae (*middle*), and W Ursae Majoris (*bottom*), the prototypes of eclipsing variables of types EA, EB, and EW, respectively. After Tsesevich [25.1].

tometric class EA are widely separated binaries composed of main-sequence components. Algol-type light curves show a pronounced primary minimum and—depending on the system geometry, inclination, and luminosity ratio of the components—a more or less deep secondary minimum. Depending on the ratio of the radii, eclipses may be total, in which case a constant level of brightness is observed for some time before and after the center of minimum. Between eclipses, the luminosity changes only slowly, as tidal deformations are virtually negligible for these widely separated stars, but the reflection effect (heating) is still noticeable.

β Lyrae also is not at all typical of the class EB. This extraordinary system is semi-detached, containing a massive, but spectroscopically unidentifiable, secondary component. Many EB systems, however, are close, yet detached, binaries with two main-sequence components nearly filling their Roche lobes and showing strong interactions. Tidal deformation and reflection cause a pronounced, continuous variation of brightness with orbital phase, which is distinctly noticeable outside of eclipses owing to the changing direction of view onto the stars.

W UMa, however, is a typical EW-type system. Both components fill their Roche lobes and form an over-contact configuration. Strong interactions cause the large brightness variation between the minima, which here, in contrast to β Lyrae systems, are about equally deep owing to the nearly equal luminosities of the components.

An analysis of the light curves of eclipsing variables which seeks to determine system parameters requires a realistic physical model and a suitable mathematical method of solution. The light variation is determined by so large a number of free parameters that analytical solutions of the problem are generally not feasible. An exception is the method of Fourier analysis proposed by Kopal [25.2,3], which, owing to its complexity, is rarely used. Purely numerical methods, such as those employed by Wood [25.4], Hill and Hutchings [25.5], Wilson and Devinney [25.6], or Wilson [25.7], are generally preferred. Kallrath and Linnell [25.8] have combined the Wilson–Devinney model with another mathematical algorithm (SIMPLEX) which yields a rapid and unambiguous convergence.

Except for the Wood method, which assumes tri-axial ellipsoids for the shapes of the components, the Roche model is the physical basis for calculations. In addition to tidal and rotational deformation, the mutual irradiation of the components (which gives rise to the *reflection effect*), limb darkening, and gravitational darkening are allowed for. The radiation flux in the direction of the observer is computed as a function of system parameters and orbital phase, and the theoretical light curve thus obtained is compared with observations. Corresponding iterative variations of parameters are obtained by fitting the computed light curve to the observed one by a least-squares solution.

25.3.3 Detached Systems

More than one-half of all stars are members of binary and multiple systems. Visual (i.e., optically separated) binaries with known absolute orbits and particularly spectroscopic binaries which simultaneously show eclipses are the primary source of knowledge on stellar masses and radii. Moreover, the two components of a binary system are of equal age, and thus offer the opportunity to test the theories of stellar

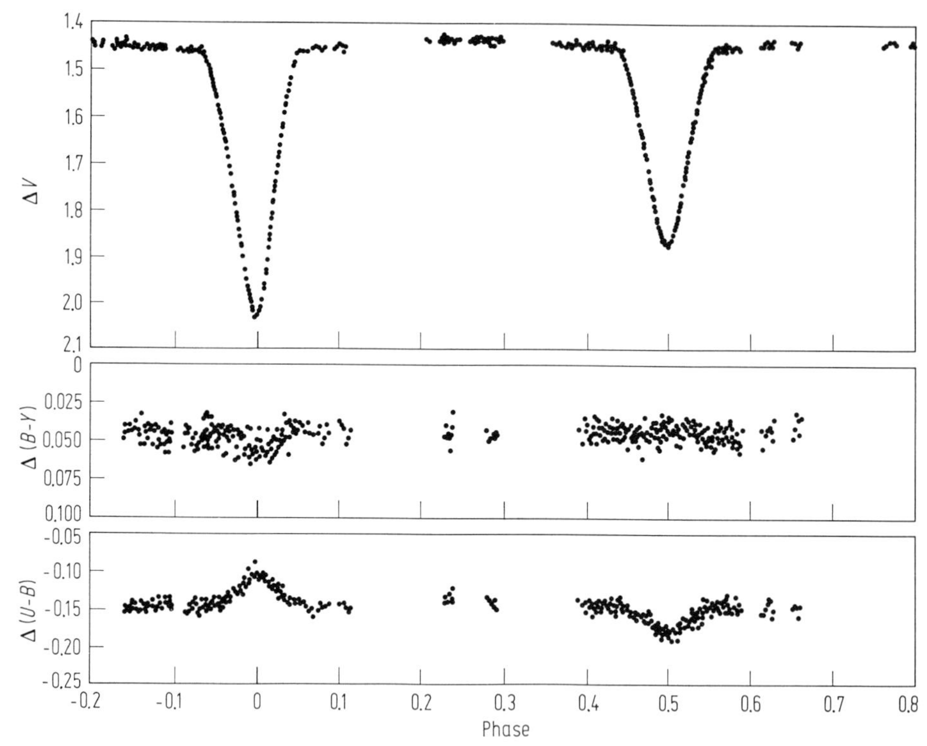

Fig. 25.4. Photoelectric measurements of the detached Algol (EA) system V451 Ophiuchi with Strömgren *uvby* filters. *Top: v* light curve; *center and bottom*: color indices $(b-y)$ and $(u-b)$. All measurements are relative to the comparison star HR 6976. After Clausen, Gimenez, and Scarf [25.10].

evolution from, for example, the observed luminosity ratio. The evolution of the individual binary components is analagous to that of single stars only when the system is sufficiently separated that the structure and evolution of one component is not affected by the other. The structure and evolution are, however, mutually influenced for most close binaries.

Widely separated Algol systems (EA) have nearly spherical components, as each star is little influenced by its companion. The stellar radii are thus small compared with the separation of the centers and with the extension of the inner Roche surface (see Sect. 25.3.1). In such systems, eclipses occur only when the orbital plane is nearly parallel with the line of sight, which means that the inclination is close to 90°. As an example, Fig. 25.4 shows the photometric light curve of V451 Ophiuchi. In this object, the component radii are 15% and 20%, respectively, of their separation. The tidal deformation in these stars is minimal (the flattening is less than 1%) and the reflection effect small. The light curve thus shows nearly constant intensity outside of eclipses.

It should again be emphasized that many eclipsing variables with EA-type light curves are detached systems, whereas the prototype Algol is an interacting semi-detached binary. The formal inclusion of a particular system into one of the various classes has traditionally been based on the shape of the light curve, while the physical structure of the system was a priori unknown, and was deduced only by more detailed analyses of photoelectric measurements. The light curves of close detached and of semi-detached systems do not differ appreciably in shape but show a smooth transition from one to the other, so that distinction between detached and semi-detached systems is not always possible on the basis of the light curve alone.

25.3.4 Ellipsoidal Variables

Detached systems with tidally deformed components nearly filling their Roche lobes may also, even if the components do not undergo mutual eclipses, exhibit a sinusoidal, orientation-dependent light variation. In the case of a partial eclipse, which occurs at inclinations significantly distinct from 90°, the amplitude of the ellipsoidal variation may be larger than the depth of the eclipse minimum. With the exception of widely separated systems, the orientational variation shows up also in all interacting eclipsing pairs between eclipses (near quadrature) as a more or less pronounced, broad "hump," which—in the case of EB and EW systems—contributes a substantial part of the strong continuous variation.

The orientational light variation is caused by the continuously changing aspect along the line of sight. Owing to tidal deformation, the projected stellar surfaces vary in size periodically with the orbital phase, so the effectively radiating areas are largest at orbital phases $\varphi = 0.25$ and 0.75 (quadrature), and smallest near conjunctions $\varphi = 0.0$ and 0.5. Hence, the brightness of the system, which is proportional to the radiating area, varies in a sinusoidal shape with a period one-half that of the revolution. The ellipsoidal light variation is enhanced by the mutual irradiation of the components (reflection effect), which plays a significant role in very close systems.

25.3.5 RS CVn Systems

The stars of RS CVn type are regarded here as a subclass of eclipsing variables, but one with peculiar properties, presumably conditioned by evolution. The recognition of these stars as a separate eclipsing class follows largely from the papers of Oliver, Catalano, Rodono, Bopp, and particularly Hall. RS CVn stars, which are characterized by F- or G-type primary components and periods of from 5 to 12 days, are actually detached systems which should show little interaction. Both the strongly perturbed light curve as well as the Ca II emissions in the spectrum indicate very strong photospheric and chromospheric activity in these stars. There is a noticeable "hump" in the light curve, which migrates with time toward decreasing phase, and is interpreted by Hall as being due to starspot activity. The continuous phase shift is caused by one binary component of inhomogeneous surface brightness rotating nearly, but not precisely, synchronously with the orbital period.

The RS CVn are also "radio stars," emitting radio radiation which is easily observed in the decimeter and meter range; measurable X-ray emission also occurs. This may also indicate enhanced coronal activity.

RS CVn systems are sometimes considered contracting, young objects, but most authors believe that these pairs contain at least one evolved component, a subgiant usually of spectral type K3–K5 IV. In some systems, both components are subgiants. It is likely that the rather rapid increase in radii of these stars early in their evolution towards becoming giants causes their peculiar properties. YY Gem (an M + M main-sequence system of period 20^h) is sometimes included in this class, as well as some W UMa stars, and even Capella (G + K giants with $P = 104^d$). Enhanced spot activity is the presumable cause of some of the similarities between these diverse objects. Nevertheless, the inclusion of such evolutionarily quite diverse objects (with the possible exception of Capella) is rather misleading. The authors share the opinion of Popper and Ulrich [25.11] that while there is no harm in stating that a particular variable has an "RS CVn syndrome," the actual RS CVn class should be restricted more narrowly.

25.3.6 Semi-detached Systems

Algol is a particularly good example of an interacting, semi-detached system. This class was in fact first recognized in connection with the so-called *Algol paradox*. Observations showed that the properties of certain types of eclipsing binaries conflicted with straightforward predictions of evolutionary theory. From about 1940 to 1950, it was found that the low-mass components of Algol systems had much larger luminosities and radii than normal main-sequence stars of the same mass, and hence appeared to be in an advanced evolutionary state. However, the higher-mass companions of these subgiants are still on the main sequence, in spite of the incontrovertible rule that more massive stars evolve faster than less massive ones. These observed results thus

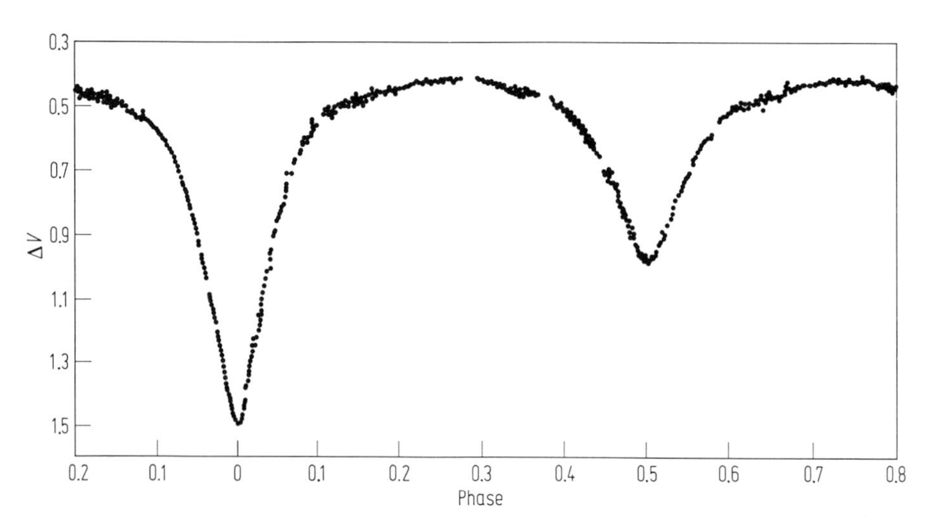

Fig. 25.5. Photoelectric *V* light curve of the semi-detached EB-type system TT Aurigae. The measurements are relative to the comparison star HD 32989. After Wachmann, Popper, and Clausen [25.13].

seemed to contradict the fact that binary-star components are of common origin and of equal age.

Crawford [25.12] recognized that the originally more massive components in such systems had in the course of evolution lost a large fraction of their mass. In Algol systems, this matter was transferred to the companion star or partly lost from the system. Therefore, the mass ratio was inverted and the now lower-mass components are farther evolved than the more massive ones, while still remaining on the main sequence.

A typical example for a semi-detached light curve is shown in Fig. 25.5. It shows the photoelectric light curve of TT Aurigae. The system consists of two early B-type stars; the lower-mass component is cooler, fills its Roche surface (contact component), and has lost so much matter to the companion during the mass-transfer phase that the mass ratio has actually been reversed. This star was thus originally of higher mass than the companion and is at a more advanced evolutionary state.

Semi-detached systems often show signs that mass exchange is currently still taking place. For instance, high-resolution spectroscopy in the optical and UV ranges show the existence of plasma rings, disks, or circumstellar envelopes rotating around the hotter and less-evolved star which accretes matter. Another indication of mass transfer can be derived from the change in the orbital period, which may occur in a secular, continuous, or discontinuous fashion; for example, U Cephei showed a period change of about 20 seconds in 100 years.

25.3.7 Contact Systems

Over 400 eclipsing contact systems are known, but only 10% of these are well studied. EW-type light curves have primary and secondary minima of about equal depth. This shows that both components are similar in surface brightness and temperature, but they in fact are (usually) a pair of stars with different masses and radii. An extreme example is a W UMa-type system, AW UMa, with a mass ratio 1:12 and a ratio of radii 1:3.5; its shape is sketched in Fig. 25.6 and the corresponding light curve in Fig. 25.7. Spectroscopic studies of these short-period objects (typical periods of 5 to 15 hours) normally show two very similar stars of the lower main sequence, but the spectral type is always compatible with only one of the components' size or mass.

The large amount of continuous light variation outside eclipses indicates strong tidal deformation of the two stars. They have a common surface at an equipotential surface located between the inner and outer critical Roche surfaces and which envelopes the entire system. The two stellar nuclei embedded in it may actually have different photospheric temperatures, but the energy flow rising into the common envelope is distributed by a mechanism which has thus far not been clarified, so that the shell has nearly homogeneous temperature. The cause of this effect is often assumed to be currents and turbulent mixing (convection) in the envelope.

Contact systems are divided into two different subclasses: W UMa systems and contact pairs of early type. There are two distinguishing criteria: first, spectral classification separates the W UMa systems distinctly from the early (OB) systems, as they belong to the middle and lower main sequence (F–K, in exceptional cases A). Second, the early-type pairs have larger absolute dimensions and also, owing to the

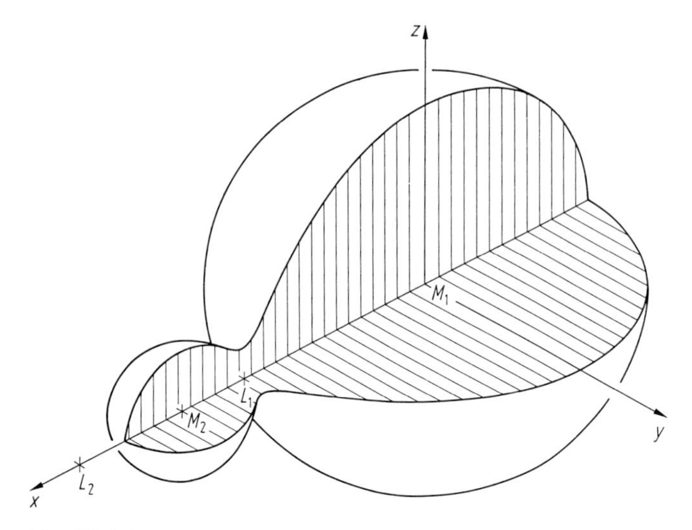

Fig. 25.6. Schematic representation of the shape of the W UMa-type system AW Ursae Majoris. The contact system has an extreme mass ratio of 1:12. After Mochnacki and Doughty [25.14].

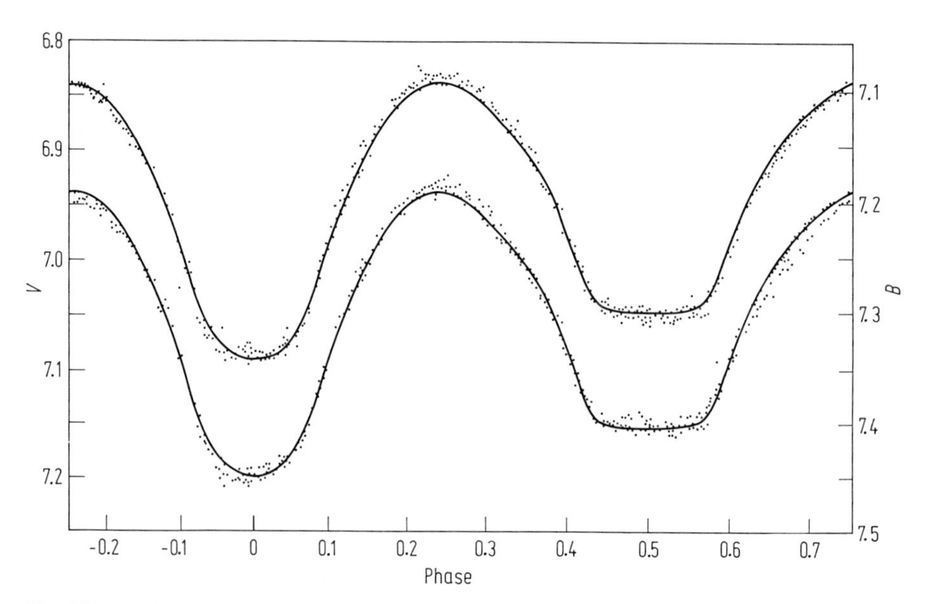

Fig. 25.7. Photoelectric B and V light curves of the EW-type contact system AW Ursae Majoris. The secondary minimum is a total eclipse. After Wilson and Devinney [25.15].

larger distance between the components, longer orbital periods (as a rule, over 1 day) than do the short-period W UMa systems (around 0.2 to 0.8 days).

W UMa systems are further subdivided into *A-types*, where the primary minimum corresponds to the occultation of the more massive, larger component by the smaller companion (transit minimum), and *W-types*, where the primary minimum is an occultation minimum when the smaller star is totally eclipsed by the larger one. Since the smaller star has a relatively higher surface brightness and higher temperature than the more massive one, the total brightness of the system during occultation is less than at transit.

Most contact systems contain unevolved main-sequence stars, but there do exist some evolved objects, such as the W UMa-type system ε CrA or the early-type contact system SV Cen. The outer shells of W UMa stars are convective; also the components rotate considerably faster than do the corresponding single stars with the same spectral type. The dynamo effect thereby gives rise to chromospheric and coronal activity in the outer atmospheric layers, as plasma flows along closed magnetic loops originating within the stellar interior. This leads to the emission of lines and continuum in the UV and X-ray regions.

A detailed discussion of observations and theoretical models for contact systems may be found, for instance, in Rucinski [25.16].

25.4 Pulsating Variables

A major group of physically variable stars are the objects which show periodic, or at least cyclic, brightness changes between a minimum and a maximum level, and usually without a well-defined "rest" magnitude. Sometimes, these brightness oscillations have a regularity which is comparable in precision to that of a quartz clock (δ Cephei), sometimes of much lower precision, but still fairly regular (Mira Ceti), and sometimes barely identifiable as cyclical (RV Tauri). Such stars are generally called *pulsating variables*. The name indicates that periodic brightness changes are caused by vibrations of the stellar body, where the stars expand and contract, usually in the radial direction, with periodic repetition. The brightness variations are merely one aspect of these changes, albeit the most easily observable one. The radial velocity and surface temperature vary with the same period, and this supplies an unambiguous proof of the pulsation process, namely, the periodic change in volume. In addition to the radial pulsations, other kinds of vibrations may occur. For instance, surface waves moving around the star can cause a periodic change in brightness of the stellar disk, or torsional-type deformations of the stellar body may occur. These *non-radial pulsations* will be only briefly mentioned; the following discussion concentrates mainly on radial pulsations.

Several important representatives of pulsating variables had already been discovered in the 17th and 18th centuries, but the idea that a star could pulsate was only gradually accepted. Although a simplified theory of pulsations can be found in some remarkable papers by A. Ritter dating back to 1879, including a basic connection with the star's pulsation period and mean density ($P\overline{\rho}^{1/2} \approx$ const), the idea of pulsations was not generally accepted. An unfortunate superficial similarity of observed radial velocity curves with those of spectroscopic binaries led to a binary-star model, which was wholly abandoned only in the 1920s. Even today, classification and terminology are somewhat confused. Astronomers often speak quite generally of "Cepheids" or "classical Cepheids" when actually referring to the whole class and not only a specific δ Cephei-type variable star. The term "Cepheid parallax" is widely used and indicates the use of pulsating variables as distance indicators.

25.4.1 The Physics of Radial Pulsations

As has been mentioned, all the important parameters of pulsating stars change as they pulsate: the observed brightness variations with time result from periodic changes in radius and effective temperature.[4]

From the observed spectral type as well as from the color, changes in the surface temperature can be estimated while the rate of change in radius dR/dt is derived from the observed radial velocity curve. The latter, upon integration, yields the difference $R_{\mathrm{max}} - R_{\mathrm{min}}$. From these data then follows the change in luminosity (total radiation per second), which for most pulsating variables is smaller than the brightness change in the visual or photographic ranges. This difference is particularly striking in the very cool, long-period variables of the Mira type. In the course of their cycle, the radiated total energy changes by at most a factor 2, while the change in the visual brightness may reach 7 or 8 magnitudes, or factors of about 1000. The explanation is found in the fact that in these extremely low-temperature stars the maximum intensity of radiation lies far beyond the visual range in the infrared, and also that at the lower temperatures additional strong absorptions from molecules enter into the spectrum. Dust formation also could play a role.

The previous comments suggest that in Mira-type stars the lowest surface temperature corresponds with a brightness minimum; this also holds for all other radially pulsating variables. Figure 25.8 demonstrates the change of various stellar parameters during one pulsation cycle of δ Cephei. In particular, the brightness change follows the temperature variation rather closely, whereas the maximum radius (see the ΔR curve in Fig. 25.8) does not occur at the time of maximum light but rather halfway between maximum and minimum on the descending light branch. This is to be expected since the luminosity depends on the *fourth* power of the temperature, but only on the *square* of the radius. It is to be kept in mind that the effective temperature constantly declines while the spectral type changes from A to F to G to K, or from about 10 000 K to about 5000 K. The longer the period, the larger also will be the brightness amplitude (see Fig. 25.9).

The rather unexpected picture that stars as gaseous spheres can perform radial oscillations becomes more plausible when considering that these structures have a kind of "spring suspension" connected with their hydrostatic equilibrium. At each point in

---

4 The effective temperature is the temperature of a blackbody which radiates the same total amount of energy from the same surface area. The total amount of radiation emitted per second (i.e., the luminosity) by a star is $L = 4\pi R^2 \sigma T_{\mathrm{eff}}^4$, where $\sigma = 5.67 \times 10^{-8}$ J m^{-2} K^{-4} s^{-1} is the Stefan–Boltzmann constant.

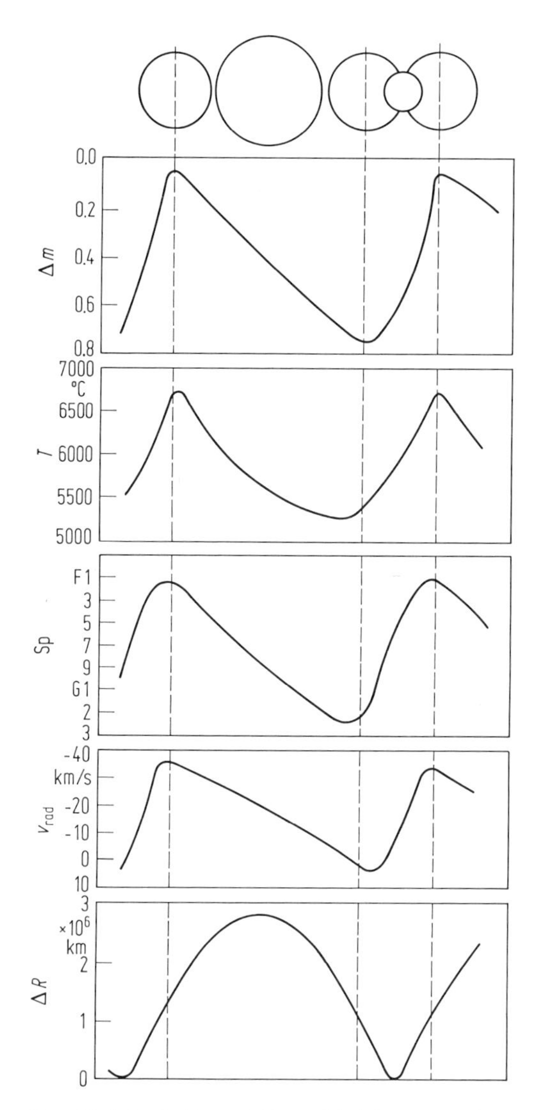

Fig. 25.8. Periodic variation of brightness, surface temperature, spectral type, radial velocity, and radius during one pulsation cycle of δ Cephei.

the stellar interior, the gravity directed inward is compensated by the outward pressure. Were this not the case, stars would change rapidly and dramatically. A difference of only 1% between gravity and pressure would lead to an expansion or contraction of the Sun by 5% of its radius on a time scale of less than one hour. This expansion or contraction, however, does not last (supernovae excepted); "elastic" oscillations will occur instead. This process can be explained as follows.

In hydrostatic equilibrium the pressure in the star increases monotonically from surface to center. The stellar temperature and density likewise increase monotonically

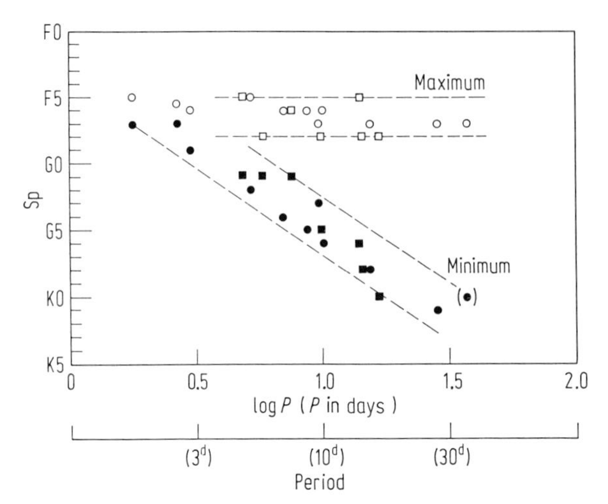

Fig. 25.9. Differences in spectral type between maximum brightness (*open circles* and *squares*) and minimum brightness (*filled circles* and *squares*) for 18 δ Cephei stars with different pulsation periods. The luminosity class is Ib in almost all cases. After Code [25.17].

inward. The increase in these quantities is such that short-term perturbations cannot grow. A local expansion of gas, for instance, moves the gaseous material into regions of lesser gaseous pressure, so that gravity predominates locally. This halts the expansion, and initiates an inward motion toward the original state of equilibrium. The inwardly moving gases overshoot the original position and subsequently encounter higher pressures which now overcome gravity. This pressure halts the motion of the gas and then accelerates it in the reverse direction, again radially outward. The repeated overshooting of the level of equilibrium causes the process to recur *periodically*, so the star will oscillate. The interaction of variations in temperature and radius during a cycle of pulsation follows is such that the instants of maximum temperature (and brightness) on the one hand, and of maximum radius on the other, are separated by a phase difference of about 90° (1/4 period), as can be seen in Fig. 25.8. This phase shift is an important criterion for the pulsation theory.

If there were no energy exchange between the system of moving gas and the surrounding gas (i.e., an adiabatic process), these vibrations would continue indefinitely. This is analogous to oscillations of a simple pendulum which, in the idealized case of no irreversible loss of mechanical energy by friction and air resistance, also continue without limit.

The comparison with the simple pendulum is not as far-fetched as it may at first seem.[5]

As in the case of a pendulum, a certain amount of mechanical energy is always dissipated into the surrounding environment. This damps the motion and continuously diminishes its amplitude until the pulsations cease at the equilibrium state. Most stars

5 A more complete, yet non-technical, presentation may be found, for instance, in Kippenhahn and Weigert [25.18], in German.

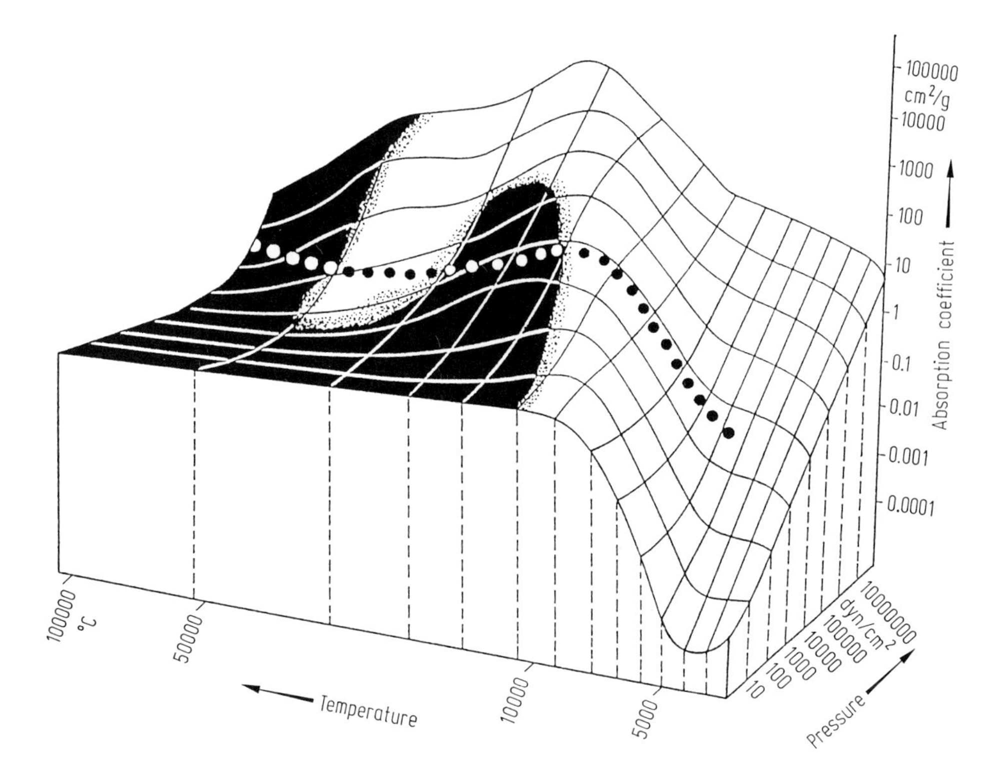

Fig. 25.10. The behavior of opacity (absorption) in the outer regions of stars. This three-dimensional model graphs absorption vertically as a function of both temperature and pressure (or density). The string of dots corresponds to the structure in a δ Cephei star from the surface (*right end*) into deeper layers. The opacity first increases inward, and then decreases again (the "opacity" or "kappa mountain"). Pulsations may be triggered in the white regions, while the dark regions, in which opacity decreases inward, have a damping effect. After Kippenhahn and Weigert [25.18].

are stable against small perturbations, insofar as all vibrations are damped around the equilibrium point.

In pulsating variables, on the other hand, such vibrations are evidently not damped, which leads to the basic question as to what maintains such undamped oscillations. This problem has engaged astrophysicists for over 30 years. On the one hand, it became clear that such vibrations cannot be related to possible periodic variations of the energy production in the interior. On the other hand, Eddington had as early as the 1920s suggested the following explanation.

If the vibrating layers receive some energy at the appropriate phase, this "surge" of power could counteract the damping effects. As an analogy, the notion of a simple pendulum could again be adduced, this time using a pendulum clock: at the moment of release, the pendulum receives a small push from the pressure exerted by the driving force, namely the weight or the spring. In pulsating variables, however, a certain interaction of the gaseous matter with radiation plays the crucial role. Energy flows constantly as radiation through the entire interior from center to surface, and it is continuously absorbed and re-emitted by the stellar matter. The rate of absorption by the gas, its resistance, so to speak, against the radiation flux, is called the *opacity*. Should the opacity then also increase during the compressional phase of the oscillation at increased temperatures and densities, then the corresponding layer could absorb more radiant energy, whereas in the expanding phase, less energy would be absorbed. Since absorbed radiation is linked with an outwardly directed impulse, this interaction of the radiation and oscillating gas layers would produce the desired "push." At the time of the original suggestion, it was believed that the opacity in stars decreases monotonically from the surface to the center. This would mean that the opacity decreases through compression and increases through expansion, which would produce exactly the opposite of the desired effect. Eddington had abandoned his basically correct idea because of this contradiction. In the 1950s it was recognized that the run of opacity not far below the surfaces of some cool stars may be substantially different than was formerly assumed. Here, the ionization of helium, the second most abundant element, becomes important. It may cause a reversal of the formerly postulated, monotonically decreasing opacity; the latter first increases downward and later decreases toward the center. Under these conditions, the described "valve" mechanism may operate quite efficiently in the manner first emphasized by the Soviet researcher Zhevakin. Compression thus does increase the opacity in the outer stellar layers, and this causes a "negative energy dissipation," as Zhevakin referred to it. It compensates the damping, and re-triggers the vibration continuously.[6] Pulsating variables thus in this sense are *not* stable against small perturbations. This explanation of undamped radial pulsations via the "kappa mechanism" (so named because the coefficient of opacity is denoted by the symbol κ) was adopted and further developed by theorists.

As an example, Fig. 25.11 shows the results of numerical calculations by Christy for the start and the development of pulsations on a suitable star model. The similarity with observed light curves is rather convincing. The theory of radial pulsations links the occurrence of pulsational instability with the internal structure of the star. Most stars do not show the instability, and the ability of theory to explain reasonably well where in the HR diagram pulsating stars are expected is remarkable. Figure 25.12 shows those regions where the various types of variables are found. Pulsating variables occur predominantly in the region surrounded by a dashed line (instability strip) with the following four most prominent types:

- RR Lyrae stars with periods under 1 day;
- W Virginis stars with periods between 1.5 and 25 days (both these and RR Lyrae types being old, metal-poor Population II stars with masses around 1 $\mathfrak{M}_\odot$);
- δ Cephei stars with periods between 2 and 40 days, young, massive stars (with 10 to 20 $\mathfrak{M}_\odot$) of Population I;
- Mira stars or long-period variables with periods of about from 150 to 500 days, stars of the older Population I.

6 The role of this "thermodynamical push" in overcoming damping may become more understandable when considering that the pulsation energy is very small, by four orders of magnitude, compared with the total thermal energy of the star.

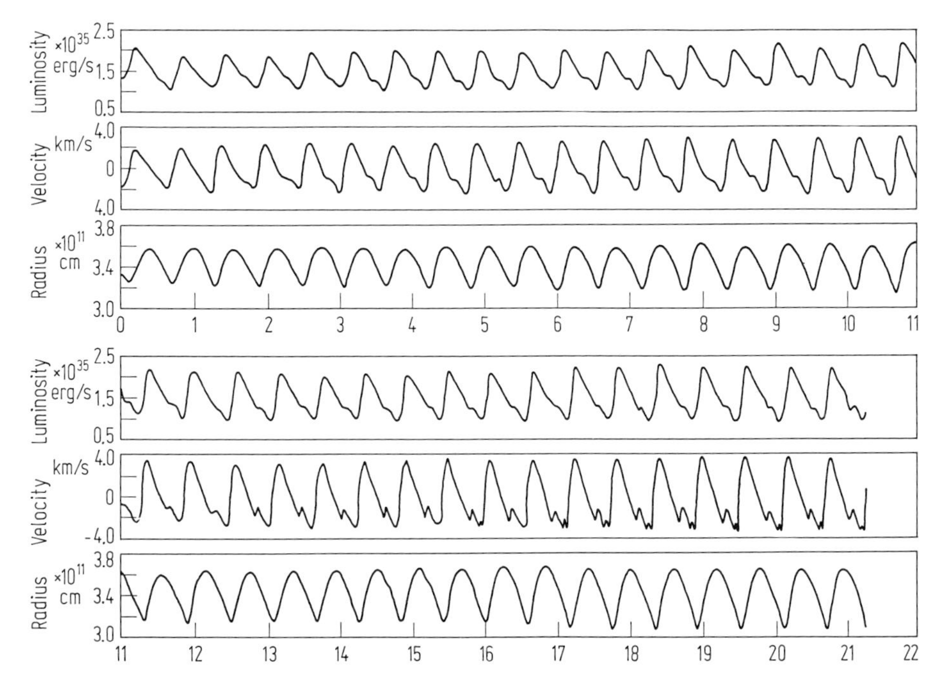

Fig. 25.11. Computer simulation of pulsations in an RR Lyrae star showing the run of luminosity, velocity, and radius of a near-surface layer for the first 20 pulsations after the instability has been introduced. After R.F. Christy [25.19].

The diagram illustrates that one can indeed identify the strip of instability or pulsations. After the main-sequence phase, the evolution of these stars advances above the main sequence in the giant and supergiant regions. Depending on the mass, these stars may reach the instability strip and leave it again after 10^6 to 10^7 years. Massive stars may traverse the strip repeatedly.

Most pulsating variables are giants or supergiants which are suffering substantial mass loss, primarily in the form of a strong stellar wind with a low terminal velocity ($\approx$ 10 km s^{-1}). Pulsations enhance the mass loss, which is then called a *pulsationally conditioned stellar wind*. This may explain why the masses for δ Cephei stars computed from pulsation theory are distinctly less than would be expected by comparison with supergiants of similar luminosity and temperature. Enhanced mass-loss rates in the instability strip are also suggested by line profiles in the spectra of some Cepheids. Unfortunately, the observed mass-loss rates are rather uncertain.

25.4.2 The Period–Luminosity Relation; Cepheids as Distance Indicators

Cepheids and other pulsating stars have achieved a particular significance in astronomy because of their use as distance indicators or "standard candles." This means that the absolute magnitude (i.e., luminosity) of such an object can be found by directly and

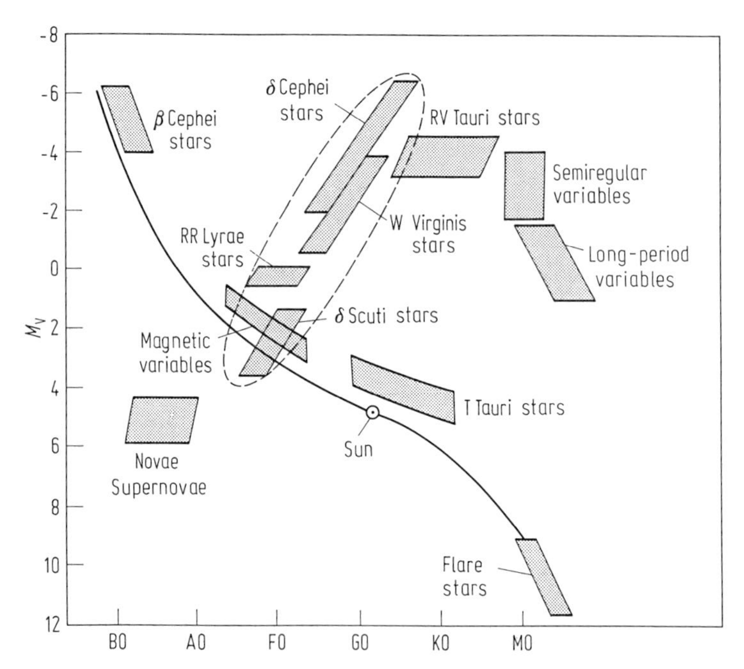

Fig. 25.12. Position of various classes of variables in the Hertzsprung–Russell diagram. The instability strip, which contains most pulsating variables, is bounded by the dashed line.

relatively easily observed properties. The direct observation of the apparent magnitude then leads to the distance. The term *distance modulus* and the addition of the interstellar absorption $A(r)$ are explained in Sect. 24.4.

Novae are in a sense also potential standard candles because, as will be seen, the luminosity of a nova can be estimated from the shape of the light curve. The information supplied by the light curves of Cepheids through the correlation between period and luminosity is much more precise, so that pulsating stars are presumably the most dependable distance indicators. Since δ Cephei-type stars are also supergiants of high absolute luminosity, they can be observed in nearby galaxies. In fact, Hubble in 1923 proved conclusively that M31, M33, and some other "nebulae" are in reality extragalactic star systems. Using the 2.5-m telescope at Mount Wilson, he discovered Cepheids in these galaxies, and determined their distances by way of the period–luminosity relation.

Actually, this important relation includes only RR Lyrae, δ Cephei, and W Virginis stars; the correlation for Mira variables is different and weaker. The period–luminosity relation had been discovered before radial pulsations became generally accepted as the cause of the characteristic light variations: Henrietta Leavitt at the Harvard Observatory had noted in 1912 that the apparent brightness (at both maxima and minima) of Cepheids in the Small Magellanic Cloud was systematically higher for stars with longer periods. The relation was found to have the form $m = a + b \log P$, and it soon became evident that a similar relation holds for absolute magnitudes, since all stars of the SMC have approximately the same distance from the solar system and hence the same distance modulus. The slope of the line in the $(m, \log P)$ or $(M, \log P)$ diagram is simply determined and yields results which are about the same for the SMC, the LMC, M31, M33, and the Milky Way. To find the distance modulus, however, the distance modulus $m - M$ for the Magellanic Clouds would have had to be determined independently of the pulsating stars, and this was not possible. If, on the other hand, the accurate distance had been known, in principle, for just one Cepheid (in practice for several), then the zero point (the quantity a) of the period–luminosity relation could have been stated and the relation then used for distance measurements in the nearer extragalactic realm. This turned out to be a difficult task. Not one pulsational variable, for instance, is so close that its trigonometric parallax could be measured. Indeed, for almost 40 years a substantial error in the zero point of the relation gave incorrect values for all extragalactic distances. Only in 1952 was the error corrected by W. Baade. As a consequence, the accepted distance to the Andromeda Galaxy (M31) was suddenly doubled. The French astronomer H. Mineur had much earlier pointed out a possible error in the zero point.

The statistical method used until then turned out to be impracticable because the proper motions were too small. Another important source of error was in failing to distinguish between δ Cephei- and W Virginis-type stars. The old, metal-poor W Vir types are not as luminous as the younger, massive δ Cephei types. Besides, the zero point must be found to an accuracy of 0.1 mag in order to determine distances to 5%. The zero point could also be derived from pulsation theory, but this would not reach the requisite accuracy. Even today the uncertainty of the zero point has not been reduced to a satisfactory level. The primary current methods are the following: δ Cephei-type stars often occur, in contrast to RR Lyrae stars, in open clusters. The distance of the open cluster can be found from multi-color photometry, and this yields a fairly reliable distance for those Cepheids which are cluster members. The zero point was derived in this fashion by Sandage and Tammann in 1969 from 15 Cepheids in open clusters. Another method is based on an analysis of radial velocities (which gives the change of stellar radius) together with the variation of spectral type (the Baade–Wesselink or Barnes–Evans methods). Sometimes the Cepheid may have a hot subdwarf companion whose distance can be estimated. A recent survey paper by Feast and Walker [25.20] reached the following comprehensive results for absolute magnitudes:

$$< M_B > = -1.16 - 2.24 \ \log P, \qquad (25.9a)$$

$$< M_V > = -1.35 - 2.78 \ \log P, \qquad (25.9b)$$

and in the near-infrared

$$< M_H >= -2.14 - 3.42 \ \log P, \qquad (25.9c)$$

where P is expressed in days. The $<>$ symbol indicates that the brightness is averaged over the light curve. (Usually the light curve is expressed in intensities rather than in magnitudes, and the intensity averaged over the period.) These formulae give $m - M =$ 18.5 or $r = 50$ kpc for the LMC, which is perhaps a little too low. A particularly strict period–luminosity relation is exhibited in the infrared; Fig. 25.13 gives the correlation

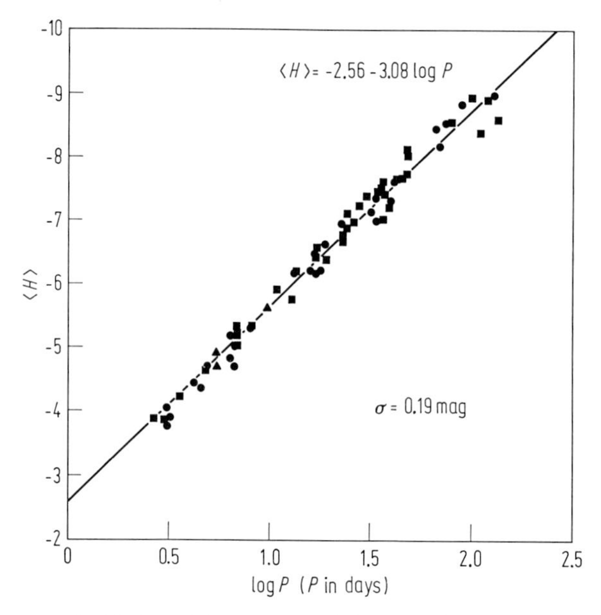

Fig. 25.13. Period–luminosity relation for Cepheids in the infrared (H-band at 1.6 μm). The graph includes data from Cepheids in the Milky Way and in the Magellanic Clouds. After Welch et al. [25.21].

for the H-band by Welch and coworkers [25.21], which is somewhat different from the previous solution, but again yields a distance modulus of 18.5 magnitudes for the LMC.

All of these formulae refer to the "simple" period–luminosity relation. A careful analysis of the data also reveals a small dependence on color. Thus, a period–luminosity-color relation can be established, even a potential dependence on the metallicity of the variable included. This complete period–luminosity relation has the following form:

$$M = M_0 + \alpha \log P + \beta(B - V)_0 + \Phi, \tag{25.10}$$

where $(B - V)_0$ is the color index corrected for interstellar absorption and Φ the *metallicity index*. These additional dependencies are not very strong, and thus in many cases it suffices to employ the simple period–luminosity relation.

The W Vir stars follow a somewhat different relation to both slope and zero point, although the scatter around the median line is somewhat larger. According to an earlier compilation by Dickens and Carey [25.22], W Vir-type stars are fainter than "classical" Cepheids by 1.5 mag for periods of $2\overset{\mathrm{d}}{.}5$, and by 2.5 mag around 20^{d}. For RR Lyrae-type stars, an absolute magnitude that is nearly period independent, but which exhibits a much larger scatter, is found: $\overline{M}_V = +0.6$.

25.4.3 Periods and Light Curves

Period values and changes in the period and in the shape of the light curve are of particular importance in pulsating variables, as they supply basic information on the structure of the stars and serve to test various stellar models.

Changes in the periods of pulsating stars are analyzed with the aid of an O–C diagram; for a short description of the method, see Sect. 8.5.5.6 in Vol. 1. For multiple-period pulsating variables (see below), Fourier-type and similar analyses of the light curve are also often employed; these methods give a "spectrum" of the possible pulsation periods.

Classical δ Cephei stars show nearly constant periods as well as quite stable light curves. As an example consider the prototype δ Cephei with $P = 5^{\mathrm{d}}.366$: the observed maxima indicate a continuously decreasing period since the discovery in 1785. A recent comprehensive study of the material to 1980 by Szabados [25.23] leads to $\Delta P = -0^{\mathrm{s}}.089$ per year (see also Fig. 25.14).

These data lead—after extrapolating with a constant or near-constant rate of change—to a *characteristic time* $P/\Delta P$ of about 5 or 6 million years. During such a long span of time, the period should change substantially. This time interval is somewhat longer than the theoretically computed transit times through the instability strip, but at least it is on the same order of magnitude. Hence, it is possible that the period change indeed represents the current evolution of δ Cephei. However, period changes in other δ Cephei-type stars appear to be more complex and ambiguous. CV Mon, with a very similar period of $5^{\mathrm{d}}.379$, shows apparently discontinuous behavior with abrupt period changes (Fig. 25.15), whereas the period of $5^{\mathrm{d}}.441$ of SW Cas has remained constant to within $0^{\mathrm{s}}.2$ over 65 years of observation.

The light curves of a few δ Cephei stars with shorter periods, between $2^{\mathrm{d}}.1$ and $3^{\mathrm{d}}.7$, exhibit a remarkably high scatter. This peculiarity, first recognized by Oosterhoff, can be explained by multiple periodicity of the light curve.

The W Vir stars, distinguished from classical Cepheids by their Population II membership, show somewhat less-stable periods and light curves. Abrupt period changes and disturbances in the light curve are more frequently encountered. In the surprising case of RU Cam ($P = 22^{\mathrm{d}}.1$), the pulsation disappeared almost completely for two years during 1964–66. Today, the star pulsates as it did until 1963, with only a slightly different period.

Typical light curves of two δ Cephei stars are compared in Fig. 25.16 with that of the prototype W Vir. Figure 25.17 shows other δ Cephei-type light curves arranged according to period. The systematic change in the shape of the light curve with increasing period is referred to as the *Hertzsprung progression*. Very characteristic changes, the interpretation of which is still controversial, are also found to occur among the short-period pulsating variables, especially the RR Lyrae stars, which, like W Vir stars, are also members of Population II. For about one-fifth of RR Lyrae stars, a characteristic change, named the *Blazhko effect* after its discoverer, is found. The light curve changes continuously and oscillates with a *beat period* of 10 to 100

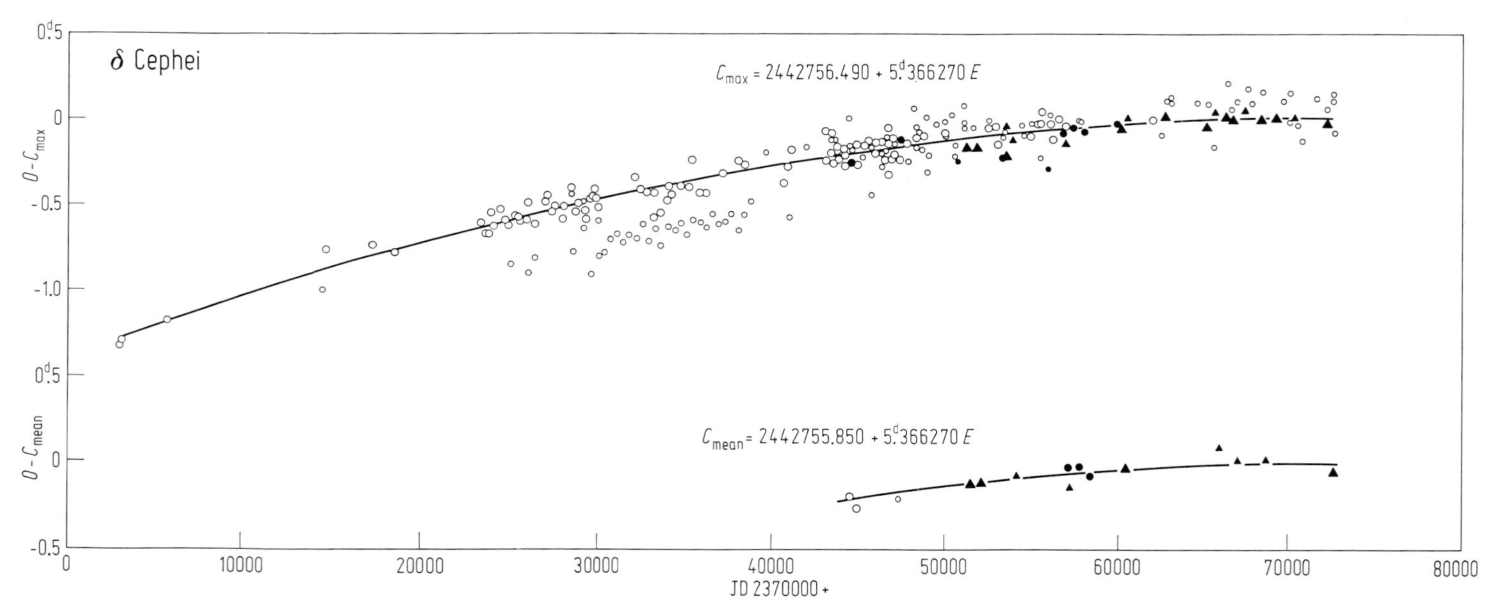

Fig 25.14. O–C diagram for epochs of maximum brightness (*top*) and mean brightness (*bottom*) of δ Cephei for the time since the star was discovered to be variable. A parabolic fit to the O–C values is convincing. After Szabados [25.23].

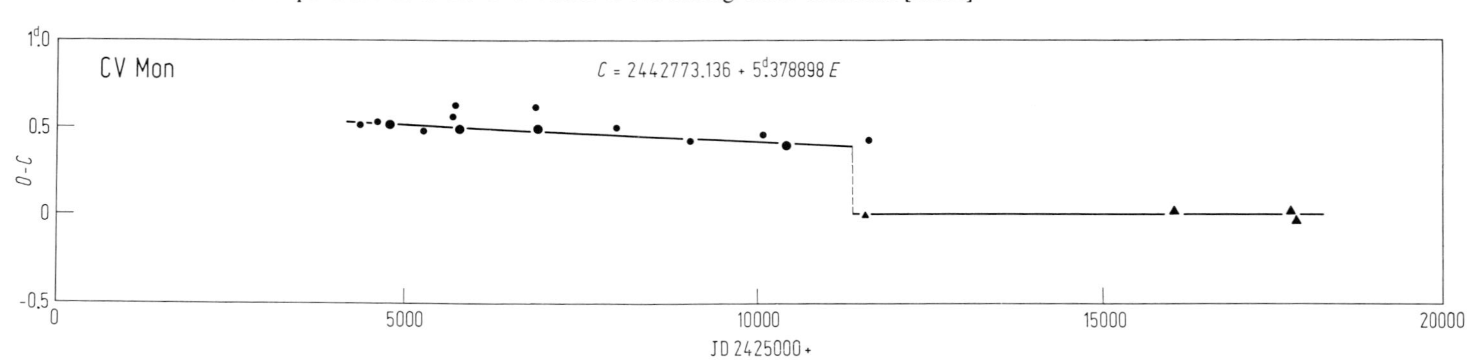

Fig. 25.15. O–C diagram for the δ Cephei-type star CV Mon. A discontinuous period change was observed in 1958. After Szabados [25.23].

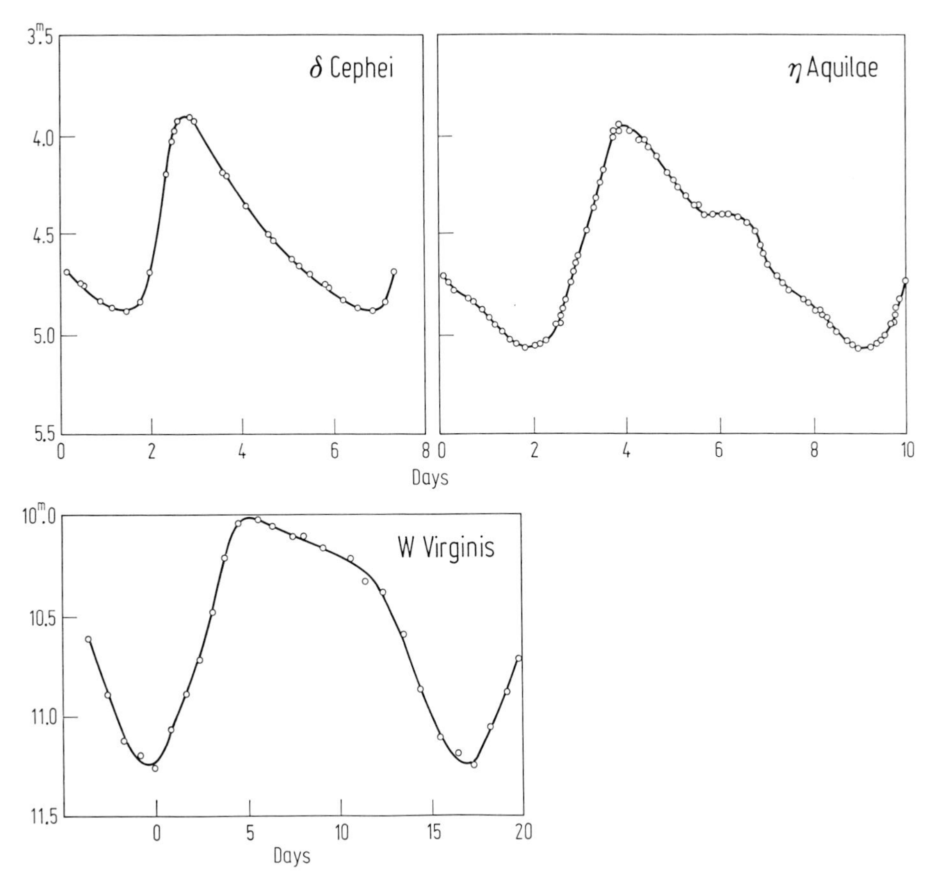

Fig. 25.16. Typical light curves of the two δ Cephei-type stars (*top*) and the prototype W Virginis (*bottom*). After Campbell and Jacchia [25.25].

days around a mean curve of type RRab.[7] This secondary period and any related in the shape of the light curve are explained by the interference of two periodic light curves with nearly, but not quite, identical periods (i.e., beats). Among the brighter RR Lyrae stars, RR Lyr, AR Her, AC And, and RW Dra are examples which show the effect especially clearly. As an illustration, an earlier study of RW Dra by Balázs and Detre [25.24], which had clearly documented the Blazhko effect, is quoted. The regular cycle of light variations (see Fig. 25.18) has a length of $31\overset{d}{.}5$, or exactly 67 periods of the normal light variation; the period is given by these authors as $11^{h}17^{m}$ (= $0\overset{d}{.}47014$). Interpreted as a beat period, the cycle of light curve changes can be generated by superposing a neighboring period of $0\overset{d}{.}46323$ onto the basic period

7 The shape of a light curve in a case classed as RRab is the usual Cepheid form with steep ascent to maximum and a slower descent. The former distinction between a and b depending on the height of maximum has been dropped. RRc-type light curves are more symmetric and almost sinusoidal.

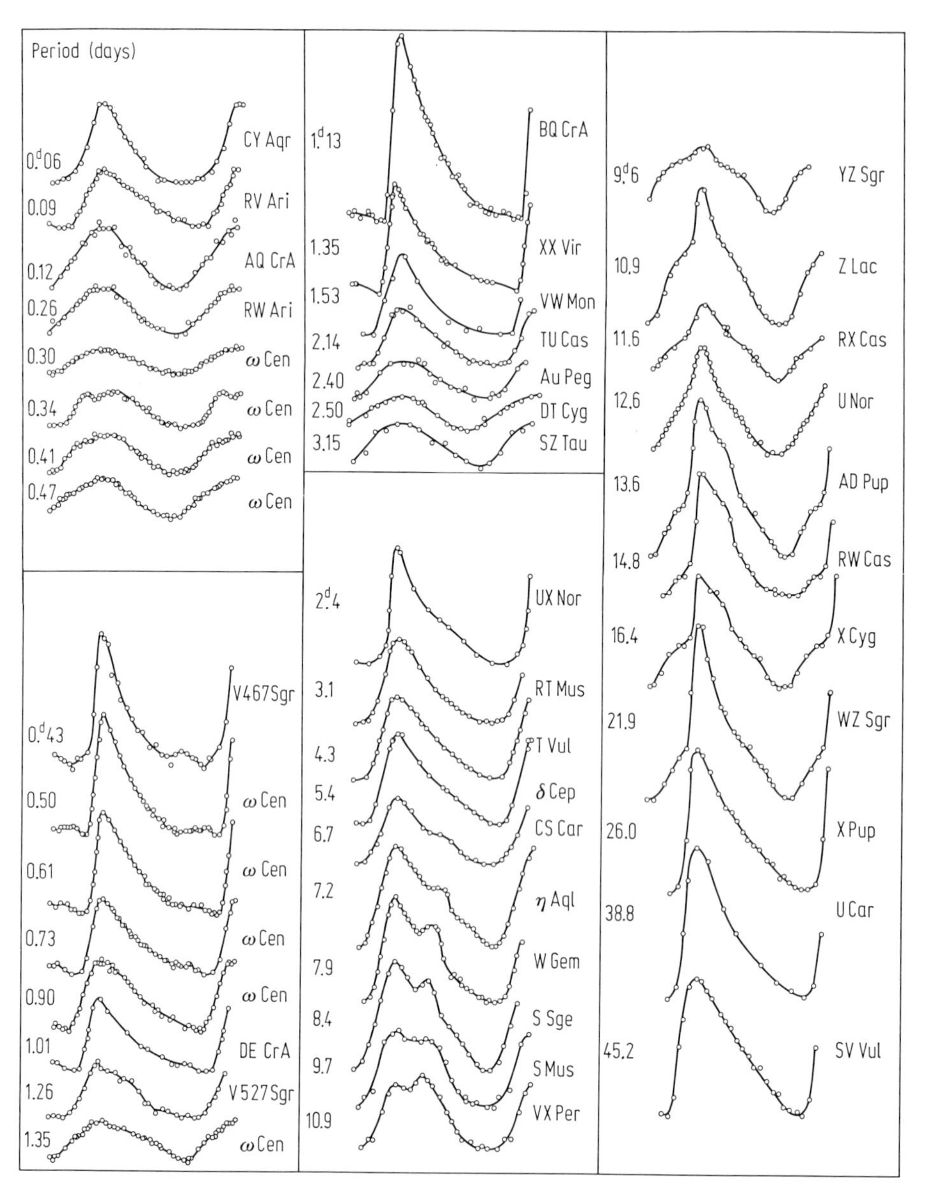

Fig. 25.17. Light curves of δ Cephei-type stars. One can clearly see how the shape of the curves change systematically with the pulsation period (the Hertzsprung progression). After Campbell and Jacchia [25.25].

→

Fig. 25.18. Illustration of the Blazhko effect: a periodic oscillation of the shape of the light curve of RW Draconis around the mean light curve (*dashed*). The curves graphed are observed at four different phases ($\Psi = 0.10, 0.30, 0.55$, and 0.90) of the 41-day secondary period. After Balázs and Detre [25.24].

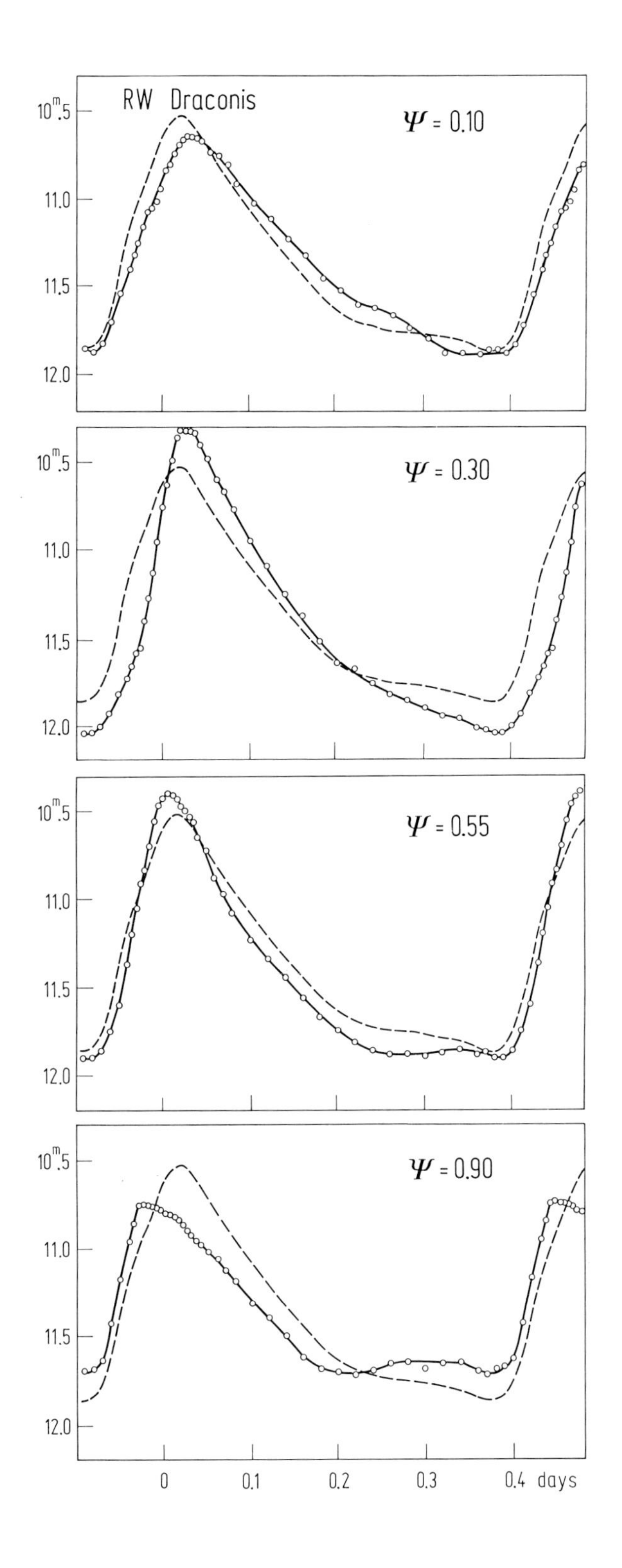
RW Draconis
Ψ = 0.10
Ψ = 0.30
Ψ = 0.55
Ψ = 0.90
10.5
11.0
11.5
12.0
0
0.1
0.2
0.3
0.4 days

mentioned ($68 \times 0\overset{d}{.}46323 = 31\overset{d}{.}5$). In cases where beats occur, the periods obey the "synodic" relation

$$\frac{1}{P_B} = \frac{1}{P_1} - \frac{1}{P_0}, \tag{25.11}$$

where P_B is the beat period. The light curves have a time-varying shape; the height of maximum, for instance, will vary depending on whether an equal-phase (high maximum) or counter-phase (low maximum) occurs in the period superposition. In AC And, the height of maximum may vary by a factor of two.

Some RR Lyrae stars (e.g., RZ Cep) show evidence of abrupt period changes. According to Prager, the period of about $7\overset{h}{.}5$ of that star shortened in 1901 August by $3\overset{s}{.}98$, but increased in 1916 November by $4\overset{s}{.}37$, and again by $1\overset{s}{.}84$ in 1923 December. This kind of apparently irregular change of period is still somewhat controversial in the literature (see also below, the "instantaneous" elements).

The *Mira-type stars* do not show very strict periodicity. The duration of successive periods (or perhaps cycles?) may easily differ by 5 to 10 days (about 2% to 4%). Valid mean periods can be deduced from longer time spans (10 to 20 years). Also, the light curves almost never repeat precisely: Mira Ceti, for instance, varies on the average between apparent magnitudes +3.0 and +9.0, but occasionally reaches +2 at maximum and +10 at minimum.

As early as 1667, Bouillard (Bullialdus) had found a fairly good mean period of 333^d for Mira Ceti, the prototype of this class. More recent discussions, for instance in the renowned bibliography *Geschichte und Literatur des Lichtwechsels veränderlicher Sterne*, give $331\overset{d}{.}6$ for the mean period. It is doubtful that the small discrepancy constitutes a secular change of period over 400 years. However, in two long-studied Mira-type stars, R Hydrae and R Aquilae, distinct decreases in the period to the amounts 30 and 50 days per century, respectively, were found. The future will show how these cases "develop." Guthnick and Prager introduced the term *instantaneous elements* to indicate elements (periods and zero epochs) valid only between abrupt changes. For Mira itself, Prager [25.26] had given a sequence of no less than 12 instantaneous elements. On the other hand, in 1898, Guthnick had suggested a harmonic analysis for the period pattern of Mira Ceti, with four significant sine terms. So far, none of these harmonic expressions for the period has been definitely proven.

This subsection will be concluded with some comments on Mira-type light variations. Upon converting a Mira light curve from a magnitude (logarithmic) scale into a "linear" curve, it may be seen immediately that brightness variations occur predominantly (up to 75%) near the maximum. This maximum corresponds to the phase of a rising shock wave reaching the surface. Thus, in contrast to classical Cepheids, changes in the light output of Mira-type stars are not necessarily caused by pulsations alone. The lesser stability of Mira-type light curves and the larger scatter of maximum epochs may also be explained by this fact.

In the context of periods of pulsating variables, Sterne and Campbell [25.27] have made an interesting suggestion which applies in general to all variable stars. They assumed that most O–C curves of variable stars represent not so much real changes as "error accumulations" due to statistical processes. Small, irregular period variations, as well as measuring errors, can accumulate and then mimic a systematic O–C curve. This mode can actually generate O–C curves which are very similar to those of some long-period or semi-regular variables. Presumably the observed deviations from computed maximum and minimum times for a number of variables represent nothing more than such statistical fluctuations. The existence of real, *physical* changes of periods cannot, however, be denied.

25.4.4 Non-radially Pulsating Stars

Under the class of *non-radially pulsating variables* are three main subclasses, which may contain some objects pulsating in radial as well as in non-radial modes. This class of not always well understood variables will receive here only a brief description.

To illustrate such oscillations, we borrow from a recent monograph of Unno et al. [25.28]: "Stars are like musical instruments and generate various oscillation modes and keys." When such a musical instrument is "played," that is to say, when the equilibrium in a β Cephei-type star is disturbed in some way, numerous (theoretically, infinitely many) vibrations are generated. In reality, only the simpler forms will appear. The mathematical treatment of the phenomenon proceeds in a fashion analogous to that of acoustic vibrations, and the separate modes of oscillations are characterized by *eigenfunctions* described by integer indices. For these, the letters l, m, and k are often used. Simple radial oscillations of the entire surface layer of a star are expressed by $l = 0$, while $l = 1$ represents dipole oscillations (when the hemispheres vibrate separately), $l = 2$ the quadrupole oscillations (with four separately vibrating spherical segments), and so on. The index m describes the behavior of the various surface parts, and k is connected with the nodal points. Skipping further refinements, two frequently occurring cases may help exemplify the matter: $l = 2, m = \pm 1$, and $l = 2, m = \pm 2$ (see Fig. 25.19). Both cases represent quadrupole oscillations, i.e., vibrations of a surface which has been divided into four parts. When $m = \pm 1$, these parts oscillate antisymmetrically, meaning that "northern" and "southern" hemispheres are in opposite phase, whereas $m = \pm 2$ indicates equal phases of the hemispheric vibrations. Both modes correspond to a revolving wave, as these surface patterns rotate around the axis of a star with a period different from that of the stellar rotation. The sign of m indicates if the vibration pattern moves with or against the direction of the rotation of the star. In the case $m = \pm 2$, the radial motions are small or absent, so that the phenomenon consists of regions of different brightness traveling around the star.

Such oscillations evidently affect the overall brightness of the star but in particular the line profiles and hence the measured radial velocity, in a quite complicated fashion. It is not surprising that the periods of β Cephei-type stars also show a complex picture.

25.4.4.1 δ Scuti Stars. Like the other non-radial pulsators, δ Scuti stars have short periods (under $0\overset{\mathrm{d}}{.}3$) and normally very small amplitudes (a few 0.01 mag). With spectral types A and F, they are not far removed from the RR Lyrae stars in the HR diagram. Their luminosity classes range from the main sequence to giants. As δ Scuti stars are not metal-poor and often occur in open clusters and in binary systems, they clearly do not belong to Population II. An interesting exception is the star SX Phe

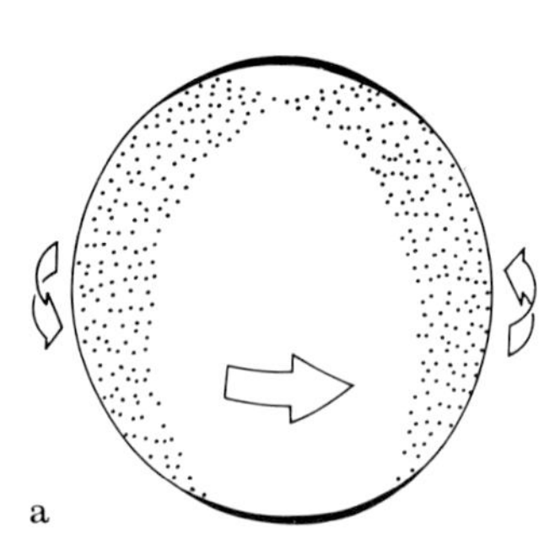

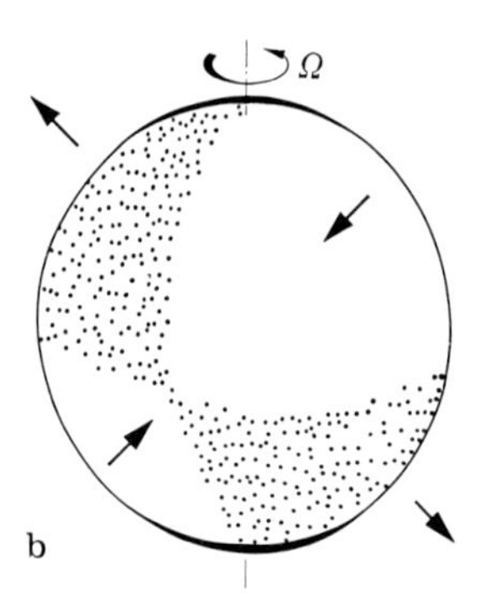

Fig. 25.19a, b. Two possible modes of rotation for non-radial pulsations (quadrupole oscillations). Here, $l = 2$ with **a** $m = \pm 2$ and **b** $m = \pm 1$. The sign of m indicates if the vibration pattern moves in the same or opposite sense of the direction of the rotation of the star. After Unno et al. [25.28].

($P = 0\overset{\mathrm{d}}{.}055 = 79^{\mathrm{m}}$), which distinctly shows low metallicity and subdwarf features. Perhaps this star represents an unexplored class of Population II δ Scuti stars.

The brilliant Vega (α Lyr) has also been proposed for membership in the δ Scuti group; see, for instance, Breger [25.40]. If confirmed, the variability of Vega, albeit only a few 0.01 magnitudes, makes it questionable whether the star could serve for defining the zero point of the stellar magnitude scale.

About 20% of all δ Scuti stars show rather large light-curve amplitudes of up to 0.7 mag. Their light curves may then look deceptively similar to those of some RR Lyrae variables. Multiple periodicities occur quite frequently in δ Scuti stars; in addition to the basic (radial) oscillation of period P_0, in many cases the first overtone of period P_1, and sometimes even the second overtone P_2, can be clearly identified. The ratios P_1/P_0 are near 0.77.

25.4.4.2 β Cephei Stars. The β Cephei stars, also called *β Canis Majoris stars*, form a small, sharply distinguished group of pulsating early-type giants. Their pulsations appear to proceed primarily in non-radial modes. They are located in the HR diagram far away from the "classical" instability strip. Their spectral types are between B0.5 and B2, and luminosity classes are quoted as IV (subgiants) and III (giants). Indeed, β Cephei stars lie about one magnitude above the main sequence, where the core-hydrogen burning presumably ends and the evolution toward the giant region begins for stars with masses in the range 10 to 15 $\mathfrak{M}_\odot$. Whether or not the pulsational instability is related to this fact has yet to be clarified. The list of brighter (mostly naked-eye) β Cephei stars contains 18 objects. In recent years, even α Virginis (Spica), which is also a spectroscopic binary, has been counted among them as a result of a pioneering interferometric study by Hanbury-Brown. The small amplitudes (under 0.1 mag) render photometric studies difficult; the radial velocity variations usually range between 20 and 100 km s^{-1} and are more readily measured. The equatorial rotational velocities are not very large, usually well below 100 km s^{-1}.

25.4.4.3 ZZ Ceti Stars. These stars are white dwarfs of spectral type DA (D = degenerate is the prefix indicating white dwarf). About 20 variables in this class are known; they lie in a narrow $B-V$ range between +0.16 and +0.20. The periods are "ultra-short." Multiple periodicities of about 100 to 1000 seconds occur almost without exception. The light variations are, with rare exceptions, of very small amplitude, for instance 0.012 mag for ZZ Ceti. The multiple periodicities cause the light curve to change from one cycle to the next. Color measurements of two members of this class yield a somewhat larger mean variation of up to 0.28 mag, which suggests that the light variations are caused solely by temperature variations at nearly constant radius.

At the high densities of white dwarfs, short periods are to be expected. The approximation of Eq. (25.9) gives at $\overline{\rho} \approx 5 \times 10^5$ g cm^{-3} a period $P \approx 11$ s. The actual values are 10 to 100 times longer, and can be reproduced with more detailed modeling and allowance for the internal structure of white dwarfs. The color range mentioned and the corresponding temperatures for 10 000 K to 13 000 K rather closely match the region where the extrapolation of the Cepheid instability strip meets the white-dwarf sequence. The observed complex period pattern and the multiple periodicities in ZZ Ceti variables, however, are far from clarified; the oscillations of these stars are most certainly not simple radial pulsations.

Photometry of ZZ Ceti stars is extremely difficult, and the fact that it has been successfully achieved is a brilliant accomplishment of observational technique. Most early observations were carried out using the "fast photometer" developed at the McDonald Observatory (Texas), notably by Robinson, Nather, Warner, McGraw, and others. It should also be mentioned that the degenerate components of cataclysmic binaries similarly show fast oscillations, a property possibly related to ZZ Ceti variability. The periods, however, are much shorter and the amplitudes still less. Some examples follow:

Object		Period (s)	Amplitude	Observer
DQ Her	(old nova)	71.1	$0\overset{m}{.}04$	M. Walker
Z Cam	(dwarf nova)	16–19	0.01	E.L. Robinson
Z Cha	(dwarf nova)	27.7	0.03	B. Warner

More than 10 such systems are known; see Sect. 25.6.2.

25.4.5 R Coronae Borealis Stars

This small but remarkable group of about 30 known members shows an uncommon type of light variation: normally the brightness is nearly constant, showing small variations of at most 0.1 or 0.2 mag, but it is so at irregular intervals (a few years on the average) interrupted by deep minima which may reach 7 or 8 magnitudes. The decline in such minima is quite steep, occurring within a time span of about 4 to 6 weeks. The ascent is slower, and may be accompanied by irregular fluctuations or secondary minima. The prototype of these variables (R CrB) had already been discovered in 1795. Observations spanning 180 years, many of which had been made by amateurs, show how totally irregular these minima are.

R CrB stars are supergiants of spectral types late F or early G. There is one important peculiarity characteristic of the whole group: they are hydrogen-poor (with very weak Balmer lines) and carbon-rich. The ratio H:He:C by numbers of atoms is 10^6:80 000:300 in the solar atmosphere but is more like 1000:80 000:1000 in R CrB stars. It has to be assumed that these stars are in a far-advanced evolutionary state and have suffered extremely high mass losses. They have all shed their original outer atmospheric layer, and now consist of a helium core in which the helium-burning into carbon and oxygen occurs, and a thin hydrogen-poor surface layer resembling an extended atmosphere. The products of helium burning obviously have reached the photosphere where the spectrum originates. This scenario explains why these stars, despite their high luminosites, have masses of only about 1 $\mathfrak{M}_\odot$; their original mass could easily have been 5 to 6 $\mathfrak{M}_\odot$. (Evidence for the present low masses is given by estimates of the gravitational acceleration at the surface from the spectrum, and by attempts to model occasionally observed pulsations.) Some experts have even suggested that these stars are just evolving into helium stars or into the central stars of planetary nebulae.

The pronounced minima of R CrB stars are generated—as is generally accepted—by the ejection of circumstellar matter, probably discharged in the form of jets into various directions. Recent studies, for instance by Feast or by Fernie and coworkers, have rendered this interpretation very convincing. The carbon-rich atmosphere could also explain the large depth of the minima, since the opacity of the ejected shell is particularly high when the carbon at larger distances from the star no longer occurs as gas but in the form of "graphite flakes" or as a soot-like, amorphous condensate. The intense infrared emission seems to support this picture: dust and soot particles are heated and re-emit the absorbed energy at long wavelengths.

In spite of their light variations' arising from two entirely different causes, the R CrB stars are discussed here along with the pulsating variables, since they are located in the HR diagram in the instability strip or close to it, and some of them also show pulsations (e.g., R CrB, UW Cen, S Aps, and very clearly RY Sgr). The periods are always about 35 to 40 days. The variations in brightness and color of RY Sgr in Fig. 25.20 illustrate the semi-regular nature of the pulsational variation.

25.4.6 Pulsating Variables in the Galaxy

Most pulsating variable stars are stars in late evolutionary phases and their distribution and motion in the Galaxy supply important hints regarding the history of the entire system. Consider first the "classical" δ Cephei-type stars, which are rather massive ($\mathfrak{M} \approx 8$–$12\ \mathfrak{M}_\odot$) and young stars. Their low age is revealed by the fact that they are strongly concentrated toward the galactic plane and are often found in open clusters. The brightest objects even show a distribution resembling the spiral structure of the Galaxy. Their young age does not at all conflict with their evolved status; they remain, for instance, only 10 to 20 million years on the main sequence, and then evolve into giants or supergiants. The corresponding time scales would be even shorter if the

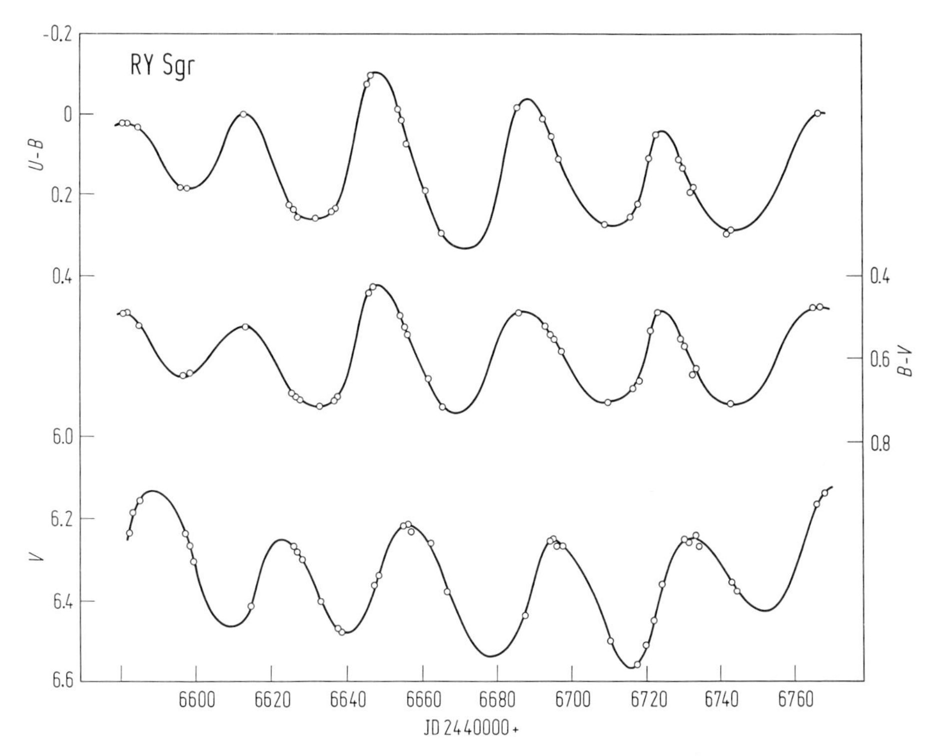

Fig. 25.20. Brightness and color changes in the R CrB-type star RY Sgr during 1986 (*top*: $U - B$, *middle*: $B - V$, *bottom*: V). The semi-regular pulsational variation, with an average period of 36 days during the time interval monitored, is evident. After Lawson et al. [25.29].

original masses had been larger (say 10–20 $\mathfrak{M}_\odot$) before intense mass loss set in. δ Cephei stars are thus massive stars which have reached, after a rather short stay on the main sequence, the supergiant phase of evolution, where they make at least one traversal across the instability strip.

Until the 1950s, the W Virginis stars were held to be members of the δ Cephei class. It is now known, however, that they represent a wholly different type of radial pulsator, despite a certain similarity of their light curves and a rather slight difference in the distribution of periods (from 1 to 25 days compared with 2 to 50 days for Cepheids). The characteristic hump on the descending branch of W Vir light curves is shown in Figs. 25.16. These stars, however, are much less thoroughly studied than δ Cephei stars, and many questions remain unanswered. The principal differences, compared with classical Cepheids, are shown in the masses (0.6 to 1.0 $\mathfrak{M}_\odot$), the much higher age, the lower metal abundance, and also in their cluster membership as well as their spatial distribution and motion within the Galaxy. Remarkably enough, over 40 W Vir stars have thus far been found in globular clusters, and this fact led to their formerly being called *Population II Cepheids*. It seems, however, that this class lacks the homogeneity of the classical Cepheids (with extreme Population I properties) or the class of RR Lyrae-type stars (extreme Population II). Perhaps there are two separate groups of different origin and age included under this category: a subdivision distinguishes between long- and short-period stars, the latter often being termed *BL Herculis-type stars*. The boundary between the two subclasses is set at 10 days, although some authors have suggested 4 days.

The W Vir stars in the Magellanic Clouds, particularly those in the LMC, display a distinct period–luminosity relation; see Sect. 25.4.2. For the variables in clusters, Harris [25.30] deduced a period–luminosity and a period–luminosity–color relation, respectively, for the Magellanic Clouds and for globular clusters. Both are in reasonable agreement, but more investigations are urgently needed.

The RR Lyrae stars are characteristic representatives of the old Population II and may be regarded as extant stellar "fossils" from an earlier time. Many of them represent the first star generation in the Galaxy at a time when it contained very few "metals," i.e., elements heavier than hydrogen or helium. The metallicity of these objects reaches hardly 1% to 5%, or maximally 10% to 20% of the solar value. Their distribution matches that of globular clusters, as halo objects with a strong concentration toward the galactic center and moving in eccentric and highly inclined galactic orbits. This connection is made particularly conspicuous by the frequent occurrence of RR Lyrae stars in globular clusters (which have typical ages of 10^{10} years). These stars were hence originally termed "cluster variables." The incidence of RR Lyrae stars varies strongly from one cluster to another, and surprisingly enough shows little correlation with the metallicity (age?) of the cluster. But Oosterhoff found that it is the predominant subtype of light variation, whether RRab or RRc, which is correlated with the cluster metallicity. More than 100 such variables have already been found in each of the clusters Messier 3 (M3), ω Centauri, and M15, and it has been conjectured that the field RR Lyrae stars may actually be escapees from globular clusters.

These variables show metal-poor, A or F spectral types, but differ fundamentally in structure from main-sequence stars of similar types. Their energy source comes from hydrogen-burning in an outer shell plus helium-burning (via the triple-α process) in the inner core. Accordingly, they are about 50 times more luminous than the Sun, even though their mass, which is deduced from pulsation and evolution theories, is only 0.6 $\mathfrak{M}_\odot$.

RR Lyrae stars are located in the HR diagram (see Fig. 25.12) along the so-called *horizontal branch*, which is a rather sharply defined region located between $B - V = +0.20$ and $+0.44$, and corresponding spectral types from about A7 to about F5. This position explains the absence of a period–luminosity relation. The fact that RR Lyrae stars are totally absent even in old open clusters like M67, NGC752, or NGC7789 is in complete accord with their advanced age. These variables are practically never found in binary systems, which distinctly confirms a known trend for Population II stars.

25.4.6.1 Mira Variables. Leaving the class of RV Tauri stars (with peculiar light curves and longer periods than those of Cepheids) for the discussion of semi-regular variables, the next type of pulsating variables in order of length of period are the Mira-type stars, sometimes called *long-period variables*. The light curve of the Mira variable R Dra is reproduced in Fig. 8.8. Vol. 1.

The shortest periods for Mira stars are around 100 days, but most cases are found between 150 and 500 days, and occasionally up to 2 years. As already mentioned in discussing the period, the pulsations are not as regular as those of the classical Cepheids, but an "average" behavior is always well defined. Mira variables have very distinctive spectra, which reveal them to be among the coolest stars, with types M6 to M9 at minimum and M3 to M7 at maximum. In general, shorter periods are correlated with earlier spectral types, but this relation is not very tight. About 10% of Mira-type variables belong to the classes S or C.[8] Remarkably, these stars always show emission lines, and this is so fundamental a property that essentially all Me-type giant stars can be assumed to be pulsating variables. This fact provides an expedient method, first introduced by E.C. Pickering and W. Fleming at the Harvard Observatory, for identifying Mira-type variables. The presence of emission lines, which occur in particular at certain phases in the light curve, may be explained by the light variation in the spectrum manifesting not only surface pulsations but also the effect of the outward-moving shock waves.

The large amplitude of the light curves and the emission lines, which expedite the detection of Mira stars, as well as their relatively high luminosity, contribute to the large number of known specimens. At present, about 6000 Mira stars, or about one-quarter of all stars contained in the new *General Catalogue*, have been recorded. Miras represent the most frequent type of variability (unless eclipsing binary stars are considered variables).

The Mira-type giants have absolute visual magnitudes around -1.5 to -2.5, and corresponding bolometric magnitudes will reach -4 or -5. Assuming, as an order of magnitude estimate, an effective surface temperature of 3000 K, then $M_{\mathrm{bol}} = -4.5$ leads to an enormously large stellar radius of about 1.5 AU. If this structure has a mass of 1 $\mathfrak{M}_{\odot}$, the corresponding mean density and the previously mentioned rule of Eq. (25.9) give periods around 500 days. Therefore, the stellar parameters are consistent to within an order of magnitude. In practice, one must also consider that a relatively large fraction of stellar mass, perhaps as much as one-half (at least 0.5 $\mathfrak{M}_{\odot}$) is concentrated in a compact, dense core which, according to current interpretation, is composed of carbon and oxygen, and has originated in an earlier stage of post-main-sequence evolution via helium and carbon burning, i.e., by the nuclear reactions 3 He $\rightarrow$ C and subsequent C + He $\rightarrow$ O. The objects observed are thus in the *second* giant stage, characterized by hydrogen- and helium-burning in alternate shells. This corresponds in the HR diagram to a position on the steeply sloped *asymptotic giant branch* or AGB. The enormous envelope surrounding the tiny core—which is barely larger than a white dwarf—loses mass at a substantial rate, and could be completely lost within a few 10^6 years.

A rather surprising result is that the galactic distribution and the statistics of motions of Mira variables indicate a fairly broad spread in population membership and hence in the ages of these objects. Actually, Miras are found in all galactic population subclasses, with the exception of the extreme Populations I and II. Those Mira

8 Type C (which has now subsumed the former R and N classes) is distinguished by a high carbon abundance; S-type stars show other chemical peculiarities (e.g., zirconium instead of titanium). These are side branches off the temperature sequence at the types K and M.

variables with the shorter periods of under 200 days may be counted among the older stars, the Population II, but with no appreciable underabundance of metals being detected. This inhomogeneity among Mira variables remains a serious problem whose investigation is urgently desired.

25.4.6.2 OH/IR Stars. A new class of variable stars closely related to the Mira group has recently been discovered: the *OH/IR stars*. It has been suggested that they be named *infrared Mira stars*. The discovery and exploration of these objects provides one of the most interesting chapters in the study of variable stars. The coding OH indicates that these stars were first noted as "molecular" radio sources through their 1612 MHz (18.6 cm) microwave radiation of the OH radical. The intensity of the radiation points to the existence of a special "pumping" mechanism in the circumstellar envelope. It is called the *maser effect*, and consists of a complex interaction of stimulation and emission in the electron shells of atoms and molecules. The maser, by the way, had been used as a very efficient amplifier in radio astronomy even before the discovery of "natural" masers in space. Infrared measurements, particularly those made from the *IRAS* satellite, also showed strong radiation sources at the same positions, but as yet no corresponding objects had been found in the visible range. Since many Mira stars also show maser effects, it seemed reasonable to assume that late-type stars of this new class, owing to strong mass loss, develop a massive circumstellar dust envelope at some distance from the star. This shell can absorb the short-wavelength portion of the radiation almost entirely, heating itself to 500 to 700 K and then again radiating intensely in the infrared range.

More recent studies (e.g., Sargent and Baud [25.31]) showed the infrared stars "behind the dust envelope" to be substantially more luminous than Mira-type stars; they lie at the top of the asymptotic giant branch. The absolute bolometric luminosities reach magnitudes in the range -4 to -8. It was expected and, by the successful observational studies of the Max Planck Institute for Radio Astronomy in Bonn (Engels [25.32]), confirmed that these stars also pulsate. A light curve in the infrared L band at 3.7 μm is shown in Fig. 25.21. The periods, which range up to 5 years, are longer than for Mira variables, and the masses may also be higher: they are expected to range from 2 to 6 $\mathfrak{M}_\odot$, or even as high as 8 $\mathfrak{M}_\odot$.

The infrared Mira-type stars are evidently in a stage of very rapid evolution, and at this point there seems only one direction indicated for them: when, after suffering a nearly catastrophic mass loss the "last" envelope has been ejected, only the hot stellar core remains as the central star of a planetary nebula.

25.4.7 Semi-regular and Irregular Variables

Following the class of regularly pulsating stars just discussed is the large class containing *semi-regular* and *irregular variables*. These two subclasses will also be treated within this section devoted to the pulsation phenomenon, since semi-regular variables certainly pulsate—although not in strictly cyclic fashion. Even in irregular variables, the light variation originates at least partly in pulsation-like changes in the volume

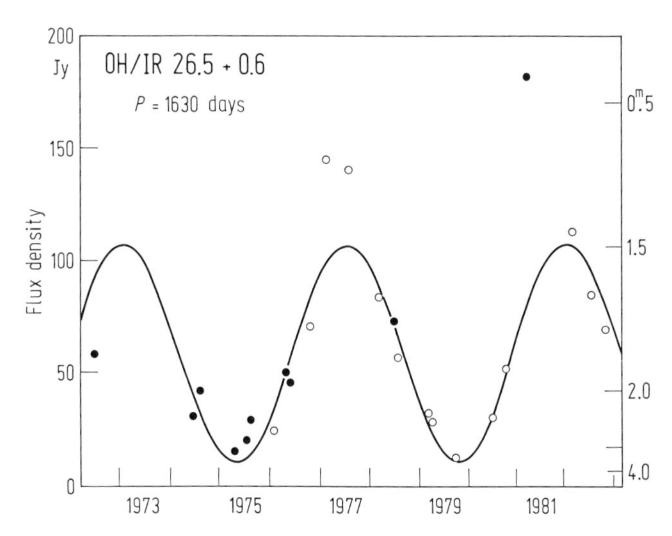

Fig. 25.21. Period variation of infrared radiation (*L*-filter, 3.7 μm), of the star OH/IR 26.5 +0.6 with a period of about 4.5 years. After Engels [25.32].

and temperature, which occur apparently erratically. It would be wrong to consider the class of irregular variables a collection of difficult-to-classify stellar oddities.

25.4.7.1 RV Tauri Stars. Before getting into a discussion of the semi-regular variables proper, there is a rather small group with less than 100 known members and of ambiguous classification, namely the *RV Tauri stars*, which must be mentioned. For some time these stars were thought to be typical representatives of the semi-regular variables, even though they show enough regularity to be considered a subgroup of "regular" pulsating variables. In fact, both by their periods of from 30 to 150 days and by their spectral types ranging from G to K supergiants, they occupy a place somewhere between Cepheids and Miras.

The characteristics of the light curves for RV Tauri stars are twofold. In most of these variables, for instance the often-observed R Scuti (Fig. 25.22), a deep and a less deep minimum appear in alternate succession; in some cases, it is not even clear if perhaps the periods should be halved. The other peculiarity features a long-term modulation which is apparently sporadic, appearing about once or twice in 10 years: the amplitudes of the deep minima decrease and simultaneously the overall brightness diminishes distinctly. Much is unclear regarding these fascinating stars. Perhaps they do not form a uniform group; some of them undoubtedly belong to Population II, as occasionally they are found in globular clusters, and many of them show underabundances of heavy elements. Some of them, however, show strong metallic lines and are thus certainly much younger, as a study undertaken by Preston and coworkers [25.33] in the early 1960s has shown. Also found at minimum, particularly in R Scuti, are strong TiO bands. The spectral classification then becomes ambiguous: the TiO bands correspond to M0–M2 types superposed on metallic lines of a much earlier G- or K-type spectrum.

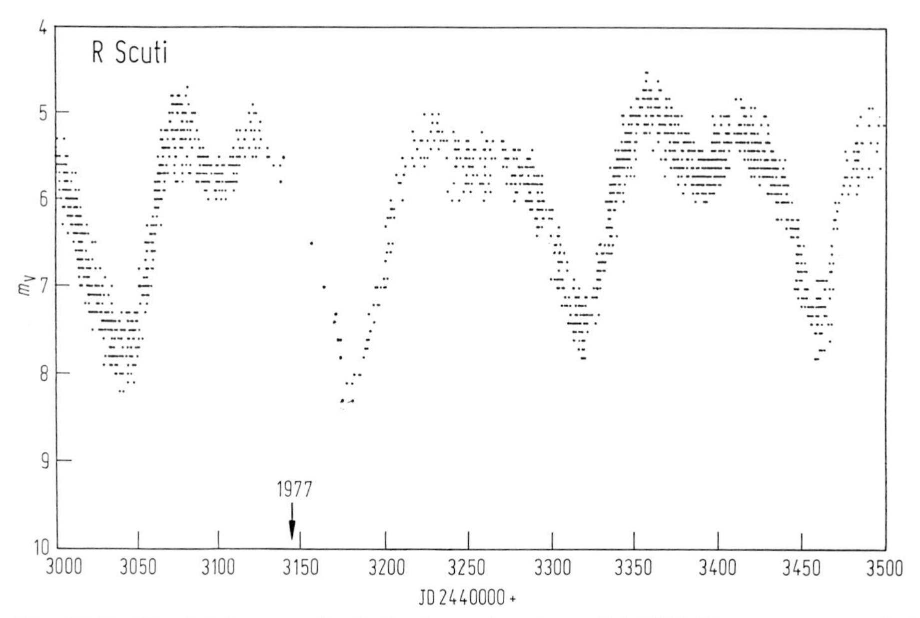

Fig. 25.22. Visual light curve for R Scuti covering the period 1976–77 serves as a typical example for the small class of RV Tauri stars. The observations were provided by the AAVSO.

Detailed spectral studies are difficult owing to the faintness of these stars; only two objects, R Scu and U Mon, reach 5th magnitude at maximum light. A thorough spectroscopic study of RV Tauri stars by researchers at the Ohio State University has, nevertheless, produced some promising first results.

25.4.7.2 Semi-regular Variables. The *semi-regular variables* are late-type giants and supergiants, with spectral type M dominating, and for which a kind of average period or cycle can be defined. These lie mostly between 100 and 1000 days, and can be broken by temporarily occurring light curve irregularities. The number of stars classified as semi-regular (SR) is substantial; about 3000 objects listed in the new *General Catalogue* fall under this heading. Together with those irregular variables, which also are late-type giants and supergiants, they form a broad class of what formerly was often termed *red variables.*

An object is placed into one of the subclasses of semi-regular variables based on the following definitions:

SRa: "fairly regular" light curve with amplitude usually less than 2 magnitudes;
SRb: less regularity, but periodicity still discernable;
SRc: like SRa,b but the variable is a supergiant, presumably quite massive;
SRd: collective class for semi-regular variables with spectral types earlier than M, S, and C.

The distinction between groups SRa and SRb is ill-defined and perhaps somewhat arbitrary. Classification under SRc requires more detailed study, and at least one useful spectrum.

The most famous representative of the many SR objects is the bright, red supergiant α Orionis (= Betelgeuse). This star was previously considered typical of irregular variables, but its more correct classification is SRc. The slow brightness variations of about 0.2 to 0.3 mag seem to indicate a cyclic pattern with about 1200 days as the length of the cycle. High-resolution speckle interferometry once detected a giant "spot" on the surface. The brightness variations, however, are more probably connected with pulsation-type changes of the radius; the few interferometric measurements of the star's diameter are not very consistent, varying from $0\farcs04$ to $0\farcs06$. It was recently reported that Betelgeuse is a binary star system, but it is possible that this duplicity has no relevance for the evolution of the red supergiant.

25.4.7.3 Irregular and Peculiar Variable Stars. This necessarily inhomogeneous category comprises primarily late-type giants or supergiants which vary slowly and with only small amplitudes (normally only a few 0.1 mag) and without a detectable periodicity. As with the semi-regular variables, here also the spectral type M is dominant. Formerly coded "I," they have been, since the third edition of the *GCVS* [25.34] (see Sect. 25.2.1), characterized by "L," with the following subdivisions:

Lb: irregularly variable giants, mostly of type M, but occasionally of types G and K;
Lc: virtually all late-type supergiants with irregular fluctuations in brightness and radius, perhaps also in color.

The class Lc is closely related to the semi-regular subclass SRc, and possibly its continuation to still less regular variations.

In addition to the irregular red variables, there are several other types of irregularly varying stars which definitely do not belong to class L, including some noted, mysterious objects. These stars are mostly early-type stars of high luminosity and with light curves often reminiscent of eruptive variables. They are sometimes listed as prototypes of very rare classes, and will be mentioned here in conjunction with the irregular stars. In our classification scheme (Sect. 25.2.2) they are also mentioned under D.3.

S Doradus is an extremely bright member of the Large Magellanic Cloud, an A5-type supergiant with absolute magnitude $M \approx -10$. At about 10-year intervals, it becomes fainter by about 1 magnitude for a few months; eclipses in a very peculiar system had for a while been suspected, but this is most certainly not the correct explanation for the light variations. In 1964 the star dimmed by 2.4 magnitudes, then became 1 magnitude brighter, and has since oscillated around the new magnitude. Occasionally, the often-observed (and perhaps related) stars P Cygni and η Carinae are counted among the S Doradus-type stars, but the rationale is doubtful. These stars behaved like very slow novae: P Cygni in 1600 reached the maximum brightness of $+3$, η Carinae in 1843 a spectacular -1. The corresponding absolute magnitudes must have been unusually large. They were in fact near -11 for P Cygni, and around -12.5 or even higher for η Carinae, whose distance is better known. Presently, P Cyg and η Car have magnitudes of about $+6$ and $+7$ respectively, which makes them still fairly bright stars. They are presumably related to the *Hubble–Sandage objects*, which are extremely bright, often irregularly variable, single stars in neighboring galaxies. What is observed here is possibly the early evolution of very massive stars (with over 100 $\mathfrak{M}_\odot$ and perhaps as much as 150 $\mathfrak{M}_\odot$) which cannot arrive a "normal" equilibrium state, and so suffer very strong mass loss.

Finally, γ Cassiopeia is a slowly variable B0 supergiant with very strong emission lines and apparently high mass loss (see Sect. 25.7.3).

25.5 Rotating Variables

25.5.1 Magnetic Variables (α_2 Canum Venaticorum Stars)

In the roughly A-type spectral range, i.e., from late B- to early F-types, numerous stars are found with—sometimes quite pronounced—spectral peculiarities:

1. Metallic lines have a strength unusual for A-type stars (*metallic A-* or *Am*-type stars).
2. Lines of certain elements usually not showing high intensity dominate the spectrum: Hg, Mn, Cr, Eu, Sr, Zr, Y, and also Si, in particular some lines in the vicinity of 420 nm (*peculiar A-* or *Ap-type stars*).

It is estimated that 15 to 20% of all A-type stars belong to the subclass Am, and from 2 to 5% to subclass Ap. Many Ap stars, especially the later-type, cooler members of the sequence (the *Si-Cr-Eu-Sr-stars*), reveal themselves as variables in a threefold fashion:

- Strong, variable magnetic fields occur, with strengths of several 10^3 to 10^4 gauss;
- Spectra are variable, both with respect to occurrence and strengths of lines as well as to radial velocities;
- There are smallish brightness variations of several 0.01 up to 0.2 mag, with the polarization also varying.

These changes usually follow one and the same period, but in a few cases the photometric and spectroscopic variations follow exactly one-half of the magnetic period. Apart from rare exceptions, the periods are in the range 5 to 9 days. The light curves are normally approximately sinusoidal, but sometimes more "triangular."

Brightness changes in the Ap stars, and especially the prototype α_2 Can Ven, are probably a manifestation of stellar rotation. The decisive role presumably is played by the strong magnetic field, which, in a fashion not entirely understood, causes a concentration of certain elements at the stellar surface. The magnetic axis does not usually coincide with the rotation axis, and the configuration is termed an *oblique rotator*. Owing to the rotation of the star, the Earth-based observer would then find that the magnetic field and the distribution of the "peculiar" elements vary. The strength of line absorption in the photosphere (blanketing) is likely to contribute, which may explain the small brightness variations. Variations of several 0.1 mag may be caused by additional starspot activity.

This theory of the oblique rotator was suggested and elaborated on by Deutsch, Mestel, Babcock, Böhm-Vitense and others, and is not discussed here in detail. It requires a "redistribution" of radiation emitted by the stellar surface, so that the energy output is reduced in certain spectral regions and enhanced in others. In this respect, the finding by Molnar, that the brightness variation in the far-ultraviolet is larger and runs in a phase counter to that in the optical range, may be important. The near-UV, on the other hand, is a sort of "neutral" range, with scarcely any changes observed.

It has sometimes been suggested that Am stars, which may be related to Ap stars, could show a similar light variation. The very small amplitude, however, is an impediment to obtaining a convincing proof of such variability.

25.5.2 BY Draconis Stars (Spotted Stars)

The differentiation between spotted stars and flare stars (UV Ceti stars; see Sect. 25.6.5) remains somewhat problematical. The variability of both types is due to activity in the near-surface layers of late-type main-sequence stars. According to present knowledge, the difference is that in BY Dra-type stars the observed phenomena originate primarily in the photosphere, but in flare stars the activity is located mainly in the chromosphere and possibly in the corona.

In the case of starspots, the dominant feature is a temperature-dependent diminution of radiant intensity. The spot areas are, as on the Sun, about 500 to 1000 K cooler than the surrounding stellar photosphere. At photospheric temperatures of only 3500 to 4000 K, this means a local reduction of surface brightness in starspot regions by 55 to 70%. However, these spots must be far larger than the solar ones, covering at least 30 to 40% of the visible surface in order to cause measurable photometric effects of some tenths of a magnitude.[9] It may be further assumed that the large spot regions have, as on the Sun, lifetimes of several stellar rotations, and thus in most cases generate nearly periodic light curves.

After early attempts to explain the light curve of Algol and other eclipsing variables by starspot formation, the idea of observable activity was again taken up about 40 years ago when Kron [25.38] suggested that the apparently periodic disturbances in the light curve of YY Gem (Castor C) could be interpreted in this way. Recognition of other objects as spot stars soon followed: BY Dra, which was chosen as the prototype and very carefully studied (Chugainov, also Krzeminski and Kraft, and later Oskanyan and coworkers), CC Eri (Evans and Bopp); a particularly extensive and homogeneous series of data on XY UMi was generated in the work by Geyer.

Although other hypotheses have been advanced, it is indeed very difficult to interpret a pronounced, smooth minimum lasting about one-half period other than by the transit of a large starspot or spot region across the stellar disk. The spot area at first appears distinctly foreshortened owing to projection, then becomes apparently larger

9 For comparison, sunspots seldom cover more than 0.3% of the surface, even at times of maximum activity. Even when allowance is made for the substantially smaller radii of K- and M-type stars, activity zones similar to those on the Sun would cover only 1 to 5% of the stellar surface. The activity of spot stars is therefore of a much higher level.

in size as it approaches the central meridian, and thereafter diminishes in apparent extent as foreshortening increases. The light curves also sometimes show strong changes and even the occasional disappearance of variability, which could be explained by the termination of an activity center, or by motions of spots. Should a spot move across the pole, it would change its "longitude" by 180°, and thus cause a characteristic jump in phase. Such jumps have actually been observed in BY Dra as well as in CC Eri. Yet a detailed "cartographic" analysis of a stellar surface "picture" from the spot light curve does not always lead to an unambiguous solution, and one is sometimes left with the impression that the demanding analysis had promised more than it actually delivered. In any event, one important qualitative result seems to have stood the test of time: strong starspot activity occurs most frequently in the polar regions of spotted stars, in contrast to the case of the Sun, where the spots prefer lower and middle latitudes.

Spotted stars also show flare-like eruptions, but it is worth noting that no relation has thus far been found between starspots and flare activity. This is not compatible with the notion of extended, active regions which on the Sun exhibit a wide range of activity, and which stretch from layers immediately below the photosphere up into the inner corona. The direct proof of chromospheric activity, which comes from the H- and K-lines of calcium in emission, also is still somewhat uncertain. Saar and Linsky [25.39], however, were recently able to detect magnetic fields of strength 2500 gauss.

Also worth noting is that the overwhelming majority (nearly 80%) of known BY Dra stars as well as flare stars occur in moderately close binary systems. This could be interpreted as a selection effect, but it is also possible that the duplicity in a yet unclear fashion causes the activity, or at least contributes to it (tidal effects?). In all RS CVn-type binary stars discussed in Sect. 25.3.5, starspot activity has been convincingly proven from the light curves and from distorted line profiles in the spectrum. Some authors assign the RS CVn stars as a subclass of the BY Dra types, but the dissimilar evolutionary states of these stars make the different arrangement in the present text preferable.

An amateur observer with photoelectric equipment can contribute much to useful research on BY Dra stars. The essential requirements are high accuracy and extended series of observations, over at least a few months, or better still over one or two years.

25.5.3 Radio Pulsars

The inclusion of *pulsars* in the system of variables documents the departure from the outdated definition: "a variable is any star which changes in brightness." Radio pulsars show variability almost exclusively in the decimeter and meter ranges of the electromagnetic spectrum. Only in a few cases (one of them being the famous Crab Pulsar) can pulsars also be observed in the visual range. The study of pulsars is primarily reserved for large radio telescopes with their complicated electronics, and thus is not accessible to most amateurs. Their importance warrants a brief and historically oriented synopsis of the essential scientific results on pulsars.

The discovery of radio pulsars in 1967 was wholly unexpected for most astronomers despite a remarkable prediction by F. Pacini. The team of discoverers at Cambridge

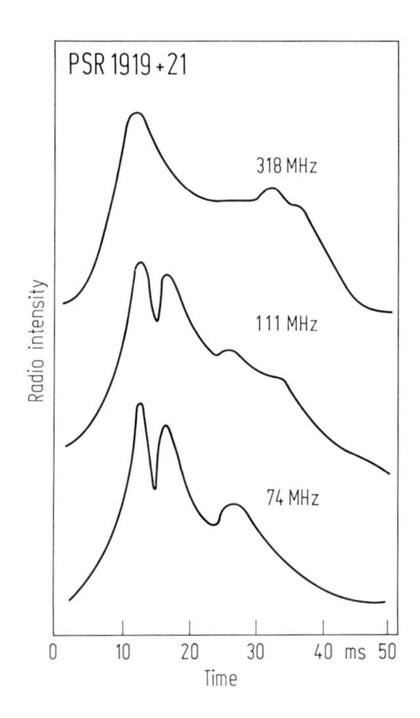

Fig. 25.23. Integrated radio pulse profiles of PSR 1919+21 at three different radio frequencies. The pulse period is $1\overset{s}{.}34$. Many radio pulsars show pulses with similar profiles. After Hankins [25.36].

University led by A. Hewish was originally interested in the scintillation of classical radio sources. For this study, a radio telescope with a short time constant (i.e., with a low inertia for recording rapidly changing signals) was needed. The high time resolution expedited the recognition of some point sources of rapid and apparently very regularly pulsing radiation. The first such source, orginally called CP 1919 (Cambridge pulsar at $\alpha = 19^h 19^m$) was found by Jocelyn Bell with a pulse period of $1\overset{s}{.}34$. Figure 25.23 shows the schematic radio pulse profiles of this object; the form is typical of many radio pulsars. The short period of CP 1919 has since been determined to astonishing precision: $P = 1\overset{s}{.}337\,301\,316\,884\,77$. This period increases by $\dot{P} = +0.1165$ ns per day. Owing to this change, the value quoted above with such spectacular precision (at least its last few digits) is valid for just a few seconds. The pulsar is now designated PSR 1919 +21, which includes the declination information. Currently over 500 radio pulsars are known, and an approximate distance from the Sun has been determined for most of these. Most of the pulsars were found in large search programs at the radio observatories of Jodrell Bank (England), Green Bank (West Virginia, USA), Arecibo (Puerto Rico, USA), and Molonglo (Australia). A recent list of 450 pulsars is given by Lyne and Graham-Smith in their book *Pulsar Astronomy* [25.35].

The characteristic feature in the curve is a main pulse which lasts about 1/10 of the period. Often, a lower interpulse occurs. The measured radio intensity goes down almost to zero between pulses. Variations in the form of the individual pulses, and even their displacements from pulse to pulse, are considerable. But after 50 to 100 periods, the intensity pattern can be averaged to yield a very stable pulse shape. The periods range from 1.5 ms (!) to about 4 s. The change of period is always positive and of the order $\Delta P/P \approx 10^{-15}$.

The clockwork-like mechanism underlying this stable, periodic pattern is evidently the rapid rotation of a compact object. Such extremely short periods are possible only for neutron stars, since in white dwarfs the rotational stability at such periods would not be guaranteed. Neutron stars have radii of 10 to 15 km and masses lying somewhere between about 1 and 2 $\mathfrak{M}_\odot$. The corresponding densities of from 10^{14} to 10^{15} g cm^{-3} lie in the range of that for atomic nuclei. These extraordinary objects also have very strong magnetic fields of up to 10^{12} gauss and extended magnetospheres of high-energy charged particles accelerated by enormous electric fields of up to 10^{12} V m^{-1}. It is the interaction of these relativistic particles with the magnetic field that generates the observed radio emission, or *synchrotron radiation*, which is strongly focused by the magnetosphere into a beam. The observed pulses occur when this beam of radiation, which rotates with the star, sweeps past the Earth (the lighthouse effect). An aperture of about 10° to 25° for the radiation beam is inferred from the ratio of pulse length to period of about 1/10 or less. Statistically, this means that four out of every five pulsars should be unobservable, as the beamed radiation does not sweep past the Earth.

The energy for this radiation is supplied by the kinetic energy of the continuously decelerating rotation of the neutron star; the periods of these objects increase without exception. After a few million years, the neutron star will rotate so slowly that it will no longer emit measurable radio radiation. Perhaps as many as 10^9 of such "silent" neutron stars populate the Galaxy, since the birthrate can be estimated as about 1 pulsar per 30 to 40 years, or even 10 to 20 years. It should be emphasized, however, that neither is there currently a satisfactory explanation for the conversion of kinetic rotational energy into energy in the magnetosphere or into radio radiation, nor is the beaming mechanism of the radiation entirely understood.

Neutron stars represent one of the final phases of stellar evolution, which implies that pulsars and supernovae should be closely linked. There exist several young supernova remnants, such as the strong radio source Cas A, which apparently do not contain a radio pulsar. On the other hand, there are only a very few young pulsars which can be unambiguously linked with supernova remnants, the Crab and the Vela pulsars[10] among them. The latter result could conceivably be explained by the substantially shorter lifetimes of the remnants in comparison with radio pulsars (see Sect. 25.6.1).

By early 1990 there were 12 pulsars known in *binary systems*, five of them in globular clusters. Five of the binary pulsars have extremely short pulse periods, in the range 1.5 to 11 ms, and these systems play an important role in the theory of close binary evolution.

The first binary pulsar, PSR 1913+16, was discovered by Hulse and Taylor in 1974, and it is still by far the best-known system. The pulse period is 59 ms, and the orbital period 7.75 hours. The binary nature becomes obvious from accurate timing of the pulses, which show both a light-travel-time effect in the arrivals and a periodic Doppler shift of the pulse frequency. This enabled pulsar observers to determine

10 The Vela pulsar (PSR 0833−45) is not to be confused with the X-ray binary Vela X-1 (4U 0900−40).

an orbit for the system with far higher accuracy than would be possible for most ordinary spectroscopic orbits. (Since only one component sends out pulses, each such system corresponds to a *single spectrum binary.*) The orbit of PSR 1913+16 is highly eccentric ($e = 0.617$), and this fact was of help to Taylor and his collaborators in verifying several important general relativistic effects, such as the relativistic advance of the periastron (i.e., the precession of the major axis of the orbit, which has, in the case of Mercury, been known since the early days of relativity) or the shrinking of the orbit (i.e., the decrease of the orbital period due to the emission of gravitational waves which carry away a small fraction of the orbital energy). This latter observation and the excellent quantitative agreement with the theory still constitute the only direct proof of the gravitational radiation emitted by rotating or orbiting systems as required by Einstein's theory.

The relativistic effects make it possible to overcome the ambiguity in the mass determination: one can derive masses for *both* stars of PSR 1913+16. The secondary, non-pulsing component is almost certainly a "silent" neutron star, and Taylor quotes the values 1.45 $\mathfrak{M}_\odot$ and 1.38 $\mathfrak{M}_\odot$ for the masses. These data are in good agreement with neutron-star masses, which have been inferred with much less accuracy from X-ray binaries, and where the values obtained seem to cluster around 1.4 $\mathfrak{M}_\odot$.

The spin-down of the pulsar rotation is considered, as was mentioned above, the ultimate energy source of the pulsar radiation. One exception of considerable importance is the group of *millisecond pulsars*, which are extremely rapidly rotating, compact objects with periods of the order of 1 to 10 ms. The earliest discovered object of this kind was PSR 1937+21, with a period of 1.56 ms. At present, there are seven objects known with periods under 11.1 ms and five of them are in binary systems. The rates of the period change as well as the magnetic field strengths are several orders of magnitude smaller than in "ordinary" pulsars. The evolutionary paths in these cases must be basically different: the clue to their physical understanding lies in their binary nature. These objects have to be closely related to the X-ray binaries, with the compact component accreting mass from its companion. The accretion rate at present is modest, but it might have been considerable at some time in the past, and it *accelerated* the rotation of the neutron star until its period reached the millisecond range. There is therefore no reason to assume that these pulsars are young objects. During the process of mass accretion the neutron star masses could increase to 1.6–1.8 $\mathfrak{M}_\odot$, and the other component might end up, for instance, as a white dwarf of 0.2–0.3 $\mathfrak{M}_\odot$.

The major difficulty with this ingenious proposal was, for a time, that PSR 1937+21, the very first millisecond object recognized, is unquestionably a single object. Yet, the remarkable discovery by Fruchter, Stinebring, and Taylor [25.37], announced in May 1988, of a very peculiar, almost bizarre object, PSR 1957+20, may provide a possible solution to the problem. This is a binary pulsar (pulse period 1.6 ms) where the secondary cannot have a larger mass than 0.020–0.025 $\mathfrak{M}_\odot$. The system (the secondary component) was found in the optical region and the faint star—which may be a very low-mass, relatively large white dwarf—shows direct evidence of the rapid, intense mass loss by the presence of a strange, comet-like nebula in its immediate neighborhood. Thus, we may be witnessing the final, catastrophic phase of the destruction of the companion star in this system.

It seems justifiable to assume in the case of PSR 1937+21 that the secondary had been completely destroyed some time earlier by the abrasive high-energy radiation from the compact object and also by Roche-lobe overflow, leaving a rapidly rotating, single neutron star.

Recent systematic searches have increased the number of binary pulsars in globular clusters to over 25, almost a dozen alone in the cluster 47 Tucanae, several of them pulsing in the ms range. These pulsars are in binary systems which have been formed, in all likelihood, through tidal captures in the dense central parts of some clusters. The interaction between the components spun up—and thus "rejuvenated"—the long silent neutron stars which are, as stellar objects, obviously very old, in fact as old as the cluster itself. (A capture which forms a binary with two "normal" components that then go through a close binary evolution until a neutron star has been formed is also possible but much less probable.)

The binary pulsars discovered to date make up only about 5% of the known pulsar population. Because they seem to be distinctly less "luminous" than single pulsars, the real proportion may be higher, perhaps 10–15%. This is still much less than the stellar binarity ratio, which is at least 60–70%. The explanation almost certainly lies in the possible disruption of a binary system after one component has undergone a supernova explosion. The components then fly apart at their previous orbital velocities, which can reach 200–300 km s^{-1}. Indeed, most pulsars are moving away from the galactic plane at comparable speeds.

25.6 Eruptive Variables

25.6.1 Supernovae

The two "new stars" in 1572 and 1604 mentioned in the historical survey were, as is now known, typical representatives of *supernovae* in the Milky Way galaxy. In reality, a supernova is quite the opposite of a new star; it is the catastrophic explosion of a much-evolved star which then becomes totally restructured or even destroyed in the process. (For comparison: the eruption of a nova involves a mass loss far less than 1% and leaves the star's structure virtually unchanged.) At the time of maximum light, supernovae may reach absolute magnitudes of -18 to -20, which translates into about 10^4 times brighter than ordinary novae. Hence, the luminosity becomes comparable with that of an entire galaxy.

During a supernova explosion, atomic nuclei of "heavy" elements (i.e., heavier than hydrogen and helium) are produced in large amounts. In particular, all heavy elements with atomic numbers larger than 209 (bismuth) are formed exclusively in supernovae. Some proportion of these heavy elements, often simply called *metals*, are dispersed by the explosion into the interstellar medium so that later generations of stars become enriched by these elements. In this way, supernovae play a considerable role in the chemical evolution of galaxies and of the entire universe.

By the end of 1992, about 830 supernovae had been discovered as telescopic objects in extragalactic systems; this number grows at a rate of at least 12–15 new discoveries annually. Most findings are owed to the Schmidt telescopes of the Mt. Palomar Observatory in California, and also to the observatories in Bern and Asiago; the amateur astronomer Rev. Robert Evans in Australia has found over 10 supernovae visually.

It has been ascertained with reasonable certainty that within historical times the following supernovae occurred in the Milky Way galaxy: SN 1006 in Lupus (declination $-41°$!), SN 1054 (remnant = Crab Nebula), SN 1181 (observed in China and Japan), SN 1572 (Tycho's Nova), SN 1604 (Kepler's nova), and a supernova in the second half of the 17th century which was not optically observed but which left the radio source Cas A as a remnant. In neighboring galaxies, supernovae appeared in 1885 in M31 and in 1987 in the Large Magellanic Cloud.

The supernova explosion results in a rapidly expanding cloud of gas with a mass in the range from a few tenths to several $\mathfrak{M}_{\odot}$. The initial velocity of expansion amounts to 10 000 to 20 000 km s^{-1}. This expanding gas collides with the interstellar medium (and possibly with pre-outburst circumstellar matter), and, by compressing and heating it, generates a strong source of radiation with a characteristic spatial structure. The bulk of this radiation lies in the *radio* domain, and, as it is generated by high-energy (relativistic) electrons gyrating in the galactic magnetic field, its spectral distribution differs markedly from that of hot stellar or interstellar gas. This so-called *synchrotron radiation* is the trademark of a *supernova remnant* (SNR), and the negligible absorption of radio radiation in space enables astronomers to detect the remnants virtually everywhere in the Galaxy. They are very sharply concentrated toward the galactic plane and show a peak in the direction of the galactic center.

A list compiled by van den Bergh [25.40] contained 135 SNRs, and about 1/3 of these are also observable in the optical and/or X-ray region. The best-known SNR is the Crab Nebula, which is associated with the "guest star" observed by the Chinese in 1054 AD. In 1968, a short-period pulsar ($P = 0.033$ s)—a rapidly rotating neutron star—was discovered in the Crab Nebula. A neutron star can be yet another result of the supernova outburst arising from the collapse of the core of a *massive* progenitor star. Contrary to some early expectations, however, one typically does not observe both the expanding nebula and the pulsar; this important circumstance has already been mentioned in connection with radio pulsars in Sect. 25.5.2. About 80% of the SNRs belong to the "shell-type" class. The Crab, with its entire volume filled by emission of mainly synchrotron radiation, is a rather rare representative of the so-called "filled-center remnants." About 15% of the SNRs are of the "combination type," with shell features and partly filled central parts; such nebulae may evolve ultimately into shell-like objects. In a Crab-like nebula, a pulsar is needed to support the strong radiation, while in the combination-type nebula the presence of a pulsar is probable but not necessary. Cas A and also the remnants of Tycho's and Kepler's supernovae are, for instance, very young, 300–400 year-old shell-type remnants (apparently containing no energizing pulsars), while the large, circular nebula in the constellation Cygnus, called the Cygnus Loop, is a typical old remnant, perhaps as old as 10^5 years.

The absence of observed pulsars in the vast majority of SNRs is not surprising, and is in fact explicable by several valid arguments. In cases where the original star was of low mass (see below), it may have been completely destroyed in the explosion, while in cases of very massive progenitor stars the formation of a *black hole*, instead of a neutron star, is to be expected. Finally, if the pre-supernova was a binary, it could have been disrupted as a consequence of the explosion, with the remnant and the generated compact object subsequently moving on different orbits in the Galaxy; thus the pulsar and SNR would in time become separated.

It is beyond controversy that the ultimate energy source of some supernova outbursts is the gravitational collapse of a star, and frequently, the burned-out *core* of a massive star. The core collapse of massive stars has been briefly reviewed by H. Bethe [25.41]. The liberation of gravitational energy can, in fact, account for the prodigious energy output observed during the supernova outburst, to an order of magnitude: 10^{43}–10^{44} J in the optical region, and 10^{44}–10^{45} J in the kinetic energy of the expanding gas. Moreover, a third form of energy output is indicated since the total gravitational energy of the collapse is near 10^{46} J, or over one order of magnitude higher. This follows easily from an estimate of the gravitational energy of the neutron star,

$$\text{gravitational energy} \quad = q\frac{G\mathfrak{M}^2}{R}, \tag{25.12}$$

where $q \approx 1$, $\mathfrak{M} \approx 1\ \mathfrak{M}_\odot$, and $R \approx 10$ km.

The theory of stellar collapse suggests that the rest of the liberated energy should take the form of neutrino emission. A very large number (about 10^{58} or more) of neutrinos and antineutrinos is generated, which, at an average of 10 or 15 MeV, represents a total energy of 1 or 2×10^{46} J. The liberated particles include not only electron neutrinos but also neutrinos linked with other leptons, the μ and τ particles. The collapse itself takes less than a second.

The neutrino events observed from the Supernova 1987A in the LMC corresponded exactly with the theoretically expected numbers, and thus favorably confirmed the picture of the progress of a gravitational collapse. This remarkable supernova event will be returned to later.

The theory of stellar collapse attempts to explain how under certain conditions part of the collapsing mass can develop nuclear densities while another part forms an outgoing shell which drives the layers of the star outward and ultimately leads to an expanding shell of the supernova. Only the neutrinos can supply information on the rapid collapse phase. The earliest optically observed phase is the originally steep increase in luminosity, which corresponds to an expanding "quasi-photosphere," an optically-thick layer radiating like an expanding atmosphere of a gigantic star—much larger than a supergiant—and, as color measurments show, steadily cooling in the process. These particular events are not unlike those of a nova eruption, except that the moving masses are four or five orders of magnitude larger and their expansion speeds 10 times higher.

Since about 1940, two main classes of supernovae have been distinguished, differing in their spectra and also in their light curves. The two classes are referred to (in a less than imaginative way) as simply Type I and Type II supernova. The Type II objects are dominated by strong Balmer lines of hydrogen, both in absorption and in emission, showing marked P Cygni profiles. At an early phase the spectra bear somewhat of a similarity to nova spectra, except that the very broad lines indicate the order of magnitude higher velocity of the expanding gas. With some difficulty, other lines, all very broad, can be identified. Type I supernovae show no trace of hydrogen and the identification of the highly broadened, overlapping emission lines seemed at first to present an intractable problem. Careful and patient studies finally revealed the presence of several metals.

In the past few years, on the basis of much-extended observational material, the need for a refined classification scheme became obvious. Among Type II supernovae, a distinction between the "plateau type" (Fig. 25.24 b) and the more "linear" light curves can be made. It appears even more important, however, to divide Type I into two subclasses according to their observed spectra. Thus, there are effectively three main classes:

Ia lacking both hydrogen and helium lines;
Ib lacking hydrogen, but showing helium lines;
II showing hydrogen (and also helium) lines.

These spectral criteria characterizing the expanding shell, in combination with the observed distribution of the supernovae among different types of galaxies and their distribution within these galaxies, have enabled astronomers to gain remarkable insight into the nature of the progenitors, the stars which have undergone the supernova explosion.

Figures 25.24 a and b show the mean light curves for the two main types (after Barbon, Ciatti, and Rosino [25.42,43]). In a somewhat later phase of development, 40 days after maximum brightness in Type I, 80 to 100 days in Type II, the light curves appear to enter a linear descent. This admits of an interesting interpretation. Since the magnitude scale is logarithmic, a strictly linear descent means an *exponential* decline of the light. Such a decline, however, is characteristic of radioactive decay. Detailed calculations of nuclear reactions occurring in supernovae make plausible that element synthesis reaches to ^{56}Ni and produces a large amount ($\approx 0.5\ \mathfrak{M}_\odot$?) of this unstable radioactive isotope. The repeated decay by electron capture and γ-photon emission leads to iron:

$$^{56}\text{Ni} \rightarrow {}^{56}\text{Co} \rightarrow {}^{56}\text{Fe (stable)}.$$

The half-lives are 6 and 77 days, respectively. The latter is nicely compatible with the observed decline of the light curve.

The calibration of absolute brightness of the supernovae is dependent on the distance scale of the galaxies, that is, on the adopted value of the Hubble constant. (On the other hand, if supernovae turn out to be useful "standard candles," an attempt can be made to derive the Hubble constant, which is a measure of the expansion of the universe, from the observed maximum brightness of supernovae.) For supernovae which appear in spiral nebulae with considerable amounts of interstellar matter, an important correction must be applied owing to the absorption of the light in the spiral galaxy itself.

Only a few years ago it was generally accepted that the brightest supernovae are of Type I, and that Type II objects are 1.5 to 2 magnitudes fainter. With the recognition of the two classes comprising Type I, the data may be subject to some revision. The following data are based on an early discussion by Kowal [25.44], as modified by van den Bergh et al. [25.45]. Assuming a Hubble constant of $H = 75$ km s^{-1} Mpc^{-1}, it

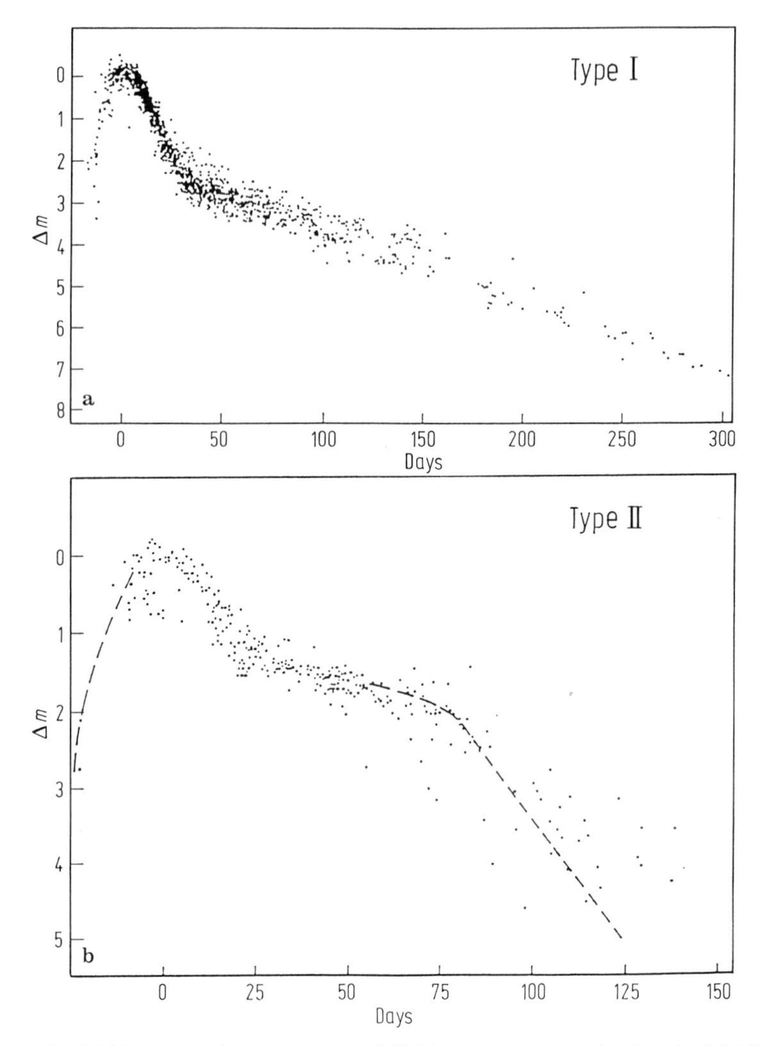

Fig. 25.24. Composite light curve of 38 Type I supernovae (*top*) and of 13 Type II supernovae (*bottom*). Note the different time scales. After Barbon, Ciatti, and Rosino [25.42,43].

is found that

$$\begin{aligned} M_V(\text{max}) &\approx -19.2 \quad \text{Type Ia} \\ &\approx -17.7 \quad \text{Type Ib} \\ &\approx -17.1 \quad \text{Type II}, \end{aligned}$$

with an estimated scatter of ± 1 magnitude in each case. Most supernova observations still make use of photographic techniques, but near maximum light the color index becomes $B - V \approx 0$. During the later development the cooling envelope becomes considerably redder.

The frequency of supernovae in galaxies is still subject to some controversy. Most statistical investigations result in an average supernova rate with very wide margins: roughly 50 to 150 years between supernova events in a bright galaxy ($L \approx 10^{10}\ L_{\odot}$), with the spirals, especially those of "late-type" (Sb, Sc), being the most productive. For our own Galaxy ($L \approx 2 \times 10^{10}\ L_{\odot}$), the supernova rate may be closer to the lower value; the six observed explosions of the last 1000 years may thus represent only a fraction (1/3 to 1/4) of the total number of events. As a crude order of magnitude estimate, about 10^4 years is obtained for the average lifespan of a SNR in the Galaxy—in any case, not an unreasonable value.

Estimating the relative frequencies of the now three main types of supernovae is a difficult task, fraught with observational bias. The above-mentioned investigation by van den Bergh, McClure, and Evans derived from a very small, but apparently homogeneous sample the approximate relative frequencies 18%, 22%, and 60% for supernovae of Type Ia, Ib, and II, respectively. (These numbers, however, are still not considered definitive.)

Of very great importance are the observed supernova "sites" in the respective galaxies. Type II supernovae occur almost exclusively in spiral galaxies, close to or in the spiral arms. Moreover, the large majority of these supernovae seems to be directly associated with giant HII regions, where star formation is going on. It is virtually certain that the progenitors are massive stars evolving very rapidly and which, after having completed nuclear evolution, reached the stage of core collapse with most of their hydrogen-rich envelopes still in place. The minimum mass of this evolutionary path is about 8 $\mathfrak{M}_{\odot}$; the most massive stars (25–30 $\mathfrak{M}_{\odot}$ or more?) may produce black holes in the supernova explosion.

Type Ib supernovae, moreover, appear amidst younger stellar populations, but are less directly associated with recent or ongoing star formation. The lack of hydrogen, as indicated by the spectrum, leads to the working hypothesis that the progenitors may be Wolf–Rayet stars which, having lost their hydrogen-rich envelopes and reduced the originally high stellar mass to a value of perhaps 4 to 8 $\mathfrak{M}_{\odot}$, are still rich in helium. The mass loss can be the consequence of a "castastrophic" stellar wind or also mass exchange in close binary evolution. Type Ia supernovae may occur in older stellar populations lacking massive stellar objects. Normally, one would expect these relatively low-mass stars (after some mass loss) to terminate their evolution as white dwarfs, without explosion. One promising hypothesis is that the progenitors are, indeed, white dwarfs, specifically, accreting CO-white dwarfs (neither hydrogen nor helium present) in a *close binary system*. Under rather specific circumstances, the mass of a white dwarf can approach—without the intervention of a nova explosion—1.4 $\mathfrak{M}_{\odot}$, that is, the Chandrasekhar limit of stability. The accreted "burnable" material provided by the non-degenerate companion can be converted to C and O in short flare-ups of nuclear burning termed *nuclear flashes*, during which no mass loss occurs. Upon approaching a critical mass close to the limit, the stability of the configuration is no longer guaranteed and the star explodes in a complicated process that most likely does *not* produce a compact central remnant[10].

---

10 For footnote see next page.

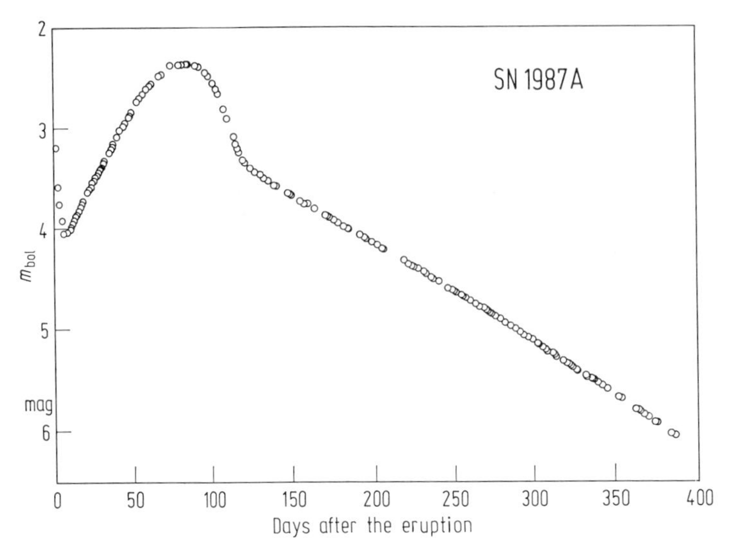

Fig. 25.25. Development of the bolometric magnitude of SN 1987A over the first 400 days after eruption (1987 February 23). A flat maximum was reached on 1987 May 20. The linear decline of brightness in this logarithmic graph (beginning about 4 months after the explosion) is caused by the radioactive decay of nickel and cobalt, which at this stage produces the energy of the supernova shell. From Biermann [25.46].

25.6.1.1 The Supernova SN 1987A in the Large Magellanic Cloud. While this chapter was being prepared, a bright supernova flared up in the Milky Way's nearest extragalactic neighbor, the Large Magellanic Cloud (LMC). This was the brightest supernova since 1604, and was visible with the naked eye. The opportunity was quite favorable to perform observations of high resolution using modern techniques and to compare them with the well developed supernova theories. As this object proved of extraordinary importance for theories of stellar evolution, a somewhat detailed discussion will be included at this point.

SN 1987A was discovered on 1987 February 24 at about 5^h30^m UT by Ian Shelton at the Las Campanas Observatory in Chile. A long exposure with the venerable 25-cm Carnegie astrograph revealed it as a 5th magnitude star. It was later discovered that almost one day earlier, on February 23 at 10^h30^m UT, the supernova had been photographed in Australia as a magnitude +6 object during its ascent to maximum, while a photograph obtained on February 22 between about 9^h and 10^h UT (also in

11 This process would bring some phases of Type Ia supernova evolution closer to that of cataclysmic variables (see Sect. 25.6.2 below). Yet the removal of burnable material before the outburst must be practically complete since recent spectroscopic studies placed a truly narrow abundance limit, indicating virtually *no* hydrogen left in the expanding envelopes of the supernovae Type Ia.

Australia) shows only the supposed pre-supernova as a star of magnitude +12. Since the neutrino observations fixed the collapse and thus the beginning of the eruption at 1987 February 23, 7^h35^m UT, the original brightness increase was very rapid, about 6 magnitudes in just three hours. Soon thereafter, the rate of increase slowed to barely more than 1 magnitude over the next day, and a very slow supernova of unexpectedly low absolute brightness developed. It was not until 1987 May 20 that SN 1987A reached a flat maximum, with $V = +2.9$ (Fig. 25.25). At the known distance to the LMC, and after allowing for interstellar absorption, this corresponds to an absolute magnitude at maximum of −15.5, or about 1.5 to 2 magnitudes below the average for Type II supernovae.

The spectrum clearly showed SN 1987A to be a *bona fide* supernova of Type II: strong, broad Balmer lines with P Cygni characteristics were seen, and gave an initial expansion speed of 18 000 km s^{-1}, albeit with some peculiarities. Its somewhat unusual behavior was possibly related to the pre-supernova object, a B3 Ia-type star with an absolute magnitude of about $M_V = -6.5$. This spectral type came as a surprise since, according to supernova theorists, the precursors were expected to be late-type supergiants. It seems, however, that the appropriate modification of theories of late evolutionary phases will not pose insurmountable problems.

The actual identification of the pre-supernova star, on the other hand, was unexpectedly more difficult. The 12th magnitude early-type star mentioned, which had been known since 1969 as an object with distinct emission lines and catalogued under the name Sk−69°202, has two fainter "companions" at separations $1\farcs5$ and $3''$. Precise astrometric and UV observations show the two neighboring stars as still existing, and make it highly likely that the progenitor was indeed the B3 supergiant. At the distance of the LMC, a separation of $0\farcs5$ gives a projected linear separation of 0.4 pc, so these objects are likely to be only optical companions.

When modeling the evolution of this star, one must also allow for the unusually low metal abundances of the Magellanic Clouds, about one-quarter that in the Galaxy. Perhaps supernovae of this kind occur normally in irregular galaxies, but are at a disadvantage from selection effects owing to their lower maximum brightness, and thus are less frequently observed.

Presumably the most important contribution by SN 1987A to the knowledge of stellar collapse and supernovae was the simultaneous observation of neutrinos which indubitably originated in the event. This was the first known observation of *stellar* neutrinos, and confirmed the previously discussed conjecture that the collapse generates a very high number (10^{57} to 10^{58}) of high-energy neutrinos (5 to 50 MeV) which contain by far the largest share in the energy balance of the event. At the given distance of the LMC of 170 000± 10 000 light years, about 10^9 to 10^{10} neutrinos per square centimeter would be expected at the surface of the Earth. Theoretically, this entire amount should be emitted in 1 second, although possible scatter of neutrinos in the dense matter of the supernova could spread the emission over several seconds. The overwhelming number of neutrinos expected to reach Earth provided the hope that they could be detected in spite of this very low effective cross section for interaction with matter. Actually, on the day of the explosion, three different institutes recorded neutrino showers, each lasting 5 to 6 seconds:

- Mont Blanc Tunnel, Neutrino Observatory,
  2^h52^m UT: 5 events;
- Kamiokande II, Japan, Proton Decay Detector,
  7^h35^m UT: 11 events;
- Irvine-Michigan-Brookhaven (IMB) Collaboration, Cleveland, Ohio, Proton Decay Detector,
  beginning at $7^h35^m42^s$: 8 events.

The solar neutrino detector in South Dakota, USA, could not have recorded this shower, mainly because it lacks adequate time resolution.

The simultaneity of the showers recorded in Japan and in the U.S. showed that the showers definitely originated in the collapse of the LMC supernova. The Cerenkov detectors used also permitted an estimate of the direction. The Italian-Soviet team in the Mont Blanc Tunnel recorded a possible event $4\frac{1}{2}$ hours earlier than the other groups, but this result is considered controversial, and most experts now feel that it was merely a statistical fluctuation. A suggestion by some astrophysicists of the Max Planck Institute in Munich that the collapse may have proceeded in a two-step fashion, first to a neutron star, and then into a black hole, was not generally accepted.

These neutrino observations provide a very important application for particle physics, as they make possible, in principle, a determination of the rest mass of the electron neutrino. Should the neutrino have a non-zero rest mass or (rest energy m_0c^2), then it necessarily travels with a speed below that of light. If this is the case, then the following relation for highly relativistic particles holds between neutrino energy and speed:

$$E = \gamma m_0 c^2, \tag{25.13}$$

where

$$\gamma = \frac{1}{\sqrt{1-(v/c)^2}} \quad \text{(Lorentz factor)}.$$

Even before the SN 1987A event, it had been suggested that this relation between neutrino energy and speed be used to determine the neutrino mass from the differences in the time of arrival after the eruption. The low-energy neutrinos will arrive somewhat later than the high-energy particles, as higher E at given m_0c^2 also means larger γ, so that v will be closer to c. This interesting method failed in the case of the LMC supernova, owing primarily to the statistically insignificant material, as Fig. 25.26 shows. On the basis of the 11 and 8 neutrino events, it can be stated only that the assumption of "higher" masses ($m_0c^2 > 10$ eV) does not give any good fit. Smaller rest energies of around 5 eV or less are more compatible with the data, and even $m_0 = 0$ is not excluded. In any event, the neutrino data from SN 1987A argue against the hypothesis that neutrinos could increase the mean density of the universe through a rest mass substantially different from zero. Clarification of this question will probably have to wait until a much closer, *galactic* supernova occurs.

SN 1987A has been recently "revisited" by R. McGray, on the pages of the 1993 volume of the Annual Rev. of Astronomy & Astrophysics [25.46a].

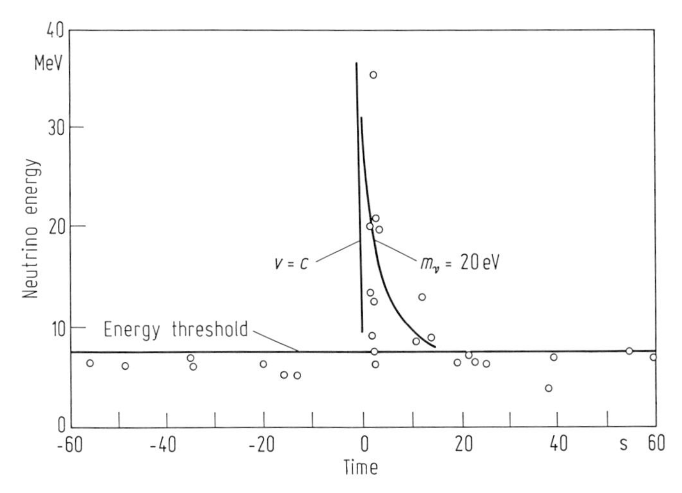

Fig. 25.26. Relation of energy versus travel time compared with neutrino events of supernova SN 1987A registered with the Kamiokande II detector in Japan. As an illustration, the arrival time as a function of particle energy is graphed for a hypothetical neutrino mass of 20 eV, chosen so as to reasonably match the group of three events at about 20 MeV. The *vertical line* at $t = 0$ shows the arrival of the light signal, the *horizontal line* is the sensitivity threshold of the detector.

25.6.2 Cataclysmic Variables

The naming convention for this type of eruptive variable has been in use for about 20 years; it derives from the Greek "kataklysmos," which means catastrophe or inundation, and refers to the physical process causing the eruption of these objects. The sudden outburst of brightness and related peculiar spectroscopic and photometric properties can be explained by the evolution-governed mass transfer between the two components of a close, interacting binary system. The analysis of such systems forms a modern chapter of astrophysics, and was advanced, particularly in the last 15 to 20 years, through the development of space-based UV and X-ray astronomy, and through high-time-resolution photometry. Owing to the large number of cataclysmic variables and to the opportunity to study important processes of interaction, these systems are significant sources of information for scrutinizing theories of evolution of close binaries.

25.6.2.1 Phenomenological Classification. The primary symptom of almost all cataclysmic systems, nova-like variables excepted, are brightness outbursts observed irregularly or only once and which liberate large amounts of energy within very short timespans. The immediate and most conspicuous effect of such eruptions is an increase of visual brightness by 2 to 19 magnitudes within hours to months, and a subsequent, slower decline of the brightness to the original level (before eruption) in timespans of days to decades, depending on the class of objects.

The diversified patterns of cataclysmic variables are classified according to physical characteristics. The primary parameter here is the energy freed during an eruption

Table 25.1. Classification of the cataclysmic variables.

Subclass	Δm_{vis}	E_{vis} (Joules)	T_n	t_2 (days)	N
Novae					
Na				2–80	
Nb	7^m–15^m	10^{37}–10^{38}	10^5 y:	80–250	~ 200
Nc				> 250	
Recurrent Novae	7^m–9^m	10^{36}–10^{37}	10–100 y	3–60	6
Dwarf novae					
U Gem	2^m–6^m		15–500 d		
Z Cam	2^m–5^m	10^{31}–10^{32}	10– 50 d	1–2	~ 300
SU UMa	2^m–8^m		10–200 d		
Novalike variables	–	–	–		~ 70

over the entire spectral range, expressed by the overall bolometric magnitude, or in a limited range, given usually by the visual or photographic magnitude. To distinguish the various subgroups, it must be ascertained whether the observed eruption was a one-time event or whether the eruptions repeat regularly in certain cycles, or irregularly. Another feature of classification is the time scale in which the outburst proceeds. In particular, the time interval between reaching visual maximum brightness and that instant at which the brightness has declined by 2 or 3 magnitudes (the t_2 and t_3 times, respectively) is often a characteristic scale for novae. Besides these general distinguishing features, more precise classification includes individual photometric and spectroscopic properties, which can now be determined with the aid of highly sensitive detectors in photometry with high time resolution, and by spectroscopic measurements in X-ray, UV, visual, and IR ranges with adequate dispersion. So, the cataclysmic variables have been subdivided into *fast* (Na), *slow* (Nb), *very slow* (Nc), *recurrent* (Nr), and *dwarf novae* (DN), with the subgroups of U Gem, Z Cam, SU UMa systems, each named after its prototype, and nova-like variables (Nl), which also contains several subgroups (UX UMa, VY Scl, AM Her, DQ Her, and AM CVn systems). Table 25.1 summarizes the phenomenological classification of cataclysmic variables, giving as the distinguishing criteria of the subclasses the following observable parameters: the *eruption amplitude* Δm_V in the visual range (in magnitudes), the energy E_V (in J) emitted over the entire outburst in the form of visual radiation, the typical *cycle length* T_n with which outbursts recur, the time t_2 (see above), and the number N of objects known in each group. The data in the table have been compiled from Robinson [25.47], Warner [25.48], Duerbeck and Seitter [25.49], and Wade and Ward [25.50]. The groups of novae and dwarf novae show a distinct correlation between the energy of outburst (amplitude) and the length of cycle between consecutive eruptions. This is known as the *Kukarkin-Parenago relation*; cf. Warner [25.51] and Antipova [25.52].

A list of particularly well-studied cataclysmic variables is found in Table 25.2. In addition to the visual and blue magnitudes at maximum and minimum, eruption cycles T_n, orbital periods of the binary systems $P_{\rm orb}$, and masses $\mathfrak{M}_1$ and $\mathfrak{M}_2$ of the primary and secondary components are also given (see also Sect. 25.3.5). Coordinates and references to published papers and to charts of the individual systems are compiled in the *Catalogue of Cataclysmic Binaries* by Ritter [25.53], from which the data in Table 25.2 are extracted. Examples of eruption light curves are presented in Figs. 25.27 a–f. Detailed surveys on the structure and properties of cataclysmic variables are found, for example, in the articles by Cordova and Mason [25.54], Robinson [25.47], Wade and Ward [25.50], and Warner [25.48].

25.6.2.2 Photometric and Spectroscopic Properties

25.6.2.2.1 Novae. Novae (from the Latin nova stella = new star) received their name from the erroneous assumption that the sudden appearance of a newly formed star had been observed. In reality, the nova precursor is normally too faint for visual and even photographic observations, so that even now relatively little information has been obtained on the properties of precursors (pre-novae) and the remnants (ex-novae and post-novae). In the known cases, they are hot blue stars with small, sporadic brightness fluctuations. Often, a nova event is detected only during or after it has attained the visual brightness maximum, because the ascent to maximum by 7 to 14 or even more magnitudes proceeds within very short timescales of from a few hours to days.

The schematic shape of a nova light curve is seen in Fig. 25.28. The classes Na, Nb, and Nc may be characterized by typical t_3 times (again, the time for brightness decline from maximum by 3 mag), by some 10, 100, or 1000 days, respectively. Measured light curves of the fast nova V1500 Cyg (= Nova Cygni 1975), of the two slow novae DQ Her (= Nova Herculis 1934) and PW Vul (= Nova Vulpeculae 1984.1), and also of the very slow nova HR Del (= Nova Delphini 1967) are graphed in Figs. 25.27 a–d. It has already been shown that the light curve shapes of individual novae may be very different. The schematic Fig. 25.28 indicates conspicuous differences occurring primarily in the transition state following the first, rapid decline after maximum. The curve there may follow a more or less smooth path, or it may show periodic fluctuations of various amplitudes, or, in slow novae, a deep brightness minimum may occur. More homogeneous are the phases of rapid ascent to maximum, often interrupted by a brief "pre-maximum halt" about 2 magnitudes below maximum, of the first, rapid descent after maximum, and of the asymptotic final descent to the original brightness with accompanying small intensity fluctuations. The maximum itself is sharply pronounced in fast novae, while in slow novae it may be several hundreds of days broad and have a complex shape (see Fig. 25.27 d).

Amplitudes of most novae are around 11 to 12 mag; extremely high values were observed in CP Pup, with over 16 mag, and in V1500 Cyg, with over 18 mag. These statistics include a selection effect which favors small amplitudes, as the magnitudes of most pre-novae are not known and particularly so for the large-amplitude cases. The mean absolute brightness at maximum is given by Duerbeck [25.55] as $M_V = -6.4$ for slow novae and $M_V = -9.4$ for fast novae. The *Catalogue and Atlas of Galactic Novae* (Duerbeck [25.56]) compiles coordinates and charts of a large number of ex-novae.

Table 25.2. Selected well-studied cataclysmic variables.[a]

Object	Type [b]	m_{max} [c]	m_{min} [c]	T_n (days)	P_{orb} (days)	$\mathfrak{M}_1$ [d] ($\mathfrak{M}_\odot$)	$\mathfrak{M}_2$ [d] ($\mathfrak{M}_\odot$)
RX And	DN (ZC)	10.9	14.9	5–20	0.211 54	1.14 (0.33)	0.48 (0.03)
AR And	DN (UG)	11.0	17.6	25	0.0938		
HV And	NI (AM:)	15.0B	16.8B		0.055 99:		
VY Aqr	DN (UG:)	8.0B	17.5B		0.22:		
AE Aqr	NI (DQ)	9.8	11.6		0.411 654	0.9 (0.1)	0.7 (0.1)
FO Aqr	NI (DQ)		13.5		0.167 71		
UU Aql	DN (UG)	11.0	16.7	71	0.140 49:		
V603 Aql	Na	−1.1	11.9		0.138 154	0.66 (0.27)	0.29 (0.02)
V794 Aql	NI (UX)	13.7	20.2B		0.23	0.88 (0.39)	0.53 (0.07)
V1315 Aql	NI (UX)	14.4	16.1		0.139 690	0.9	0.4
TT Ari	NI (VY)	9.5	16.3		0.137 551	0.8:	0.4
T Aur	Nb	4.1	15.1		0.204 378	0.68:	0.63:
SS Aur	DN (UG)	10.5	14.8	40–75	0.1828	1.08 (0.40)	0.39 (0.20)
KR Aur	NI (VY)	11.3	18B		0.162 80	0.59 (0.17)	0.35 (0.02)
V363 Aur	NI (UX)	14.2	15.0		0.321 242	0.86 (0.08)	0.77 (0.04)
Z Cam	DN (ZC)	10.5	14.8	19–28	0.289 840	0.99 (0.15)	0.70 (0.03)
AF Cam	DN (UG)	13.4	17.3	75	0.0525		
SY Cnc	DN (ZC)	11.1	14.5	22–35	0.380	0.89 (0.28)	1.10 (0.05)
YZ Cnc	DN (SU)	10.5	15.5	6–16	0.0864	0.39 (0.12)	0.27 (0.02)
AC Cnc	NI	13.8	15.4	204	0.300 478	0.82 (0.13)	1.02 (0.14)
AT Cnc	DN (ZC)	12.7B	16.2B		0.238 691		
AM CVn	NI (AC)	14.1	14.2				
HL CMa	DN (ZC)	11.7	14.5		0.2145	1.0:	0.45 (0.10)
BG CMi	NI (DQ)	14.3	14.7		0.134 79	0.8 (0.2)	0.38
OY Car	DN (SU)	11.4	17.3	25–50	0.063 121	0.90 (0.04)	0.100 (0.005)
QU Car	NI	11.1	11.5		0.454		
HT Cas	DN (SU)	10.8	18.4	30–35	0.073 647	0.60 (0.16)	0.20 (0.03)
V425 Cas	NI	14.5	18		0.149 64	0.86 (0.32)	0.31 (0.02)
BV Cen	NI	10.5	13.3	150	0.610 116	0.83 (0.10)	0.90 (0.10)
V436 Cen	DN (SU)	11.3	15.5	32	0.062 501	0.7: (0.1)	0.17:

Continued next page

Table 25.2. (continued)

Object	Type [b]	m_{max} [c]	m_{min} [c]	T_n (days)	P_{orb} (days)	$\mathfrak{M}_1$ [d] ($\mathfrak{M}_\odot$)	$\mathfrak{M}_2$ [d] ($\mathfrak{M}_\odot$)
V834 Cen	NI (AM)	14.2	16.0		0.070 497		
WW Cet	DN (ZC)	9.3	15.7	31	0.175 78	0.50 (0.14)	0.35 (0.03)
Z Cha	DN (SU)	11.9	17.2	82	0.074 499	0.54 (0.01)	0.081 (0.003)
TV Col	NI (DQ)	13.6	14.1		0.228 600		
GP Com	NI (AC)	15.7	16.0		0.032 31		
SS Cyg	DN (UG)	8.2	12.1	24–63	0.275 130	1.20 (0.10)	0.71 (0.06)
EM Cyg	DN (ZC)	12.5	14.4	13–46	0.290 909	0.57 (0.08)	0.76 (0.08)
V751 Cyg	NI (UX)	13.2	16:		0.25:		
V1500 Cyg	Na	2.2	17.2		0.139 613		
V1668 Cyg	Na	6.0	19.0		0.4392:		
CM Del	DN (UG)	13.4	15.3		0.162	0.48 (0.15)	0.36 (0.03)
HR Del	Nc	3.3	13.0		0.214 167	0.9: (0.1)	0.58:(0.01)
AB Dra	DN (ZC)	12.3	15.8	8–22	0.151 98		
DO Dra	NI	10.6B	15.1B		0.1658		
EF Eri	NI (AM)	13.7	17.7B		0.056 266		0.13
U Gem	DN (UG)	9.1	15.2	125	0.176 906	1.18 (0.15)	0.56 (0.06)
AH Her	DN (ZC)	11.3	14.7	7–27	0.258 116	0.95 (0.10)	0.76 (0.08)
AM Her	NI (AM)	12.0	15.5		0.128 927	0.39:	0.26:
DQ Her	Nb	1.4	17.7		0.193 621	0.62 (0.09)	0.44 (0.02)
V533 Her	Na	3.0	16.0		0.28:		
V795 Her	NI (DQ)	12.5B	13.2B		0.6157		
EX Hya	NI (DQ)	11.7	14.1	574	0.068 234	0.57 (0.27)	0.13 (0.01)
VW Hyi	DN (SU)	8.5	13.8	27	0.074 271	0.63 (0.15)	0.11 (0.02)
WX Hyi	DN (SU)	11.4	14.8	14	0.074 813	0.9 (0.3)	0.16 (0.05)
BL Hyi	NI (AM)	14.3	17.3		0.078 914		
DI Lac	Na	4.6	14.6B		0.543 773		
T Leo	DN (UG)	11.0	15.7	450	0.058 819	0.16 (0.04)	0.11 (0.01)
X Leo	DN (UG)	12.4	15.8	20	0.1644		
DP Leo	NI (AM)	17.5B	> 22		0.062 363	0.4:	0.1:
ST LMi	NI (AM)	15.0	17.2		0.079 089		

Continued next page

Table 25.2. (continued)

Object	Type [b]	m_{max} [c]	m_{min} [c]	T_n (days)	P_{orb} (days)	$\mathfrak{M}_1$ [d] ($\mathfrak{M}_\odot$)	$\mathfrak{M}_2$ [d] ($\mathfrak{M}_\odot$)
AY Lyr	DN (SU)	12.3	18.4B	8–43	0.0730		
MV Lyr	Nl (VY)	11.1	18B		0.1336		0.17:
TU Men	DN (SU)	11.6	> 16	37	0.1176	0.6	0.35
BT Mon	Na	4.5:	18.1		0.333 814		
CW Mon	DN (UG)	11.9	16.3	122	0.1762		
V380 Oph	Nl:	14.5	> 16.1		0.16	0.58 (0.19)	0.36 (0.04)
V426 Oph	Nl (DQ)	11.5	13.4	14	0.250		
V442 Oph	Nl	12.6	> 14.5	14	0.1406	0.34 (0.10)	0.31 (0.02)
V2051 Oph	DN (UG:)	13.0	17.5		0.062 428	0.44 (0.05)	0.13 (0.04)
CN Ori	DN (ZC)	11.9	16.3	8–22	0.1639	0.94 (0.25)	0.56 (0.02)
BD Pav	DN (UG)	12.4	> 16.5		0.179 30		
RU Peg	DN (UG)	9.0	13.1	75–85	0.3746	1.21 (0.19)	0.94 (0.04)
IP Peg	DN (UG)	12B	18.5B	95	0.158 208	0.8 (0.1)	0.35 (0.10)
GK Per	Na	0.2	14.0		1.996 803	0.9 (0.2)	0.25
RR Pic	Nb	1.2:	12.5		0.145 026	0.95:	0.4:
TY Psc	DN (SU)	11.7	16.3	11–35	0.071		
AO Psc	Nl (DQ)	13.3	13.6		0.149 626		
TY Psc	DN (SU)	12	16		0.080 63		
VV Pup	Nl (AM)	14.5	18.0		0.069 747	> 1.0	> 0.25
CP Pup	Na	0.2	15:		0.061 15:		
VW Sge	Nl	10.5	13.9	550	0.514 198	0.74	2.8
RZ Sge	DN (SU)	12.2	17.4B	62–93	0.067		
WY Sge	N: DN:	5.4	> 21B		0.153 634		
WZ Sge	DN (SU)	7:	15.5		0.056 688	0.8: (0.3)	0.09: (0.02)
V1223 Sgr	Nl (DQ)	12.3	16:		0.140 232	0.5 (0.1)	0.4
V3885 Sgr	Nl (UX)	9.6	10.3		0.206:	0.8: (0.2)	0.7: (0.1)
VY Scl	Nl (VY)	12.9	18.5		0.1662:		
VZ Scl	Nl (UX)	15.6	> 18		0.144 622	0.3:	0.4:
LX Ser	Nl (UX)	14.5	16.5		0.158 432	0.41 (0.09)	0.36 (0.02)
MR Ser	Nl (AM)	14.9	17		0.078 873		

Continued next page

Table 25.2. (continued)

Object	Type [b]	m_{max} [c]	m_{min} [c]	T_n (days)	P_{orb} (days)	$\mathfrak{M}_1$ [d] ($\mathfrak{M}_\odot$)	$\mathfrak{M}_2$ [d] ($\mathfrak{M}_\odot$)
RW Sex	Nl (UX)	10.4	10.8		0.2451	0.8 (0.3)	0.54 (0.09)
SW Sex	Nl (UX)	14.8B	16.7B		0.134 938	0.58 (0.20)	0.33 (0.06)
RW Tri	Nl (UX)	12.6	15.6		0.231 883	0.44 (0.08)	0.58 (0.03)
EK TrA	DN (SU)	12.0	> 17	231	0.0636		
SU UMa	DN (SU)	11.2	15.0	5–33	0.076 35		
SW UMa	DN (SU)	9	17.0B	459	0.056 81	0.71 (0.22)	0.10 (0.01)
UX UMa	Nl (UX)	12.7	14.1		0.196 671	0.43 (0.10)	0.47 (0.03)
AN UMa	Nl (AM)	14.5:	18.9B		0.079 753		
CH UMa	DN (UG)	10.7	15.9	204	0.3448:		
CU Vel	DN (SU)	10.7	15.5	113	0.0769		
IX Vel	Nl (UX)	9.1	10.0		0.1220:		
KO Vel	Nl (AM:)	16.7	17.7		0.071 83:		
TW Vir	DN (UG)	12.1	16.3	15–44	0.182 67	0.91 (0.25)	0.40 (0.02)
VW Vul	DN (UG:)	13.6	15.6	14–23	0.0731	0.24 (0.06)	0.14 (0.01)
QQ Vul	Nl (AM)	14.5	15.5		0.154 522		
PHL 227	Nl (UX)	13.5			0.1356		
H0538+608	Nl (AM)	14.6	> 17		0.129		
1H0542−407	Nl (DQ)	15.7			0.258:		
1H0832+488	DN (UG)	14.9B			0.268 10		
PG1030+590	Nl	14.9	16.4		0.1361		
PG1346+082	Nl (VY)	13.0	17.5		0.017 25:		

[a] The data in Table 24.2 are taken from the 4th edition of the *Catalogue of Cataclysmic Binaries, Low-Mass X-Ray Binaries, and Related Objects* of Ritter (1987). Dubious values are flagged with a colon ":".

[b] Na = fast nova; Nb = slow nova; Nc = very slow nova; Nr = recurrent nova; DN = dwarf nova of Type U Gem (UG), Z Cam (ZC), or SU UMa (SU); Nl = novalike variable of Type UX UMa (UX), VY SCL (VY), AM Her (AM), DQ Her (DQ), or AM CVn (AC).

[c] m_{max} denotes the maximum visual magnitude of the novae, recurrent novae, and dwarf novae during the outburst, for dwarf novae of SU UMa type during superoutburst, or the maximum attainable brightness of novalike variables during an active state.
m_{min} is the magnitude of novae, recurrent novae, and dwarf novae at minimum, that is, before, after, or between outbursts, respectively; for novalike variables, the smallest observed magnitude in the inactive state is given.
Those magnitudes labeled with a B were measured in the blue ($\sim$ 430 nm) spectral range.

[d] $\mathfrak{M}_1$ and $\mathfrak{M}_2$ are, respectively, the masses (in $\mathfrak{M}_\odot$) of the more luminous and usually more massive component (white dwarf + accretion disk) and of the fainter secondary component (cooler main-sequence star); the values in parentheses give the error in the mass determination.

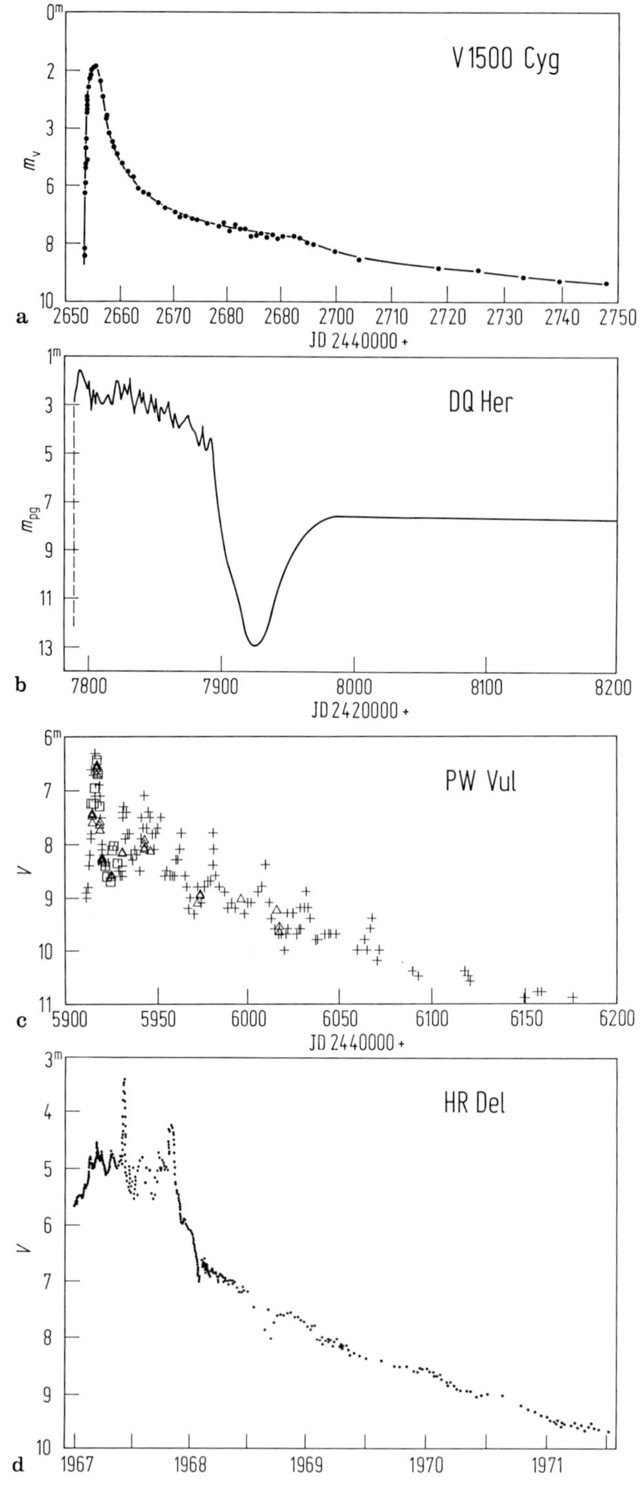

Fig. 25.27 a–f. Eruption light curves of novae and dwarf novae. **a** Fast nova (Na) V1500 Cygni (after Young et al. [25.102]); **b** slow nova (Nb) DQ Herculis (1934); **c** slow nova (Nb) PW Vulpeculae (1984): + = visual, □ = photoelectric, △ = IUE FES data; **d** very slow nova (Nc)

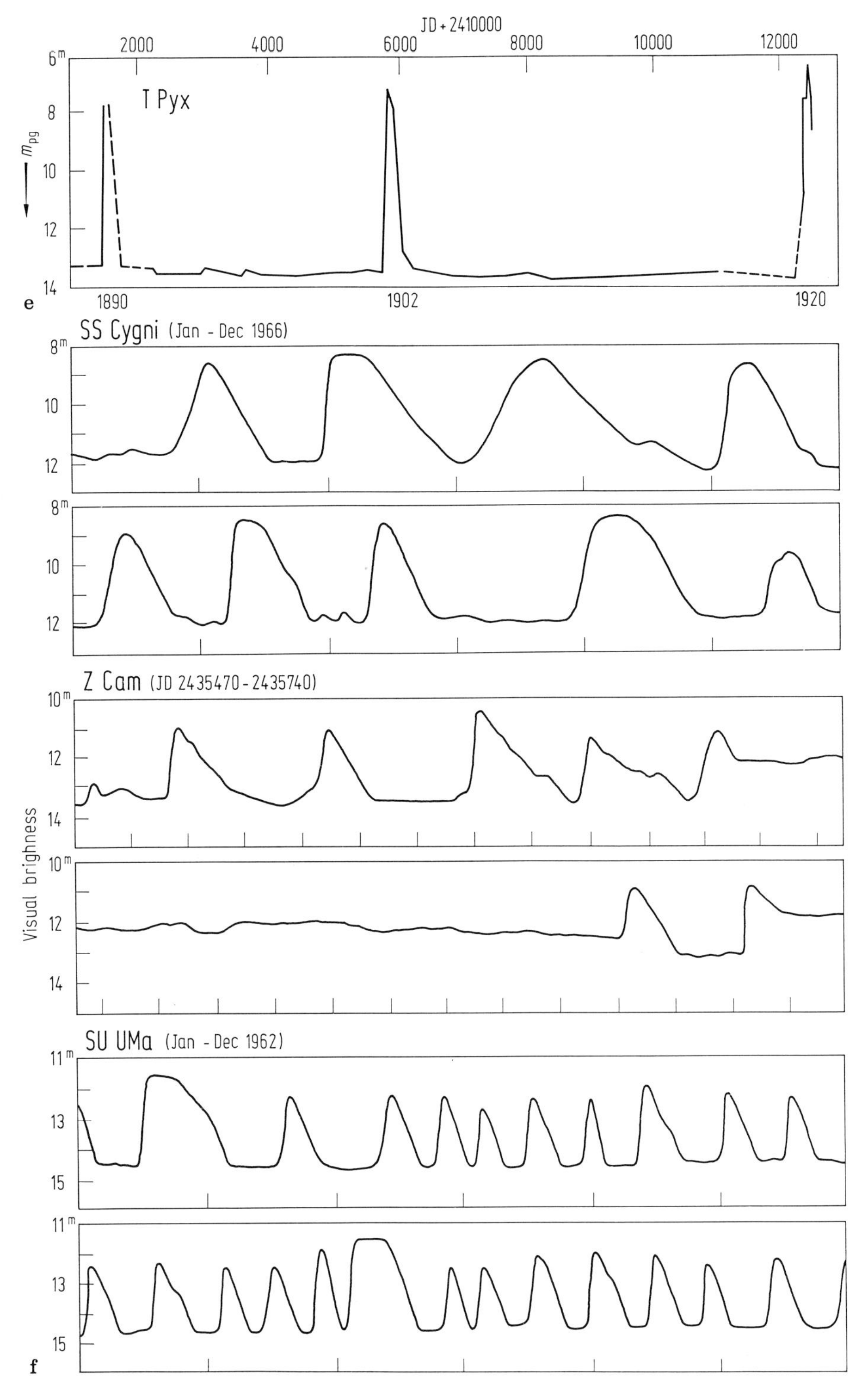

HR Delphini (1967); **e** recurrent nova (Nr) T Pyxidis (after Hoffmeister et al. [25.84]); **f** dwarf novae SS Cygni, Z Camelopardalis, and SU Ursae Majoris (after Glasby [25.103]).

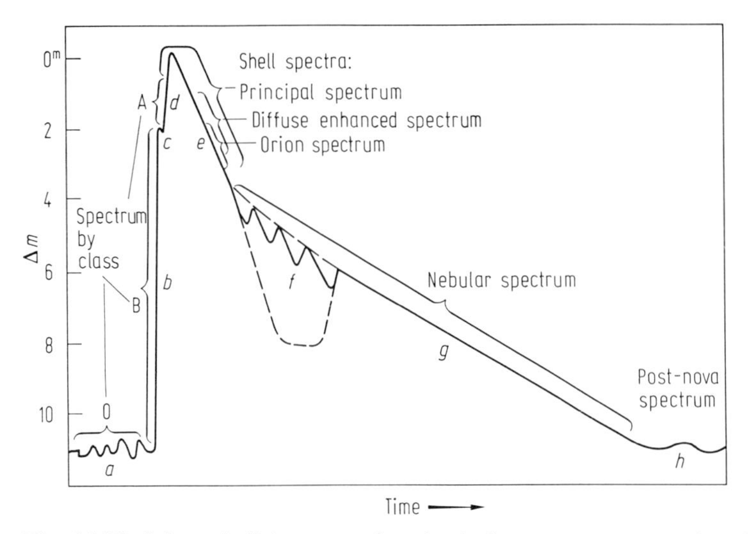

Fig. 25.28. Schematic light curve of a classical nova. *a* pre-nova; *b* and *d* ascent; *c* pre-maximum halt; *e* first descent; *f* transition stage; *g* final descent; *h* post-nova. After Hoffmeister, Richter, and Wenzel [25.60].

The distinction between novae and other classes of variables (e.g., supernovae) is made primarily by the spectral features during outburst, as has been described in detail by, for instance, Payne-Gaposhkin [25.57,58], McLaughlin [25.59], and Warner [25.48]. Pre-novae spectra are virtually unknown; the few exceptions show a strong, blue continuum of OB-type and with intense emission lines. At maximum, the spectrum resembles that of a B-, A-, or F-type star with broad absorption lines, Doppler blue-shifted by the amount of several hundred km s^{-1}, and originating in an expanding pseudo-photosphere. With increasing expansion of the shell after maximum, the spectrum changes to that of a supergiant with a middle to late spectral type. On the short-wavelength edge of the weakening, diffuse absorptions, new, sharp absorption components appear. Also, on the long-wavelength side of the broad absorption lines, symmetric with the laboratory wavelengths, broad emissions of hydrogen (Balmer lines) and singly-ionized metals (e.g., CaII, FeII) appear. While the absorption can be formed only in that column of the expanding shell projected onto the source, which moves toward the observer (hence the blueshift), the emissions are produced by photons escaping from all parts of the optically thin envelope, and this leads to a more or less symmetric Doppler broadening of the unshifted emission lines. Shortly after the appearance of the sharp absorptions (the *principal spectrum*), there appear parallel to these much more blueshifted absorptions (the *diffuse enhanced spectrum*), followed a short time later by another system of usually even broader absorption and emission lines (the *Orion spectrum* of strong HeI, CII, NII, OII lines). Near the beginning of the transition state, the nova shell, due to its continuous expansion, has rarefied sufficiently so that the plasma density approaches values typical for planetary nebulae. Absorption lines and continuum thus disappear almost entirely, and the spectral picture is created solely by strong emission lines characteristic of nebular spectra. At the beginning of this *nebular stage*, the *forbidden lines* are strongest in the visual and UV range (e.g., OIII lines at 495.9, 500.7, and 436.3 nm, and lines of multiply ionized C, N, Ne, Si, or Fe). Ionized coronal lines (up to Fe XIV) are also observed. The nebular stage may last up to several years, and then passes gradually into the post-nova stage, where the system arrives back at the original pre-nova brightness. The envelope is now entirely transparent, and the spectrum is once again generated, apart from some increasingly faint nebular lines, solely by the central binary system.

The complex appearance and the spectral variations during the course of a nova eruption can, in principle, be modeled as an explosion in the central binary system which ejects a small fraction (10^{-4} to 10^{-5} $\mathfrak{M}_\odot$) of the total system mass in the form of the nova shell. A substantial fraction of the total explosive energy (10^{37} to 10^{38} J) goes into the kinetic energy of the shell, accelerating it to speeds of several thousand km s^{-1}.

25.6.2.2.2 Recurrent Novae. Recurrent novae are distinguished from classical novae primarily, as their name suggests, by their repeated outbursts, with typical cycle lengths of a few decades, and by a smaller amplitude, averaging about 3 to 4 magnitudes less (see Fig. 25.27 e). It is generally accepted that classical novae also will show repeated eruptions, but at long time intervals of 10^4 to 10^5 years, so that only one outburst is historically documented in each case (Bath [25.83]) More detailed photometric and spectroscopic studies of recurrent novae (e.g., Barlow et al. [25.61]) lead to the expectation, however, that their secondary components—contrary to those of most classical novae—are evolved, late-type giants and not main-sequence stars. The origin and progress of the recurrent eruptions are largely similar with those of classical novae. Only six objects are known, the best known and investigated being T CrB (outbursts: 1866, 1946), T Pyx (1890, 1902, 1920, 1944, 1966), U Sco (1863, 1906, 1936, 1979), and RS Oph (1898, 1933, 1958, 1967, 1985). Simultaneous UV and IR measurements in photometry and spectroscopy have thus far been made only in the eruptions of U Sco in 1979 by Barlow et al. [25.61] and of RS Oph in 1985 by Snijders [25.62].

25.6.2.2.3 Dwarf Novae. Once generally called U Geminorum stars, dwarf novae received their name from the similarity to the nova phenomenon, but the cyclicly recurring outbursts have an amplitude several magnitudes less than in classical novae.

The structure of these systems is similarly described by the basic model of an interacting close binary, and the cause of the outbursts is here also due to mass transfer between the components. The basic difference, however, between novae and dwarf novae consists in the physical process which triggers the outburst. The responsible mechanisms will be discussed later.

As with recurrent and classical novae, dwarf novae also show a relation between the mean time interval between eruption outbursts and the eruption amplitude, the previously mentioned Kukarkin-Parenago relation (see e.g., Antipova [25.52]). Dwarf novae can be subdivided by characteristic photometric properties and by eruption patterns into three groups:

1. *U Geminorum* or *SS Cygni systems* show outbursts which raise the visual brightness by 2 to 8 magnitudes in one or two days, with a subsequent decline to minimum in a few days or weeks (see Table 25.2). Such outbursts occur quasi-periodically

and in similar shape with an average cycle length of about ten to several hundred days.

2. *SU Ursae Majoris systems* are different from U Gem stars. These systems show, after several "normal" outbursts, a *supermaximum* which is about 1 to 2 magnitudes brighter and lasts several days longer than a normal maximum (see e.g., Vogt [25.63]). An interesting phenomenon observed during supereruptions are periodic "superhumps," sharp intensity peaks superposed on the broad supermaximum, and recurring with a photometric period which generally differs by only a few percent from the orbital spectroscopic period of the binary. A notable feature of SU UMa systems is their very short orbital periods, usually less than 2 hours. Table 25.2 lists the orbital periods of many dwarf novae; for SU UMa stars, the maximum brightnesses given are mean magnitudes during supermaxima.
3. *Z Camelopardalis systems* show quasi-periodic outbursts often interrupted by longer lasting stationary brightness levels. The brightness may, shortly after time of maximum, remain constant at about 1/3 of the amplitude below maximum intensity for some days to many months, before the system then, in most cases, drops to minimum brightness.

25.6.2.2.4 Nova-like Objects. A heterogeneous group of objects partly of different physical structure and evolutionary status is comprised under the term *nova-like systems*. They are phenomenologically related to cataclysmic systems, but show no outbursts. As for dwarf novae at minimum and for ex-novae, random brightness variations from seconds to minutes, and with amplitudes of some hundredths to tenths of a magnitude are found. This phenomenon, called *flickering*, is caused by the mass transfer in the close binary (see Mumford [25.64]; Robinson [25.65]; Warner [25.48]; also Fig. 25.31).

When studied photoelectrically with high time resolution in conjunction with numerical methods of Fourier analysis, some nova-like systems and dwarf novae also exhibit coherent, periodic, or quasi-periodic oscillations with amplitudes of some thousandths to hundredths of a magnitude, and with periods in the range from seconds to minutes (Wade and Ward [25.50]). These periodic variations are interpreted as oscillations of the degenerate component of the binary. Examples are the pulsations of the ex-novae DQ Herculis and V533 Her with periods of 71 and 64 seconds, respectively (Nather [25.66], Patterson [25.67]), or of the AM Her-type system 3A 0729+103 with a 15.2 minute period (Warner [25.68]; see also Fig. 25.29).

The spectra of nova-like systems feature a continuum similar to that of hot, OB-type stars, which strongly increases toward the ultraviolet. As in dwarf novae, variable emission components are superimposed upon the broad Balmer absorption lines. Also observed, among others, are HeI, HeII (especially at 468.6 nm) CIII/NIII (464 nm), and forbidden emissions (e.g., of OIII, NeIII), which also occur in the nebular state of novae and in planetary nebulae. Moreover, in longer-period systems ($P \gtrsim 6^{\mathrm{h}}$), an absorption-line spectrum of late type (K,M) can be identified.

In recent years, detailed photometric, spectroscopic, and polarimetric studies have led to a systematic division of the various nova-like systems into the following subclasses:

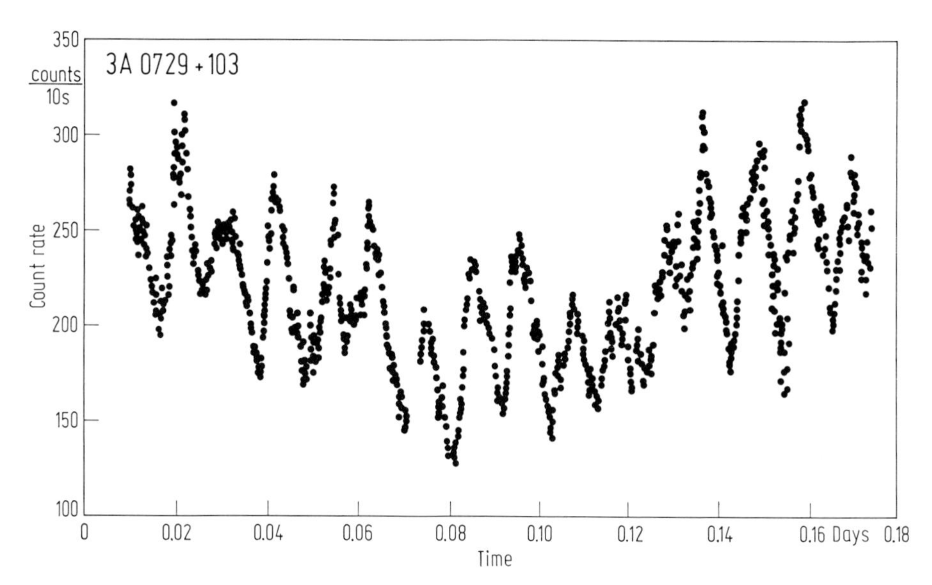

Fig. 25.29. Coherent oscillation of the intermediate polar system 3A 0729+103 with a period of 15.2 minutes. Photon counts are graphed against time. After Warner [25.68].

1. *UX Ursae Majoris systems*: these often have near-constant brightness over several decades. In addition to the usual flickering and statistical fluctuations of 0.1 to 1 mag within minutes and hours, low-amplitude coherent oscillations are in some cases observed. The spectra much resemble those of dwarf novae at minimum.
2. *VY Sculptoris systems*: also dubbed "anti-dwarf novae," these objects are found predominantly in so-called active phases which last for a few years when they have—apart from fluctuations by some tenths of a magnitude and the usual flickering—nearly constant mean brightness. These active states are sporadically interrupted by considerably shorter "inactive" phases, with the brightness diminished by 3 to 5 magnitudes. A well known specimen is TT Ari. The spectra of VY Scl stars in active phases can be compared with those of dwarf novae in eruption, or of Z Cam systems at stationary brightness, whereas the spectra in inactive phases are similar to those of dwarf novae at minimum, especially through the appearance of certain emission lines (Duerbeck et al. [25.69]; Krautter et al. [25.70]).
3. *AM Herculis systems* or "polars": their distinctive feature is a strong and time-variable circular polarization (> 10%) of visual and IR radiation. The continuous intensity distribution in the optical range can be explained by synchrotron radiation emitted by electrons in magnetic fields of the order 10^3 to 10^4 tesla. In the higher energy range, soft X-ray bremsstrahlung ($kT \approx 10$ to 100 eV, corresponding to temperatures of about 10^5 to 10^6 K), and hard X-rays ($kT > 10$ keV; $T \approx 10^8$ K) are found. The existence of locally strong magnetic fields is in some systems also documented by spectroscopic measurements during the rather short, inactive phases; the Balmer absorption lines of the degenerate component show Zeeman splitting corresponding to a magnetic strength of some 10^3 tesla (Schwope and

Beuermann [25.71]). Similar to VY Scl but contrary to dwarf novae in eruption, the prominent active phases show emission lines, particularly of hydrogen and helium in the visual range, and of multiply ionized elements in the UV. A recent survey on the properties of AM Her systems is given by Liebert and Stockman [25.72].

4. *DQ Herculis systems*, also called *intermediate polars*, have photometric and spectroscopic properties similar to polars, that is to say, long-lasting, active states are occasionally interrupted by phases of lesser brightness ($\Delta M \approx 1$ to 3), and the spectra show strong emission lines. Typical for DQ Her stars, however, are brightness pulsations with amplitudes of some tenths of magnitude (Mouchet [25.73]) and long-term stable periods ranging from about 1 minute to 1 hour. The name "intermediate polars" derives from the comparatively weaker polarization of the optical radiation, which in the case of DQ Her is under 1% as compared with 20% in polars. A general description of the properties of DQ Her systems is given by Warner [25.74].
5. *AM Canum Venaticorum systems*: the two objects AM CVn and GP Com are eclipsing binaries with the shortest known orbital periods of 18 minutes and 46 minutes, respectively. These binaries, each presumably composed of two degenerates, show a pure helium emission-line spectrum. The relation with cataclysmic variables is photometrically indicated by the chronic flickering in the visual range, and by periodic, coherent oscillations with periods of a few minutes. Superposed on the rapid fluctuations are phase-dependent, ellipsoidal brightness changes owing to the orientational light variation of the binary components deformed by tidal action (Nather, Robinson, and Stover [25.75]).

25.6.2.2.5 The Binary Model. Spectroscopic observations of cataclysmic variables have revealed them to be close binary systems. The cogent proof are periodic radial velocity variations of spectral lines of one or both components generated by the Doppler effect of the orbital motion about the center of the system, and/or the occurrence of eclipse effects as well as the spectroscopic or spectrophotometric evidence for a cool source of radiation in the red and infrared ranges.

Of the approximately 600 known cataclysmic binaries, orbital periods are known for over 100; of this subset other system parameters are also known. The periods lie, with few exceptions, in the interval between about 75 minutes and 15 hours, clustering between 1.5 and 2 hours, and also between 3 and 4 hours, while there is a conspicuous gap between 2 and 3 hours. A catalogue of objects with known orbital periods has been published by Ritter [25.53], from which Table 25.2 is excerpted.

The now generally accepted model of a cataclysmic system is schematically illustrated in Fig. 25.30. The more luminous (and usually more massive) primary component of the semi-detached binary system always is a compact white dwarf (a hot, degenerate star with radius of about 0.01 $R_\odot$), and thus about the size of the Earth, and with a mass between 0.5 and 1 $\mathfrak{M}_\odot$. The secondary component is normally a cool main-sequence star of spectral type G, K, or M which fills its maximally permitted Roche-volume (contact component).

The characteristic effects of this class of variables originate through a mass stream directed from the secondary toward the primary component. Any expansion owing

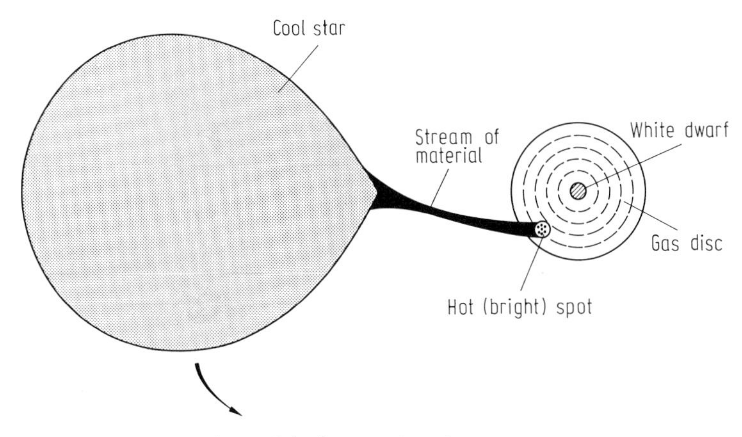

Fig. 25.30. Schematic model of a cataclysmic system.

to evolution or any contraction of the Roche surface due to loss of mass and of orbital momentum leads to a mass transfer through the inner equilibrium point into the potential well of the white dwarf. This compact object occupies only a small fraction of its Roche volume, so that the high specific angular momentum of the transferred mass makes the mass spiral into an accretion disk surrounding the white dwarf. A viscosity process thus far not understood causes mass to move inward and momentum outward through this disk, so that the matter lost by the secondary component accretes at least partly onto the surface of the white dwarf.

In the collisional region of the mass stream and outer edge of the disk, a shock front is generated and heats the surrounding plasma. The energy liberated is radiated away in the optical and UV ranges. This source of radiation is called, due to the rather high luminosity, a *hot spot*, although its extension and geometry are still largely unclear. Its existence, however, has been proven in many cases, and there is substantial support for the correctness of the basic model for cataclysmic variables. The best evidence is the occurrence of depressions of the total brightness of the system, while the hot spot is occulted by the secondary component and partly by optically dense matter in the accretion disk; this is illustrated in Fig. 25.31 for the dwarf nova U Geminorum. The discontinuous and inhomogeneous conversion of potential energy into thermal energy around the hot spot is probably part of the origin of the familiar flickering.

The mass transfer proceeds differently when the primary component is a magnetic white dwarf with a field strength of the order of 10^3 tesla. The plasma flow from the secondary component is channeled in these AM Her-type systems (polars) along the field lines of the magnetic dipole field directly toward the magnetic poles onto the surface of the white dwarf. The accretion therefore is not directed through a differentially rotating disk in the orbital plane of the binary, but through the spatial structure of the magnetic field. The magnetic poles then heat to temperatures up to 10^8 K, so that the surrounding plasma emits hard X-ray *bremsstrahlung* ($kT > 10$ keV) and thermal X-rays ($kT \approx 10$–100 eV). One more confirmation of this model is afforded by the observed, strong circular polarization of visual and infrared radiation.

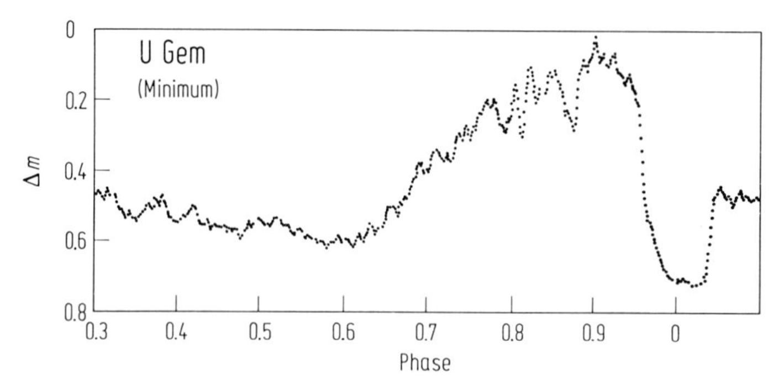

Fig. 25.31. High-time-resolution photometric light curve for the dwarf nova U Geminorum in minimum state. The deep depression in brightness at phase 0.0 is caused by the occultation of the hot spot by the secondary component. The much-reduced flickering during eclipse is also distinctly seen. After Warner and Nather [25.76].

In systems of lesser magnetic fields ($\approx$ 100 tesla) hybrid cases with both disk and column accretion are also possible, and this is presumably realized in the intermediate polars (DQ Her sytems). A detailed and up-to-date survey on magnetic cataclysmic variables has been given by Lamb and Melia [25.77].

25.6.2.2.6 Theories of the Eruption. Novae. The total amount of energy liberated in a typical eruption is 10^{37} to 10^{38} Joules. Some of this energy appears as photons, with wavelengths in all ranges of the electromagnetic spectrum, and which are radiated away within a few weeks to months. Much of it also supplies the kinetic energy of the ejected shell. For comparison, the Sun would require several thousands of years to emit this amount of energy at its present rate. Evidently, an energy reservoir this large is of nuclear origin, and its energy is freed via an explosive event.

It has been learned that thermonuclear chain reactions can and do occur near the surface of the white dwarf, essentially fusing hydrogen into helium. The mass transfer from the cool main-sequence star continuously increases the mass of the hydrogen-rich matter in the accretion disk, and with time deposits it onto the surface of the white dwarf. Typical accretion rates are 10^{-8} to 10^{-9} $\mathfrak{M}_\odot$ yr^{-1}. Owing to the steep potential gradient near the massive, compact white dwarf, the accretion of matter converts so much gravitational energy into thermal energy that the surface temperature of the white dwarf increases to over 10^8 K.

An important parameter for the beginning and progress of thermonuclear fusion is the relative abundance of heavier elements in the white dwarf atmosphere. Convective mixing within the upper atmospheric layers with the collected hydrogen-rich matter provides the C, N, and O nuclei as catalysts for the Bethe-Weizsäcker cycle, and accelerates the hydrogen burning. The manner in which the chain reaction develops also depends on the mass and the temperature of the white dwarf, which determine, among other things, the *speed class* of the nova event.

Theoretical model computations show that this process frees enough energy to explain the observed radiation in a nova eruption, including the observed expansion velocities of the ejected envelope. The computed light curves are in qualitatively and quantitatively good agreement with the observations. A survey on the recent models of nova eruptions is found in Starrfield and Sparks [25.78].

Part of the matter accumulated on the surface of the white dwarf surface and the accretion disk is blasted by the explosion into interstellar space to become the nova shell. It expands with speeds of several thousand km s^{-1}, and can be identified spectroscopically, and in some cases, by direct photographs. Famous examples are V603 Aql (Nova Aquilae 1918) and DQ Her (Nova Herculis 1934), whose shells have been followed over decades (see e.g., Mustel and Boyarchuk [25.79]; Weaver [25.80]). A systematic study of numerous nova envelopes has been made by Duerbeck and Seitter [25.81].

Dwarf Novae. The eruption of a dwarf nova liberates about 10^6 times less energy than the eruption of a classical nova. This in itself indicates that dwarf nova outbursts are caused by a wholly different mechanism. It is not entirely clear whether the eruption is triggered by the secondary or the primary component.

One class of models treats quasi-periodic, hydrodynamical instabilities in near-surface layers of the cool secondary star which arise when the mass transfer is temporarily enhanced and affects, by dissipation of a larger amount of potential energy over a short time, the flaring up of the system (e.g., Bath and Pringle [25.82]; Bath [25.83]). Other studies show that the outburst could be caused by cyclic instabilities in the structure of the accretion disk. In principle, the gravitational energy stored in the disk matter suffices to supply the energy of the eruption. It had been assumed earlier that sudden changes in the viscosity of the accretion disk may temporarily cause an increased matter flow in the radial direction, and thus an increased liberation of potential energy, and hence an outburst of brightness (Osaki [25.84]). The cause of such viscosity changes can be explained as follows. A sudden increase in viscosity, and hence of mass accretion, occurs in those parts of the disk where, at sufficiently low temperatures ($< 10^4$ K), there occurs a transition from complete to partial ionization of the plasma, and convection sets in (Meyer and Meyer-Hofmeister [25.85]; Meyer [25.86]). Future simultaneous measurements of dwarf novae before and during eruption in different spectral ranges from the UV to the IR should permit a definitive clarification of the mechanism of outburst.

25.6.3 X-ray Binaries

The first cosmic X-ray sources were found in 1963 by rocket experiments. The most intense point-source was found in the constellation Scorpius, and was, being the first X-ray source discovered there, named Sco X-1. The question of the mechanism of energy generation was answered soon afterwards. It was supposed that stellar X-ray sources are interacting, close binary systems in which transfer and accretion of matter onto a compact component causes emission of high-energy radiation, ultimately to the large potential gradient.

While in cataclysmic systems the matter is collected by a white dwarf, which, as a degenerate object, has a mass of about 1 $\mathfrak{M}_\odot$ and a diameter of about 10^4 km (density $\approx 10^7$ g cm^{-3}), the much higher luminosity (by several orders of magnitude) in X-ray sources indicates that the accreting component is substantially more compact than a

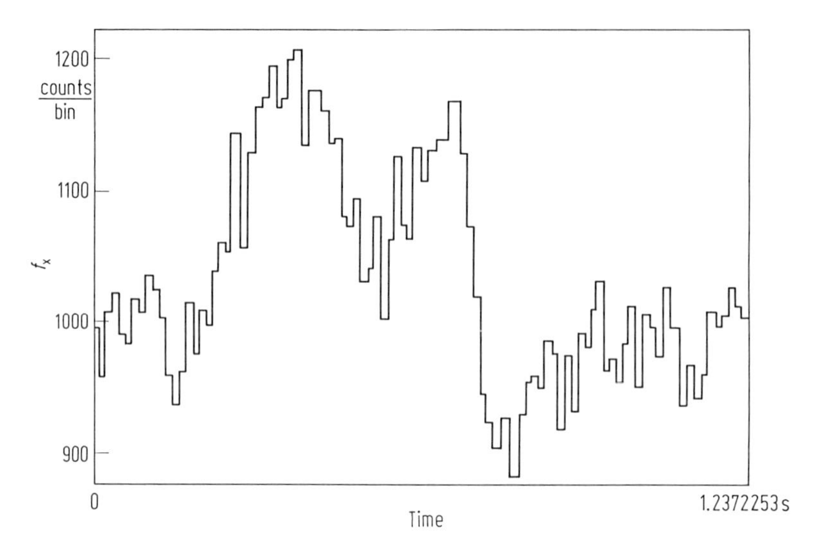

Fig. 25.32. Typical X-ray pulse from Hercules X-1, which repeats in very similar shape with a period of 1.237 2253 seconds. After Doxsey et al. [25.87].

white dwarf. It was supposed from the very beginning that a neutron star (diameter $\approx$ 10 km, density $\approx 10^{15}$ g cm^{-3}) or a black hole is present here.

This model was solidified by the discovery, made with the X-ray satellite *UHURU* launched in 1971, of the two X-ray binaries Her X-1 and Cen X-3, as well as through numerous subsequent identifications of X-ray sources with binary systems. The primary component is usually optically observable, either as an O- or B-type giant or supergiant, or as a main-sequence star of middle- to late-type, while the optically invisible secondary star is identified by an X-ray luminosity of 10^{36} to 10^{38} erg s^{-1}. The X-rays show variations on very short time scales, from seconds to milliseconds, and often exhibit pulses which occur periodically or quasi-periodically. As an example, Fig. 25.32 shows an X-ray pulse from Her X-1 which repeats in similar form every 1.24 seconds.

By contrast with pulsars, the breadth of the pulse relative to the length of the period is here substantially larger. It would therefore be somewhat problematic to try to explain the pulsed radiation by a similar "lighthouse model," in which the synchrotron radiation is emitted in a light cone rotating with the magnetic field. Appearance and disappearance of a "hot spot" near the magnetic pole of the neutron star supplies a most probable explanation. An important proof for the binary model is in the pulsation period showing a Doppler frequency variation owing to orbital motion of the neutron star about the center of mass of the system; the period agrees with the orbital period of the system, which is on the order of days. X-ray binaries with such very precisely pulsed sources offer the opportunity to estimate the masses of the neutron star components, which is of utmost importance for validating theories of the late evolutionary phases of stars. The binary character of these objects is proven in several cases also by the periodic occultation of the X-ray source, when a temporary disappearance of the X-ray pulses is observed over about 10–30% of the orbital period, and it also recurs exactly at that interval of phase where the normal primary component eclipses the X-rays for the observer.

Table 25.3 gives the characteristic parameters of some well-studied X-ray binaries with some optically identified primary components. Depending on the mass of the visible primary, low-mass (e.g., Her X-1) and high-mass (e.g., Cen X-3) systems are distinguished. The high-mass systems have giant or supergiant components of types O and B, and with masses from about 10 to 30 $\mathfrak{M}_\odot$. Low-mass systems, on the other hand, have late-type (i.e., solar-type) primary components with masses of the order of 1 $\mathfrak{M}_\odot$. Numerous other systems seem also to belong to the low-mass X-ray binaries, which appear primarily in the direction of the "galactic bulge" (indicated earlier by the prefix GX). These low-mass X-ray binaries (frequently referred to as LMXBs) show pulsations only in a few cases and, unexpectedly, eclipses are not observed either. The binary nature of these sources is, however, accepted as a certain conclusion. Another distinguishing criterion is the orientational light variation in the visual range, which occurs because of the time-varying (due to orbital motion) perspective on the tidally deformed primary star, and the different surface brightnesses presented by that star. In very close low-mass systems, both the tidal deformation of the optically visible companion and the heating of that part of its surface which points toward the X-ray component and is thus exposed to the high-energy radiation contribute more to the light curve than is the case in the more widely-separated high-mass systems. In the latter, the radiation is produced almost entirely by the high-luminosity O or B supergiant and is rather little modulated by such effects. Hence, the ratio of brightness variation relative to mean visual brightness, $\Delta m_V / m_V$, is much smaller in high-mass than in low-mass systems. Observed brightness amplitudes are here only a few hundredths to tenths of a magnitude, while low-mass systems may reach amplitudes of several magnitudes (e.g., $\Delta m_V \approx$ 1–2 mag for Her X-1 = HZ Her).

An important subclass of X-ray binaries must be at least mentioned here: the so-called *burst sources*, discovered and much studied by W.H.G. Lewin and his colleagues at MIT. The X-ray "light curve" remotely resembles the optical behavior of some dwarf novae but the time scale of bursting is much shorter, from hours to a few days (Type I burst sources). Even shorter is the time scale in the case of the still unique Rapid Burster (MXB 1730−335), characterized by time intervals from seconds to minutes between fairly regular bursts of Type II. Instabilities in the accretion flow may play the dominant role in this case.

Cases of extreme peculiarity among X-ray binaries are SS 433 (V 1343 Aql) and Cyg X-3 (V 1521 Cyg).

SS 433. Frequently referred to as a "small-scale model" of an active galactic nucleus when discovered as a relatively faint X-ray source, this object did not reveal its unusual nature immediately. This was first recognized by observing very large periodic Doppler shifts in its spectrum, sometimes called "moving lines"; the corresponding radial velocities reach no less than 26% of the velocity of light! These velocities are ascribed to a pair of relativistic jets of gas which were ejected in opposite directions from a central object. The central object is a distant binary with a period of 13.1 days, where the visible component is a strongly reddened, highly luminous star with a peculiar spectrum probably showing substantial mass loss. The other component is

Table 25.3a. Parameters of X-ray double stars. After Manchester and Taylor [25.106].

X-ray Source/ Optical ID	X-ray Luminosity (erg s^{-1})	Distance (kpc)	Orbital Period (days)	Orbital Inclination (degrees)
SMC X-1 / SK 160	1.4×10^{38}	60 ± 10	3.8927 ± 0.0010	70°
Cen X-3 / Krzeminski's Star	6×10^{37}	5–10	2.087 129 $\pm 0.000\ 007$	90°
30 1700-37 / HD 153919	6×10^{37}	1.5 ± 0.5	3.4120 ± 0.0003	90°
Cyg X-1 / HD 226868	1.4×10^{37}	2.5 ± 0.5	5.5999 ± 0.0009	27°
Sco X-1 / V818 Sco	10^{37}	0.3–1	0.787 313 $\pm 0.000\ 001$	?
Her X-1 / HD 153919	7×10^{36}	2–6	1.700 165 $\pm 0.000\ 002$	90°
Vela X-1 / HD 77581	3.2×10^{36}	$1.4 \pm 0.38.95$	90° ± 0.02	

Table 25.3b. Parameters of X-ray double stars (continued).

X-ray Source/ Optical ID	$\mathfrak{M}/\mathfrak{M}_\odot$	X-ray Source		Optical Component		m_V/ Δm_V
		Variability	Occultation	$\mathfrak{M}/\mathfrak{M}_\odot$	Spectral Type	
SMC X-1 / SK 160	2.2–4.2	periodic 0.7157 s	+	26–30	B0.5 I	+13.3 $0^m\!.09$
Cen X-3 / Krzeminski's Star	0.6–1.8	periodic 4.842 s	+	16.5–20	O6.5 II	+13.4 0.08
30 1700-37 / HD 153919	0.6	not periodic to 0.1 s	+	10	O6f	+6.6 0.04
Cyg X-1 / HD 226868	0.6–1.8	quasi-periodic to 1 ms	–	> 10	O9.7 Iab	+8.9 –
Sco X-1 / V818 Sco	~ 1	not periodic to 1 s	–	< 2	?	+13 0.2
Her X-1 / HD 153919	~ 1	periodic 1.23782 s	+	~ 2	A7–B0	~ +14 ~ 1
Vela X-1 / HD 77581	1.35–1.9	not periodic to 1 s; periodic 283 s	+	18.5–24	B0.5 Ib	+6.9 0.07

a compact object, presumably a neutron star, surrounded by a massive accretion disk. The relativistic jets emerge from the center of the disk and perpendicular to it. Owing to a precession of the accretion disk with a period of 164 days, the jets will move along a conic surface with the same period, alternately approaching and receding from Earth. The gas, perhaps in the shape of blobs or pellets, moves outward in the jets at a velocity of about 80 000 km s^{-1}. The jets show a remarkable degree of collimation, and, according to observations made with the *EINSTEIN* satellite (*HEAO-2*), the major axis of the surrounding diffuse nebula W 50 shows an alignment with the extended source SS 433. The energy budget of the source and the mechanism of the alignment still present major unsolved problems.

Cyg X-3. Radiation from this object is unique on several counts. The primary energy source may be accretion onto a compact object, but the binary motion must be only indirectly inferred from a smooth and stable, 4.8 hour modulation of the X-ray intensity. It must be assumed that the observed X-ray photons are scattered into the direction toward Earth by some optically thick medium, the position and structure of which is not yet understood. Thus, one does not "see" this X-ray binary in the same way as the other systems. Moreover, Cyg X-3 also radiates at radio frequencies in a very peculiar way, producing giant outbursts, which were first observed in 1972 September. High- and very high-energy gamma rays (up to 10^{16} eV) have been frequently observed from Cyg X-3, and a recent detection of highly energetic *neutral* particles coming from the same direction is still puzzling particle physicists.

In the early 1970s, as the existence and the detectability of *black holes* were being very intensely discussed, it was somewhat prematurely conjectured that the puzzling characteristics of peculiar systems such as ε Aurigae or β Lyrae could be ascribed to black hole components. Among the massive X-ray binaries, however, a few systems do indicate the presence of a possible black hole, notably the systems Cyg X-1 and LMC X-3. The main observational find is the large mass of the invisible component, which is far above the assumed upper limit for neutron star masses (perhaps around 2.5 $\mathfrak{M}_\odot$), along with the complete absence of X-ray pulsations. The *minimum* mass of the X-ray component can be estimated from the observed mass function (see Sect. 25.3.1) by putting $\sin i = 1$ and using a "plausible" value for the mass of the OB-type component. The values obtained for the invisible, compact component (its compactness follows, of course, from the strong X-ray radiation) are about 6 $\mathfrak{M}_\odot$ in the case of Cyg X-1 and even higher, around 10 $\mathfrak{M}_\odot$, for LMC X-3. One must assign, for instance, the disquietingly low value of 2 to 3 $\mathfrak{M}_\odot$ for the O9.7 Ib component in the Cyg-X-1 system, in order to get the mass of the invisible component in the range of neutron star masses. Yet, closer studies of color and spectrum of the B-type supergiant render such hypotheses very unlikely. The case seems to be very strong, but a general consensus on the existence of black holes in certain X-ray binaries is not yet being reached, although recent studies did add two more, rather strong candidates, the compact objects in the transient sources A0620–00 and V404 Cygni, to the list.

25.6.4 Symbiotic Stars

Symbiotic variables are spectroscopic binaries with spectra composed of an absorption spectrum of late-type and combined with emission lines implying a rather high degree of ionization (e.g. He II, C IV, N V, O IV, Ne V) and with large excitation energies (UV lines).

Some objects show the same spectral features but no variability; these may be single stars with peculiar effects. The binary nature of symbiotic stars has been proven with rare exceptions. Certain safe evidence are periodic radial velocity variations, as shown, for example, in Fig. 25.3, or the occurrence of eclipse effects. The observed periodic or quasi-periodic light variations typically have amplitudes of 1 to 3 magnitudes, and photometric and spectroscopic periods of several hundreds of days. The shape of the light curve is variable with time, and partly irregular. Substantial or longer lasting increases in luminosity often occur, and for this reason these objects were formerly also called "nova-like" variables, in contrast to the current definition. A well-known example of such an eruption is the active state in RR Tel, which lasted several years. This system increased its luminosity in the year 1944 by about 7 magnitudes. Other objects, such as AG Peg, also show brightness increases which are stretched over several decades, and these qualitatively resemble outbursts of extremely slow novae. Outside of active states, the light curves with their sinusoidal variations resemble those of Mira variables. The spectra of the late-type components are usually classified as M giants, and are compatible with that type of pulsational variability. In some objects, a strong infrared excess and radio emission indicate the presence of dense, extended circumstellar dust shells. In this context, it is being considered whether symbiotic stars may be the precursors of planetary nebulae. The preliminary model for symbiotic binaries originates from the assumption that these systems consist of a cool, late-type giant with a radius of about 100 $R_\odot$ and a hot subdwarf which is smaller and substantially hotter than the Sun. The entire system is enveloped in a gaseous shell formed by nonconservative mass transfer or by mass loss from the expanded giant component due to pulsations. The variability of the system is contributed to by the pulsations of the giant, variations in the mass accretion of the hot subdwarf, or changes of density and structure in the circumstellar envelope. In most cases, all effects may be combined. A list of selected symbiotic binaries with magnitudes and periods is given, for instance, by Hoffmeister, Richter, and Wenzel [25.60].

25.6.5 Flare Stars

Flare stars, also known as UV Ceti-type stars, are late dwarfs of spectral types K to M (mostly dMe), which are made conspicuous by their repeated, very rapid flaring. The visual brightness in these eruptions or "flares," which occur at irregular intervals of typically a few days, increases within a few seconds to minutes by from 1 to 6 magnitudes, and then retreats within a few minutes to 2 hours back to normal minimum brightness. On the average, a brightness gradient of about 0.05 to 0.1 mag per second is reached during ascent to maximum; in some cases it can be substantially larger. (In UV Ceti up to 2.8 mag per second were measured at a total amplitude of 6.5 mag.) Along with such normal flares, fast photometric methods have also revealed very rapid "spike" flares, where the entire event occurred within less than one minute. Figures 25.34 a,b show a typical example each of normal and spiked flares.

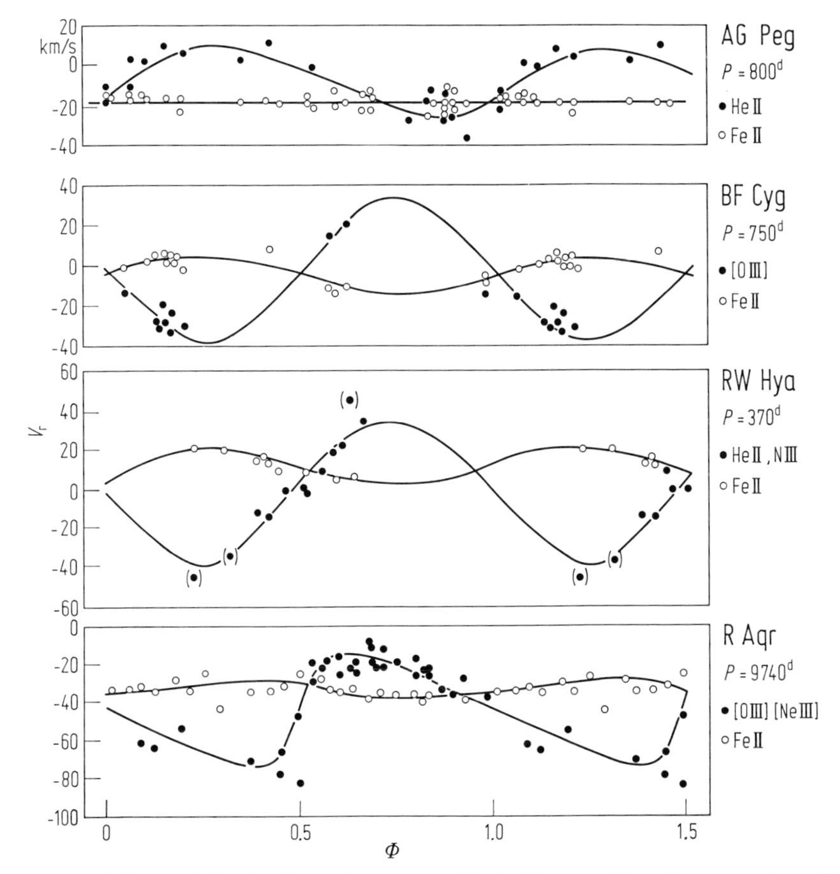

Fig. 25.33. Radial velocities of some symbiotic binaries. The diagrams show the periodic Doppler shifts of different emission lines. After Boyarchuk [25.88].

The amplitude of a flare increases with decreasing wavelength, and therefore is larger in the blue and ultraviolet than in red light. According to Moffet [25.89], the mean $B-V$ color of the flare radiation is $+0.34 \pm 0.44$, whereas the $U-B$ index of -0.88 ± 0.31 corresponds to a substantially hotter continuum source (UV excess). During eruptions, the spectral intensity distribution changes in such a way that the continuum in the blue and ultraviolet regions becomes relatively stronger. Also, there appear numerous emission lines (particularly of H and He), which are much weaker or absent at minimum.

Owing to the low luminosity of very late-type dwarf stars, which are located slightly above the lower main sequence in the Hertzsprung-Russell diagram, UV Ceti stars can be detected only in the local neighborhood of the Sun (within 20 pc). Presumably they are evolutionarily young stars which, despite an age of about 10^9 years, have not reached the main-sequence state since the evolution at a mass of only a few tenths of the Sun's mass is correspondingly slow. Similar flare activities are observed in the

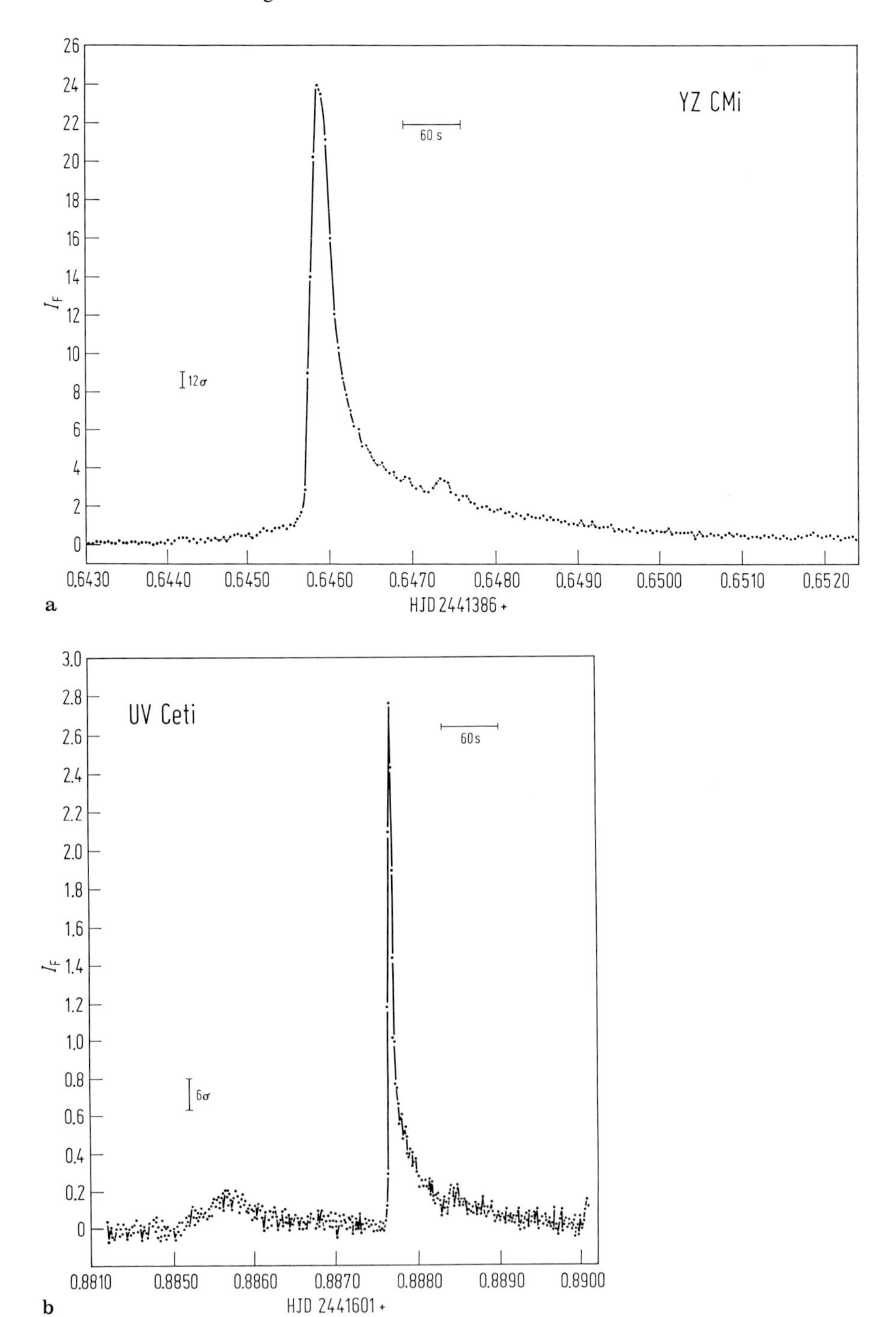

Fig. 25.34. a Typical flare of YZ Canis Minoris, the fast photoelectric photometry was made with a U-filter. **b** Photoelectric measurement in white light with no filter of a spiked flare of UV Ceti, recorded with an integration time of 1 second. After Moffet [25.89].

evolutionarily very young T Tauri stars, which are found at somewhat earlier spectral classes (F to K) in the HR diagram, but also above the main sequence.

The physical process responsible for flare activity has thus far not been made clear. A possible relation with various mechanisms of solar activity and also with eruptive phenomena within the chromosphere and corona of the Sun (solar flares, surges, prominences) has been noted. It is thus reasonable to link the flares with complex magnetic field structures undergoing rapid changes with time and space. In this connection, the emission of relativistic electrons and photon scattering by the inverse Compton effect could cause short-term enhanced radiation of electromagnetic relation.[12] Also, local time-limited nuclear fusion in the convective outer layers of flare stars could conceivably cause the eruptions.

There may be some connection between the flare activity in late-type dwarfs and solar activity, and thus the question may arise whether a kind of cycle could be expected in flare stars. Some longer series of measurements have indeed shown such cycles; long-term flare activity in AD Leo, for instance, hints at a cycle of length about 8 years.

The extended observations conducted at Mt. Wilson Observatory strongly support the assumption of solar-like stellar activity. Since 1966, O.C. Wilson and coworkers have monitored about 90 neighboring K- and M-type dwarf stars, measuring the faint, but distinctly observed emission components of Ca II in the center of the strong, broad absorption lines. These emission components are ascribed to the chromospheric activity of the stars. On the Sun the strength of chromospheric features varies with the activity cycle, and in some flare stars a cyclic variation of the emissions, and hence the presumably underlying chromospheric activity, was found with cycles lasting from 7 to 10 years. With this result, the question posed at the beginning of the chapter regarding the variability of the Sun has been returned to. Through much effort and patience, activity similar to that of the Sun has indeed been found in the flare stars.

It seems justified to consider these stars as *bona fide* variable stars. The question arises whether beyond this variable activity the total amount of radiation also varies. This would clinch the question of variability for the Sun (and any solar-type stars).

The variability of the solar constant remained an intractable problem for ground-based astronomy: the best determinations resulted in values scattered in the broad range between about 1325 and 1395 W m^{-2}. The *Solar Maximum Mission* (*SMM*) satellite contributed decisively to this problem. In 1980, the year of solar maximum, the relatively accurate value of 1368.3 W m^{-2} was found, but with some short-term variations (Fig. 25.36). Afterwards, the value of the solar irradiance slowly decreased, closely paralleling the decline of the photospheric activity, until in 1986, at the time of the solar minimum, it reached 1367.0 W m^{-2}. After the minimum, the curve seems to be on the rises again. The range of this variation is merely 0.1% or 0.001 mag, but its reality is beyond doubt. It is worth mentioning that while the total solar

---

12 The *Compton effect*—which provided an early proof of the particle nature of light—consists of a collision between a light quantum and an electron of low energy or at rest, conferring part of the photon energy to the electron. The inverse Compton effect involves a collision of a photon in the radio or visible range with a *high-energy* (relativistic) electron, in which case the photon receives additional energy at the expense of the electron.

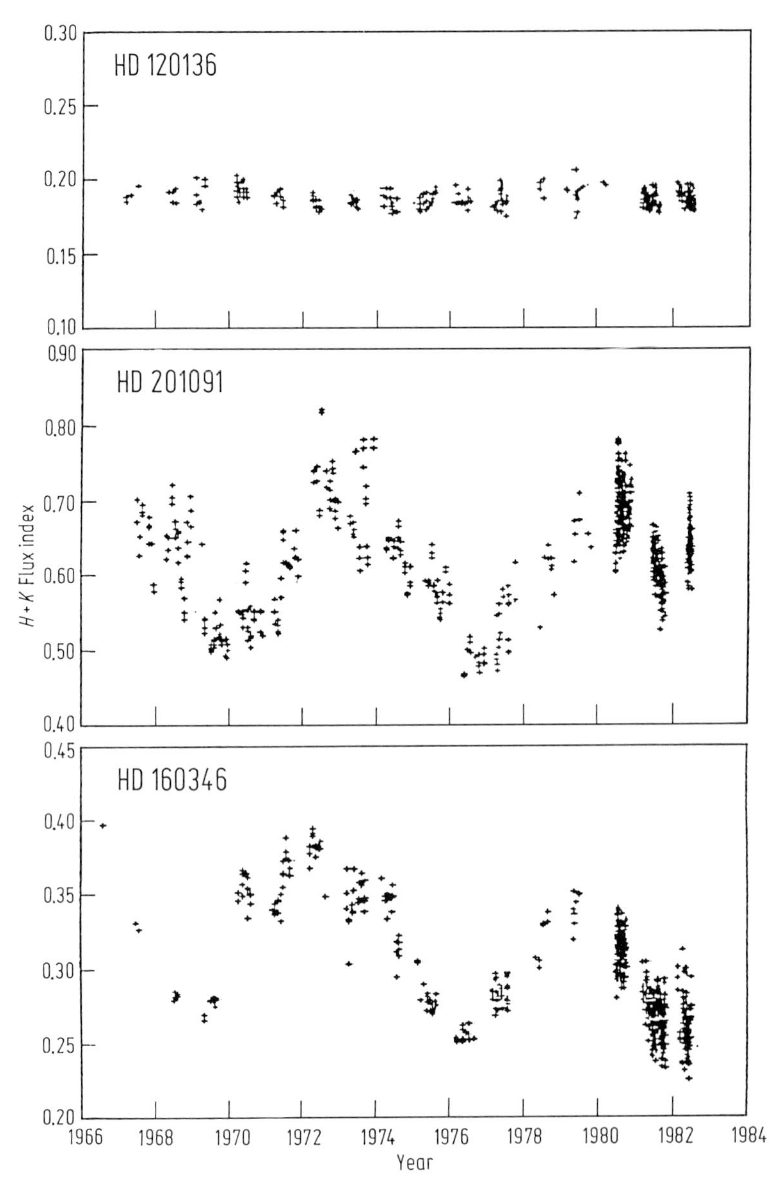

Fig. 25.35. Cyclic changes of chromospheric activity of main-sequence stars. In HD 120136 (*top*), a cycle of about 10 years is slightly hinted at, while the variations in the other two cases appear quite pronounced. Measures for the degree of activity are weak emission components superposed on the photospheric absorptions of the Ca II H- and K-lines (Vaughan [25.90]). A long-term observing program of such phenomena, introduced by O.C. Wilson, operates on the Mount Wilson Observatory. These data pose again the question introduced in the introduction, namely, what are the magnitude limits over which a star can be considered variable?

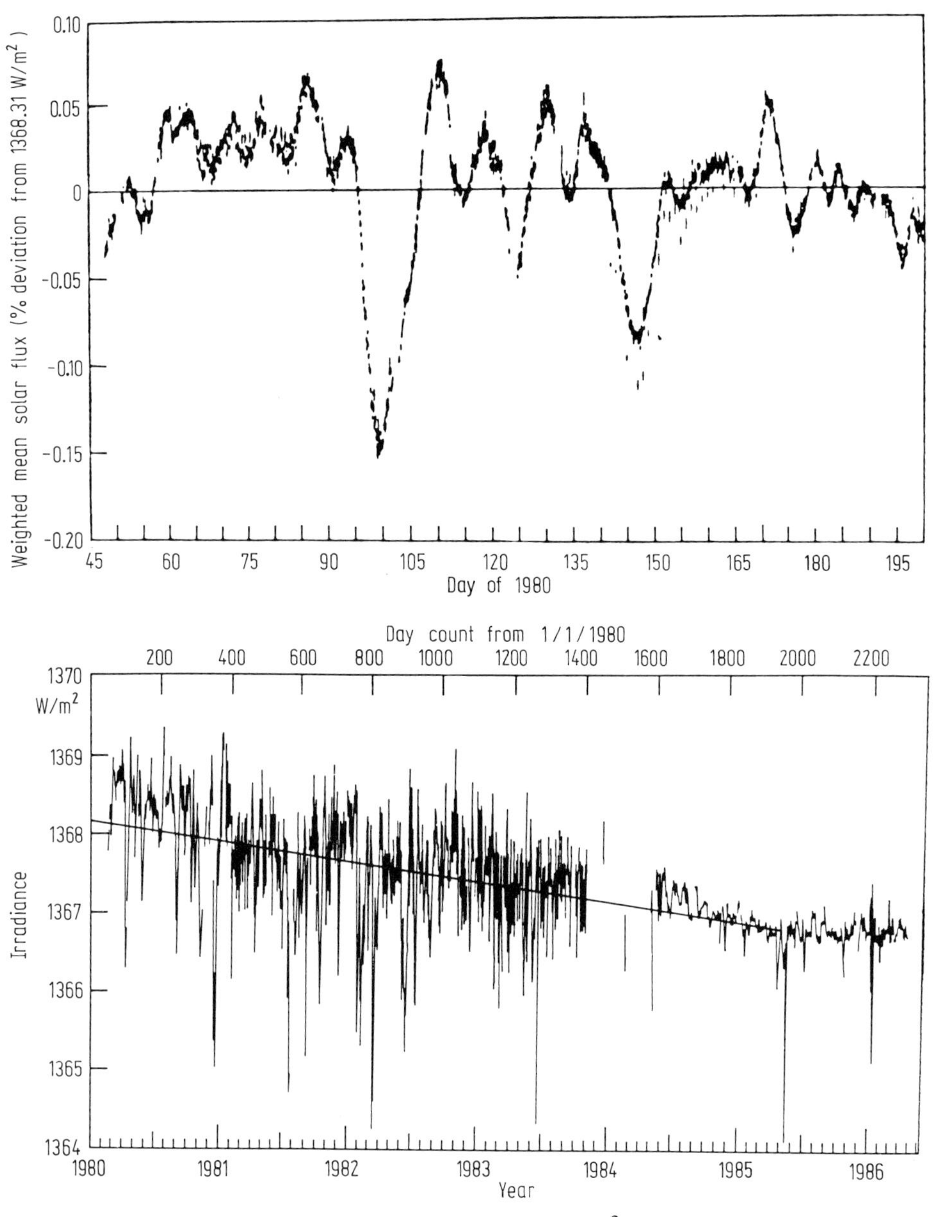

Fig. 25.36. The solar irradiance expressed as the flux (W m^{-2}) at 1 AU, monitored by the *Solar Maximum Mission* (*SMM*) satellite. The first six months of measurements in 1980, the year of the solar maximum, are shown in the *top part* of the figure. The *bottom part* depicts the extended series of measurements from 1982 to 1986 (the year of the solar minimum). Data from NASA Jet Propulsion Laboratory.

radiation (corresponding to a "bolometric magnitude") is ever so slightly decreasing with decreasing solar activity, the transit across the solar disk of large, active sunspot groups is frequently accompanied by marked *decreases* of the solar constant, up to 0.2% (note the two dips in the top part of Fig. 25.36).

Variations of this type may remain extremely difficult to detect in distant stars, even by relative measurements and in a limited spectral region, but a few bold attempts at it have nevertheless been made.

25.7 Young Irregular Variables

Variations in photometric and spectroscopic properties are shown by the evolutionarily young stars in the pre-main-sequence phase, and this fact requires that they be classed as a separate group of variables. Originally, in the years between 1920 and 1930, such objects, which display irregular brightness variations, were found in the Orion Nebula. By about 1950 many other related objects, which were located slightly above the main sequence and in the subgiant region on the HR diagram, had been identified. They showed a broad scattering of spectral types from O to M, with middle and late types having a larger share than early-type stars. Such stars are found primarily in groupings, termed *associations*, within dark or emission nebulae, often in the vicinity of massive, young O- and B-type stars (OB associations). Systematic investigations by Joy [25.91] and Herbig [25.92] concentrated on the subgroup called *T Tauri-type stars*. These objects are young, solar-type proto-stars characterized by distinct spectroscopic features. Groupings of proto-stars in star-forming regions are generally termed *T associations*, after the important class of T Tauri stars.

After the occurrence of these irregularly variable stars in extended interstellar clouds had been identified as a collective feature, the first attempts to find a cause for the photometric and spectroscopic irregularities centered upon the process of accretion of interstellar matter by normal stars. Comparison with theoretical evolutionary models soon showed that these objects had to be extraordinarily young stars, still in an evolutionary phase just prior to reaching the main sequence. The irregularity of the photometric variations and spectroscopic features is due to the complexity of the process by which newly-formed stars accrete interstellar matter.

25.7.1 Orion Variables and RW Aurigae Stars

Orion variables, also called *nebular variables*, are defined as irregularly variable stars connected with interstellar, diffuse gaseous nebulae. Among them are early spectral types (O, B, A; examples: TY CrA, XY Per, T Ori, Z CMa, BF Ori), as well as middle- to late-types (F to M; examples: SU Aur, SS Mon, AH Ori). In most cases, the absorption line spectrum is superposed by emission lines (for example, Hα), which is interpreted as evidence for circumstellar shells and for inflow and outflow of plasma (P Cygni line profiles). The irregular brightness variations cannot be unambiguously characterized. Brightness variations of the order of one or more magnitudes on time scales of several tens of days are observed against temporarily constant brightnesses, with dips lasting from one to a few days, short-term eruptions for hours or minutes, as well as quasi-periodic changes with cycle lengths in days.

RW Aurigae stars are not in principle different from Orion variables, but they are generally not found in connection with dense interstellar gas and dust clouds. Brightness variations are also irregular, including the fluctuations with amplitudes less than 1 magnitude on timescales of hours to days, and the less frequent large variations (up to several magnitudes) on a longer timescale (examples: RW Aur, T Cha, RR Tau, BO Cep, WW Vul). Detailed survey articles include those by Glasby [25.93] and Strom, Strom, and Grasdalen [25.94].

25.7.2 T Tauri Stars

Closely related to the nebular variables and the RW Aurigae stars are the T Tauri stars, named after the prototype located in the huge Taurus-Auriga complex, which consists of dark clouds and is known as a star-forming region. The spectral types of the objects with masses between about 0.3 and 3 $\mathfrak{M}_\odot$ range from F to M. In the HR diagram, they are found about 2 or 3 magnitudes above the main sequence, at absolute magnitudes between +7 and +3. Joy [25.91] first recognized them as a related class of variables.

T Tauri stars, unlike the RW Aurigae objects, are classed solely on the basis of spectroscopic features, the criterion being the presence of the following spectral lines:

- Emission lines of the Balmer series and of Ca II (the H- and K-lines);
- Fluorescence emission lines of Fe I ($\lambda\lambda$ 406.3, 413.2 nm);
- Forbidden emission lines of neutral oxygen ([O I] ($\lambda\lambda$ 630.0, 636.3 nm) and of singly ionized sulfur ([S II] $\lambda\lambda$ 406.8, 407.6, 671.6, 673.0);
- Strong absorption line of Li I (λ670.7 nm).

The intensity distribution of the continuum shows, compared with other late-type stars, a strong infrared excess caused by thermal emission from circumstellar dust. In addition to spectroscopic and spectrophotometric features and the position in the HR diagram, another characteristic mark of T Tauri stars is their position in the Milky Way. T Tauri stars are found primarily in the T associations (named after them), in dark clouds or in emission nebulae, and often in the vicinity of young, massive stars in OB associations. Examples include the Taurus-Auriga complex, the Orion Nebula, the Scorpius-Ophiuchus complex, the North American Nebula (NGC 7000), and the Pelican Nebula (IC 5067).

A significantly large subgroup ($\approx$ 40%) of T Tauri stars are the YY Orionis stars (e.g., S Cr A), with particularly clear spectral indicators for the accretion of matter from the surrounding circumstellar gas and dust cloud (the so-called inverse-P Cygni line profiles, emissions with red-shifted absorption components). Lists of known T Tauri stars and survey articles have been written by, for example, Baschek [25.95], Rydgren, Strom, and Strom [25.96], Cohen and Kuhi [25.97], and Wolf [25.98]. Mundt [25.99] and Mundt and Bastian [25.100] give a survey of YY Orionis objects.

25.7.3 Variable Be Stars

The *variable Be stars* are often called γ Cassiopeiae-type stars, and are closely related with the so-called shell stars. In the HR diagram, they are arranged between late O and early A types, near the main sequence at luminosity classes V to III. They are rapidly rotating, early-type stars, with spectroscopic indications for more or less extended rings and envelopes in the vicinity of the photosphere. They are characterized by equatorial rings, disks, and spherical envelopes which generate emission and absorption lines of various intensities, ionization and excitation energies, Doppler shifts, and peculiar line profiles, depending on physical parameters (circumstellar temperatures, densities, flow velocities) and on the geometric perspective (equatorial to polar). Such circumstellar emissions and shell absorptions are superposed upon a normal photospheric absorption spectrum distinguished by rather broad line profiles, the Doppler broadening being caused by the rapid rotation of the star. The shell spectrum often varies with time, the intensity and profiles of shell lines changing distinctly, sometimes within a few days. Examples are γ Cas and ζ Tau.

Photometric changes are detected in about 60% of the Be stars, primarily on longer scales—on the order of 10^2 to 10^3 days, and mostly in a pattern that shows irregular minima with depth $\lesssim 1$ mag relative to a long-period average brightness. (Light curves of γ Cas and XX Oph are displayed by Hoffmeister, Richter, and Wenzel [25.60].) In some cases, a correlation between brightness fluctuations (brightness decline after shell ejection) and corresponding spectroscopic features has been found (e.g., γ Cas, BU Tau).

Models for Be and shell stars generally assume the presence of circumstellar matter in the shape of equatorial rings, disks, or envelopes to explain the observed features. The mass loss in these stars may be triggered by hydrodynamical processes within the stellar interiors and atmospheres (dissipation of acoustic waves and mechanical oscillation energy). The centrifugal action in the atmospheres of the rapidly rotating stars (especially in the equatorial regions) may overcompensate gravity, and therefore is of decisive importance for the formation of temporarily stationary shells. Detailed data and other references can be found in Seitter and Duerbeck [25.101].

References

25.1 Tsesevich, V.P.: in: *Eclipsing Variable Stars*, Wiley & Sons, New York 1973, p. 1.
25.2 Kopal, Z.: *Proc. Am. Philos. Soc.* **85**, 399 (1942).
25.3 Kopal, Z.: Language of the Stars. *Astrophys. Space Sci. Library* **77** (1979).
25.4 Wood, D.B.: *Astron. J.* **76**, 701 (1971).
25.5 Hill, G., Hutchings, J.B.: *Astrophys. Space Sci.* **20**, 123 (1973).
25.6 Wilson, R.E., Devinney, E.J.: *Astrophys. J.* **166**, 605 (1971).
25.7 Wilson, R.E.: *Astrophys. J.* **234**, 1054 (1979).
25.8 Kallrath, J., Linnell, A.P.: *Astrophys. J.* **313**, 346 (1987).
25.9 Drechsel, H., Rahe, J.: *Sterne und Weltraum* **5**, 229 (1983).
25.10 Clausen, J.V., Gimenez, A., Scarfe, C.: *Astron. Astrophys.* **167**, 287 (1986).
25.11 Popper, D.M., Ulrich, R.K.: *Astrophys. J.* **212**, L131 (1977); see also Popper and Ulrich in *Proc. IAU Symp. 88*, Toronto 1977.
25.12 Crawford, J.A.: *Astrophys. J.* **121**, 71 (1955).

25.13 Wachmann, A.A., Popper, D.M., Clausen, J.V.: *Astron. Astrophys.* **162**, 62 (1986).
25.14 Mochnacki, S.W., Doughty, N.A.: *Monthly Not. Roy. Astron. Soc.* **156**, 51 (1972).
25.15 Wilson, R.E., Devinney, E.J.: *Astrophys. J* **182**, 542 (1973).
25.16 Rucinski, S.M.: in *Interacting Binary Stars*, J.E. Pringle, R.A. Wade (eds.), Cambridge University Press, Cambridge 1985, pp. 85 and 113.
25.17 Code, A.D.: *Astrophys. J.* **106**, 309 (1947); see also the tables in Duerbeck and Seitter [25.49].
25.18 Kippenhahn, R., Weigert, A.: *Sterne und Weltraum* **4**, 148 (1965).
25.19 Christy, R.F.: in *Methods of Computational Physics*, Vol. 7, p. 213.
25.20 Feast, M.W., Walker, A.R.: *Ann. Rev. Astron. Astrophys.* **25**, 345 (1987).
25.21 Welch, D.L., McAlary, C.W., McLaren, R.A., Madore, B.F.: in *Cepheids: Theory and Observations*, (IAU Colloquium No. 82), ed. B.F. Madore, 1985, p. 119.
25.22 Dickens, R.J., Carey, J.V.: *Royal Obs. Bull. Greenwich* **129**, 335 (1967).
25.23 Szabados, L.: *Mitt. Sternwarte Budapest* **76** (1980).
25.24 Balàzs, J., Detre, L.: *Mitt. Sternwarte Budapest* **8** (1939).
25.25 Campbell, L, Jacchia, L.: *The Story of Variable Stars*, The Harvard Books on Astronomy, H. Shapley, B.J. Bok (eds.), Blakiston, Philadelphia 1941.
25.26 Prager, R.: *Geschichte und Literatur des Lichtwechsels der Veränderlichen Sterne*, 2. Ausgabe, Dümmler, Berlin 1934, p. 286. (1934).
25.27 Sterne, T.E., Campbell, L.: Wash. National Acad. **23**, 115 (1938) (= Harvard Reprint No. 134).
25.28 Unno, W., Osaki, Y., Ando, H., Shibahashi, H.: *Non-radial Oscillations of Stars*, Tokyo 1979.
25.29 Lawson, W.A., Kilmartin, P.M., Gilmore, A.C., Clark, M.: *Inf. Bull. Var. Stars. No. 3085*, (1987).
25.30 Harris, H.C.: *Astron. J.* **86**, 719 (1981).
25.31 Sargent, A.I., Baud, B.: in *Mass Loss from Red Giants*, M. Morris, B. Zuckerman (eds.), Reidel, Dordrecht 1985.
25.32 Engels, D.: *Sterne und Weltraum* **23**, 243 (1984).
25.33 Preston, G.W., Krzeminski, W., Smak, J., Williams, J.A.: *Astrophys. J.* **137**, 401 (1963).
25.34 *General Catalogue of Variable Stars, Third Edition*, B.V. Kukarkin et al. (eds.), Astronomical Council of the Academy of Sciences in the USSR, Moscow 1969; continued as: *Fourth Edition*, eds. P.N. Kholopov et al. (Vols. I, II, III), N.N. Samus (Vol. IV), Nauka Publishing House, Moscow 1985–1990.
25.35 Lyne, A., Graham-Smith, F.: *Pulsar Astronomy*, Cambridge University Press, Cambridge 1990.
25.36 Hankins, T.H.: *Astrophys. J.* **181**, L49 (1973).
25.37 Fruchter, A.S., Stinebring, D.R., Taylor, J.H.: *Nature* **333**, 237 (1988).
25.38 Kron, G.E.: *Astrophys. J.* **115**, 301 (1952); see also Kron: "Starspots?", ASP Leaflet No. 257 (1950).
25.39 Saar, S.H. Linsky, J.L.: *Adv. Space Res.* **6**, 235 (1986). (1985).
25.40 Breger, M.: *Publ. Astron. Soc. Pacific* **91**, 5 (1979).
25.41 Bethe, H.: *Physics Today*, Sep. 1990, p. 24.
25.42 Barbon, R., Ciatti, F., Rosino, L.: *Astron. Astrophys.* **25**, 24 (1973).
25.43 Barbon, R., Ciatti, F., Rosino, L.: in *Supernovae and Supernova Remnants*, C.B. Cosmovici (ed.), Reidel, Dordrecht 1974, p. 115.
25.44 Kowal, C.: *Astron. J.* **73**, 1021 (1968).
25.45 van den Bergh, S.: *Bull. Astr. Soc. India* **10**, 199 (1983).
25.46 Biermann, P.: *Phys. Bl.* **44**, No. 11, 419 (1988).
25.46a McGray, R.: *Ann. Rev. Astron. Astrophys.*, in press (1993).
25.47 Robinson, E.L.: *Ann. Rev. Astron. Astrophys.* **14**, 119 (1976).
25.48 Warner, B.: in *Structure and Evolution of Close Binary Stars*, Proc. IAU Symp. No. 73, P. Eggleton, S. Mitton, J. Whelan (eds.), Reidel, Dordrecht 1976, p. 85.
25.49 Duerbeck, H.W., Seitter, W.C.: *Variable Stars*. In *Landolt-Börstein, New Series*, Group VI, Vol. 2b, K. Schaifers, H.H. Voigt (eds.), Springer, Berlin 1982, p. 197.

25.50 Wade, R.A., Ward, M.J.: in *Interacting Binary Stars*, J.E. Pringle, R.A. Wade (eds.), Cambridge University Press, Cambridge 1985, p. 134.
25.51 Warner, B: in *Cataclysmic Variables. Recent Multi-Frequency Observations and Theoretical Developments* (Proc. IAU Coll. No. 93), H. Drechsel, Y. Kondo, J. Rahe (eds.), Reidel, Dordrecht 1987, p. 3.
25.52 Antipova, L.I.: In *Cataclysmic Variables. Recent Multi-Frequency Observations and Theoretical Developments*, Proc. IAU Coll. 93, H. Drechsel, Y. Kondo, J. Rahe (eds.), Reidel, Dordrecht 1987, p. 453.
25.53 Ritter, H.: *Catalogue of Cataclysmic Binaries, Low-Mass X-Ray Sources and Related Objects* (5th ed.), *Astron. Astrophys. Suppl.* (1990).
25.54 Cordova, F.A., Mason, K.O.: in *Accretion Driven Stellar X-Ray Sources*, W.H.G. Lewin, E.P.J. van den Heuvel (eds.), Cambridge University Press, Cambridge 1982.
25.55 Duerbeck, H.W.: *Publ. Astron. Soc. Pacific* **93**, 165 (1981).
25.56 Duerbeck, H.W.: *A Reference Catalogue and Atlas of Galactic Novae. Space Science Reviews*, **45** (1987).
25.57 Payne-Gaposhkin, C.: *The Galactic Novae*, North Holland, Amsterdam 1957.
25.58 Payne-Gaposhkin, C.: *Astrophys. Space Sci. Library* **65**, 3 (1977).
25.59 McLaughlin, D.B.: *Novae, Novoides et Supernovae*, Centre Nationale de la Recherche Scientifique, Paris 1965, p. 3.
25.60 Hoffmeister, C., Richter, G. Wenzel, W.: *Verändliche Sterne* (2nd edn.), Springer, Berlin 1984.
25.61 Barlow, M.J., Brodie, J.P., Brunt, C.C., Hanes, D.A., Hill, P.W., Mayo, S.K., Pringle, J.E., Ward, M.J., Watson, M.C., Whelan, J.A.J., Willis, A.J.: *Monthly Not. Roy. Astron. Soc.* **195**, 61 (1981).
25.62 Snijders, M.A.J.: in *Cataclysmic Variables. Recent Multi-Frequency Observations and Theoretical Developments*, (Proc. IAU Coll. No. 93), H. Drechsel, Y. Kondo, J. Rahe (eds.), Reidel, Dordrecht 1987, p. 243.
25.63 Vogt, N.: *Astron. Astrophys.* **88**, 66 (1980).
25.64 Mumford, G.S.: *Sky & Telescope* **26**, 190 (1963).
25.65 Robinson, E.L.: *Astron. J.* **80**, 515 (1975).
25.66 Nather, R.E.: *Vistas Astron.* **15**, 91 (1973).
25.67 Patterson, J.: *Astrophys. J.* **233**, L13 (1979).
25.68 Warner, B.: in *Cataclysmic Variables and Related Objects*, Proc. IAU Coll. No. 72, M. Livio, G. Shaviv (eds.), Reidel, Dordrecht 1983, p. 155.
25.69 Duerbeck, H.W., Klare, G., Krautter, J., Wolf, B., Seitter, W.C., Wargau, W.: in *Proc. Second European IUE Conference*, ESA SP-157, 91 (1980).
25.70 Krautter, J, Klare, G., Wolf, B., Wargau, W., Drechsel, H., Rahe, J., Vogt, N.: *Astron. Astrophys.* **98**, 27, (1981).
25.71 Schwope, A., Beuermann, K.: in *Recent Results on Cataclysmic Variables*, ESA-Workshop, Bamberg, 1985 April 17–19, ESA SP-236, 173 (1985).
25.72 Liebert, J., Stockman, H.S.: in *Cataclysmic Variables and Low-Mass X-Ray Sources*, eds. D.Q. Lamb, J. Patterson, Reidel, Dordrecht 1985, p. 151.
25.73 Mouchet, M.: *ESO Messenger* **34**, 3 (1983).
25.74 Warner, B.: in *Cataclysmic Variables and Low-Mass X-Ray Binaries*, D.Q. Lamb, J. Patterson (eds.), Reidel, Dordrecht 1985, p. 269.
25.75 Nather, R.E., Robinson, E.L., Stover, R.J.: *Astrophys. J.* **244**, 269 (1981).
25.76 Warner, B., Nather, R.E.: *Monthly Not. Roy. Astron. Soc.* **152**, 219 (1971).
25.77 Lamb, D.Q., Melia, F.: in *Cataclysmic Variables. Recent Multi-Frequency Observations and Theoretical Developments* (Proc. IAU Coll. No. 93), H. Drechsel, Y. Kondo, J. Rahe (eds.), Reidel, Dordrecht 1987, p. 511.
25.78 Starrfield, S., Sparks, W.M.: in *Cataclysmic Variables. Recent Multi-Frequency Observations and Theoretical Developments*, (Proc. IAU Coll. No. 93), H. Drechsel, Y. Kondo, J. Rahe (eds.), Reidel, Dordrecht 1987, p. 379.
25.79 Mustel, E.R., Boyarchuk, A.A.: *Astrophys. Space Sci* **6**, 183 (1970).
25.80 Weaver, H.: in *Highlights of Astronomy* **3**, 509 (1974).

25.81 Duerbeck, H.W., Seitter, W.C.: in *Cataclysmic Variables. Recent Multi-Frequency Observations and Theoretical Developments*, (Proc. IAU Coll. No. 93), H. Drechsel, Y. Kondo, J. Rahe (eds.), Reidel, Dordrecht 1987, p. 467.
25.82 Bath, G.T., Pringle, J.E.: *Monthly Not. Roy. Astron. Soc.* 194, 967 (1981).
25.83 Bath, G.T.: In *Cataclysmic Variables. Recent Multi-Frequency Observations and Theoretical Developments* (Proc. IAU Coll. No. 93), H. Drechsel, Y. Kondo, J. Rahe (eds.), D. Reidel, Dordrecht 1987, p. 293.
25.84 Osaki, Y.: *Publ. Astron. Soc. Japan* **26**, 429 (1974).
25.85 Meyer, F., Meyer-Hofmeister, E.: *Astron. Astrophys.* **126**, 34 (1982).
25.86 Meyer, F.: in *Recent Results on Cataclysmic Variables*, ESA Workshop, Bamberg 1985 April 17–19. ESA SP-236, 83 (1985).
25.87 Doxsey, R., Bradt, H.V., Levine, A., Murthy, G.T., Rappaport, S., Spada, G.: *Astrophys. J.* **182**, L25 (1973).
25.88 Boyarchuck, A.A.: in *Non-Periodic Phenomena in Variable Stars*, Academic Press, Budapest 1969.
25.89 Moffet, T.J.: *Astrophys. J. Suppl.* **29**, 1 (1974).
25.90 Vaughan, A.H.: in *Solar and Stellar Magnetic Fields: Origins and Coronal Effects*, IAU Symp. No. 102, J.O. Stenflo (ed.), Reidel, Dordrecht 1983, p. 113.
25.91 Joy, A.H.: *Astrophys. J.* **102**, 168 (1945).
25.92 Herbig, G.H.: *Adv. Astron. Astrophys.* **1**, 47 (1962).
25.93 Glasby, J.S.: *The Nebular Variables*, Permagon Press, Oxford 1974.
25.94 Strom, S.E., Strom, K.M., Grasdalen, G.L.: *Ann. Rev. Astron. Astrophys.* **13**, 187 (1975).
25.95 Baschek, B.: *Sterne und Weltraum* **19**, 205 (1976).
25.96 Rydgren, A.E., Strom, S.E., Strom, K.M.: *Astrophys. J. Suppl.* **30**, 307 (1976).
25.97 Cohen, M., Kuhi, L.V.: *Astrophys. J. Suppl.* **41**, 743 (1979).
25.98 Wolf, B.: *Sterne und Weltraum* **10**, 506 (1984).
25.99 Mundt, R.: *Sterne und Weltraum* **16**, 234 (1977).
25.100 Mundt, R., Bastian, U.: *Astron. Astrophys. Suppl.* **39**, 245 (1980).
25.101 Seitter, W.C., Duerbeck, H.W.: in *Landolt-Börnstein, New Series*, Group VI, Vol. 2b, K. Schaifers, H.H. Voigt (eds.), Springer, Berlin 1982, p. 290.
25.102 Young, P.J., Corwin, H.G. Jr., Bryan, J., de Vaucouleurs, G.: *Astrophys. J.* **209**, 882 (1976).
25.103 Glasby, J.S.: *Variable Stars*, Constable, London 1968, p. 176.
25.104 Bath, G.T., Shaviv, G.: *Monthly Not. Roy. Astron. Soc.* **183**, 515 (1978).
25.105 Engels, D.: *Mitt. Astron. Ges.* **60**, 252 (1983).
25.106 Manchester, R.N., Taylor, J.H.: in *Pulsars*, Freeman, San Francisco 1977.
25.107 Marlborough, J.M.: in *Be and Shell Stars*, IAU Symp. No. 70, Reidel, Dordrecht 1976, p. 335.
25.108 Prialnik, D.: In *Cataclysmic Variables. Recent Multi-Frequency Observations and Theoretical Developments* (Proc. IAU Coll. No. 93), H. Drechsel, Y. Kondo, J. Rahe (eds.), Reidel, Dordrecht 1987, p. 431.
25.109 Vogt, N.: *Mitt. Astron. Ges.* **57**, 79 (1982).

26 Binary Stars

W.D. Heintz

26.1 General Overview

Binary and multiple systems of stars first appeared as a topic of astronomical research about two hundred years ago. Early micrometric observations then established orbital displacements, supporting the conjecture that most of the double stars seen are not perspective random pairings but rather are physically connected systems. Also, the notion of mutual stellar eclipses was put forward as an explanation for the periodic brightness changes of certain apparently variable stars.

Even the casual observer will certainly have encountered many fine double stars when sweeping the sky, and will have used them in various degrees of difficulty for optical and seeing tests (Fig. 26.1). As large telescopes, technologies, and theory progressed, binary stars became a rewarding target for high-precision astronomy; the literature shows that the research emphasis has shifted toward close (interferometric and photometric) systems and subsystems, and toward binaries containing low-mass, low-luminosity dwarf stars.

Visual binaries in particular are usually characterized by long periods of revolution. Their study often depends on position measurements collected over many years, but the quantities to be measured are still minute enough that attention must be paid to the effects of random and systematic errors. Many of the archived data have only statistical significance until perhaps much later when a particular star receives special attention for one reason or another; investigators then appreciate finding measurements which date far back in time. Visual double stars have always been the domain of only a few researchers, indicating that experience and perseverance do play a role.

The two components of a binary star move in elliptical orbits around their common center of mass. The center of mass is not, however, a visible point; what is actually observed is the relative orbit of one star with respect to the other, which is the sum of the component orbits and similarly elliptical. For the relative orbit, Kepler's third law states that

$$\alpha^3 = (ra)^3 = (\mathfrak{M}_1 + \mathfrak{M}_2)P^2, \tag{26.1}$$

where α is the orbital semimajor axis in astronomical unit (AU) distances, a the corresponding quantity in arcsec, r the distance of the pair from the Sun in pc, $\mathfrak{M}_{1,2}$ the masses in solar units, and P the (mutual) period of revolution in years. All stellar masses are determined by this formula. It also indicates that the great majority of binaries separated in small telescopes must have large true distances between their

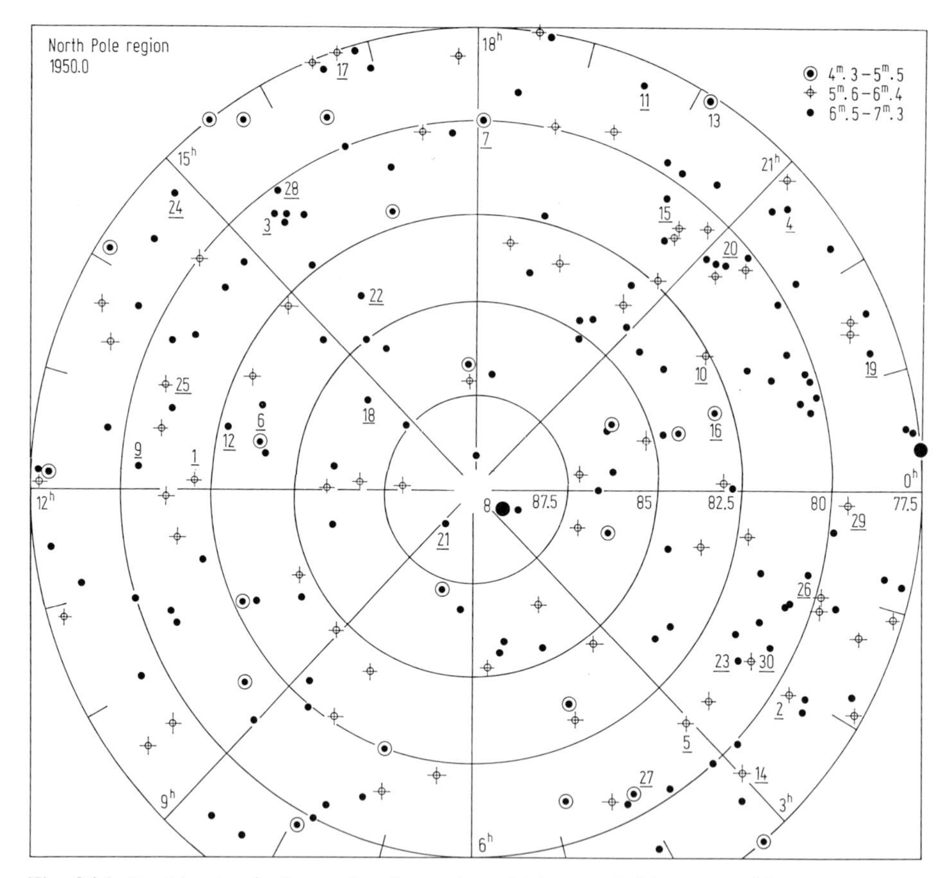

Fig. 26.1. Double stars in the north polar region which are suitable as test objects.

components, corresponding to periods of millenia and longer. Only in those binaries which are relatively close to the Sun will the orbital displacement become conspicuous over a few decades or, in a few cases with periods less than 100 years, within a few months or a year.

Apart from these genuine (physically associated) objects, there also exist *optical doubles* whose components are not related but merely seem close together owing to perspective along the line of sight; such pairs are without scientific interest. A physical binary system is identified by its curved orbital motion (in contrast to the rectilinear relative motion of independent, optical components), or by a common proper motion in the sky in cases where the relative motion is too slow to distinguish its curved or linear shape. Bright, close pairs are almost certainly physical; among stars brighter then magnitude +8 with separations under 5″, optical pairs are rare exceptions, as observations and also statistical considerations show. For separations over 20″, the optical objects dominate, but there are still many known pairs and groups of common proper motions which have much larger separations. (The term "binary" in English tends to emphasize the physical nature, in contrast to the more general "double" star.)

Binary and multiple systems are quite abundant in the solar neighborhood. It is estimated that roughly 85% of stars (excluding red dwarfs) are components of binary systems, thus leaving the truly single stars as a 15% minority, and even they are likely to have originated within groups and escaped thereafter. The triple and higher-multiple systems—also by no means rare—show components normally in very unequal mutual distances, such as a close pair with a third star ten or more times farther out, or two closer pairs circling each other in a very wide orbit. These patterns of distances are vital for the long-term stability of the systems. Near-equal component separations, for which the Trapezium in the Orion Nebula is the best-known specimen, are found almost exclusively among young stars. This type of arrangement is evidently unstable and dissolves quickly—even faster than do open clusters. The higher incidence of single stars among M-type dwarfs (perhaps 40%) is likely to have two causes: first, these red dwarfs were born with far too little angular momentum to form close binaries, and second, as very low-mass objects, they were preferentially ejected from the unstable original groupings.

Orbit dimensions and periods of binaries occur over a vast possible range. The closest pairs revolve with their surfaces practically in contact or within joint envelopes around both stars, while the widest pairs are separated so far that their gravitational bond barely exceeds that of the general gravity field of the Galaxy. Binaries substantially wider than Alpha and Proxima Centauri (separated by 13 000 AU) will not remain bound over their lifetime. (This is one of several arguments against the untrustworthy hypothesis that the Sun has a very distant, undetected companion which approaches the planetary system at long intervals for the purpose of exterminating dinosaurs.) The need to study the much different orbit scales by separate techniques is reflected in the observational subdivision into the *visual* (long-period) *binaries*, the *photometric* (eclipsing short-period) *binaries*, and the *spectroscopic binaries*, which overlap with the two other groups but contain largely non-eclipsing, short-period objects.

Mass determinations by Eq. (26.1) require that α be known in km or in astronomical unit (AU) distances, freed from the orbital inclination i and split into the relative orbits. This leaves two possibilities:

1. *visual binaries*, in which the orbit (defined by a and P) is supplemented by astrometric studies of parallax and mass ratio (by making measurements relative to several unrelated background stars in the field). The parallax usually contributes the largest relative error and may be replaced by spectroscopic measurements, if available, of the radial-velocity amplitude of at least one component;
2. *eclipsing binaries*, in which radial-velocity amplitudes of both components are needed to find the scales of relative and absolute orbits; the distance r from the Sun does not enter, so its uncertainty does no harm, but then the absolute magnitudes cannot be directly determined either.

The *mass–luminosity relation* (MLR) for main-sequence stars is:

$$\begin{aligned} \log L(\text{bol}) &= 3.8 \log \mathfrak{M} && \text{for} \quad 0.5 < \mathfrak{M} < 2.5, \\ \log L(\text{bol}) &= 2.6 \log \mathfrak{M} - 0.3 && \text{for} \quad 0.1 < \mathfrak{M} < 0.5, \end{aligned} \tag{26.2}$$

where $\mathfrak{M}$ and L are the mass and luminosity in solar units. The slope of 2.6 for low-mass stars is also approximately followed by stars with masses over 2.5 $\mathfrak{M}_{\odot}$, but here it is less well defined owing to the high incidence of evolving, overluminous stars. The evolution can be calculated through stellar models as a set of luminosity *isochrones* (loci of equal age). The change of energy generation on the main sequence is slow enough that in many cases the zero-age line (arrival at the main-sequence equilibrium) still holds. Data on distances and masses, however, show a substantial number of these stars to be incipient subgiants—even if not yet spectrally classified as such.

When components of a visual binary can be identified as main-sequence stars, and provided there is no third mass in the orbiting system, the MLR may be employed (in place of an unknown or uncertain parallax) as an additional condition to find the masses, since the latter are linked with the absolute magnitude by both the MLR and the parallax (see [26.1], p. 62). For instance, the distance 46.5 ± 1 pc found for the Hyades cluster by MLR-fitting seems to be of better accuracy than that found by other methods. A similar reasoning would apply to eclipsing components (if their spectroscopic data are deemed too weak) but the basic assumption—membership of the main sequence—is here riskier.

Bodies below a certain mass limit, which is theoretically expected to be near 0.08 $\mathfrak{M}_{\odot}$, cannot tap the resource of nuclear energy, and—lacking hydrostatic equilibrium—they never become stars proper. These objects have "substellar" masses, and are termed (with a catchy but ill-suited phrase) *brown dwarfs*. Also conjectured to exist are *black dwarfs*, which may be former stars and white dwarfs which have run out of all stored heat, or which perhaps never even reached brown-dwarf status. Since substellar objects are characterized by mass, which here does not relate to luminosity, they cannot be identified by their brightness.

The frequency of stellar masses peaks at about one-fourth of the solar mass (absolute visual magnitude ≈ 12); stars with masses around 0.1 $\mathfrak{M}_{\odot}$ are less common, and substellar masses are probably quite rare. Although not much is known quantitatively about the origin of the planetary system, its genesis (the accretion of very small bodies under the influence of a central gravitational force) is utterly different from that of binary and multiple stars, and inferences from one process as to the other are not valid. It may be true that substellar and planetary masses both form infrequently, although for quite different reasons.

26.2 Features of Visual Double Stars

The spherical position of the secondary (fainter) star relative to the primary is given by two numbers (Fig. 26.2): the polar coordinates, as obtained with a micrometer, are the separation ρ, given in arcseconds, and the position angle p counted in degrees from north through east, south, west from 0° to 360°. Photographic plates give rectangular coordinates x (positive toward north) and y (east), both in arcseconds. The conversion

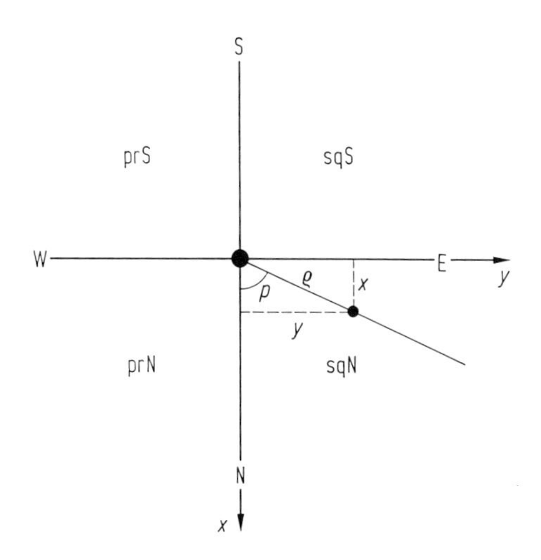

Fig. 26.2. Relative coordinates and quadrants for a binary star in the field.

Table 26.1. Relation between magnitude difference Δm of a pair and the amount $m_A - m_{tot}$, by which the primary star is fainter than the combined light.

Δm	$m_A - m_{tot}$	Δm	$m_A - m_{tot}$	Δm	$m_A - m_{tot}$
$0^m_.0$	$0^m_.75$	$1^m_.0$	$0^m_.36$	$2^m_.0$	$0^m_.16$
0.2	0.66	1.2	0.31	3.0	0.07
0.4	0.57	1.4	0.26	4.0	0.03
0.6	0.49	1.6	0.22	5.0	0.01
0.8	0.42	1.8	0.19	6.0	0.00

formulae are

$$\begin{aligned} \Delta\delta = x = \rho\cos p \quad & \text{(Difference in Decl.)}, \\ 15\cos\delta\Delta\alpha = y = \rho\sin p \quad & \text{(Difference in R.A.)}. \end{aligned} \tag{26.3}$$

Or, conversely,

$$\tan p = y/x \quad \text{and} \quad \rho = \sqrt{x^2 + y^2}. \tag{26.4}$$

The axes divide the field into four "quadrants" referred to in Fig. 26.2.

From the total magnitude m_{tot} and the magnitude difference Δm of a pair, the component magnitudes m_A and m_B can be found and vice versa. Table 26.1 relates Δm with the amount $m_A - m_{tot}$, by which the primary star is fainter than the combined light.

Example: Data for Castor are $m_{tot} = +1.54$ and $\Delta m = 0.96$. Table 26.1 gives $m_A - m_{tot} = 0.37$, so the primary is $+1.91$ and the companion $+2.87$. Equally bright components thus are 0.75 magnitudes fainter each than the combined light, whereas at $\Delta m > 3$ the companion contributes almost nothing to the total light.

Visual binaries are named with a discoverer's code and a number. Most objects accessible to small telescopes in the northern hemisphere are already numbered in the oldest systematic catalogs (before 1850), those by W. Struve (Σ) and O. Struve (OΣ); numerous other codes added since need not be detailed here. Often used is the ADS number after Aitken's compilation catalog of double stars (north of $\delta = -30°$ and known before 1927), and the coding now preferred is by the coordinates for the year 2000. Thus, Castor (α Geminorum) can be found under any of the designations Σ 1110, ADS 6175, and 07346 N3153 (coordinates $7^{h}34^{m}_{.}6$, $+31°53'$ for 2000). Within a system the components are distinguished by a capital letter, generally according to brightness: A = primary star, B = companion, C, D, ... for more distant components, if present.

The examination of stars in the Durchmusterung catalogs (to about 10th magnitude) for duplicity has been largely completed; most of the recent discoveries are very close (interferometric) pairs or stars fainter than 9th magnitude. The last compilation to appear in print was the *Index Catalogue* (IDS) in 1961. The *Washington Index* (WDS) is available in a preliminary 1984 version on tape, containing about 80 000 entries. Since completeness is a vital aspect of documentation, the compilations include tens of thousands of very wide, faint, and "unimportant" pairs which somehow got into the records; presumably, many optical pairs are among them. Recently the literature became further cluttered by the collection of many more very wide pairs, allegedly of significance for some space-mission star catalogs but without physical interest.

The color and magnitude differences seen in visual binaries basically correspond to the Hertzsprung–Russell Diagram. Main-sequence pairs may have components of equal magnitudes and colors, or unequally bright stars showing the corresponding color difference. The combination of a red or yellow giant with a white main-sequence star also occurs (with the well-known contrast illusion that the white star looks blue), and occasionally also with a degenerate companion. For larger magnitude differences the irradiation normally prevents a reliable determination of the color and sprectrum of the companion, and even its magnitude estimates may be quite uncertain. In an evolutionary sense, visual (and many spectroscopic) pairs are wide enough to be non-interacting, that is, the stars affect each other only through orbital gravity. This familiar color–luminosity picture does not hold for many eclipsing components, whose constricting equipotential surfaces force deformation, mass transfer, mass loss, and unusual evolutionary patterns.

26.3 Micrometer and Visual Observations

Screw micrometers are an 18th century invention and still used. Although for other observational applications mentioned in this volume their research function has ceased, some use on visual double stars remained. Even in this area, eye and micrometer are being replaced by more elaborate measurement technologies which improve the precision and which depend less on good seeing conditions and on the observer's skill. On the other hand, the present volume purports to address the interests of instructional and recreational astronomy, and the versatile and fairly easily built micrometer is a case in point. Precision screws as used elsewhere in industry are now machined with

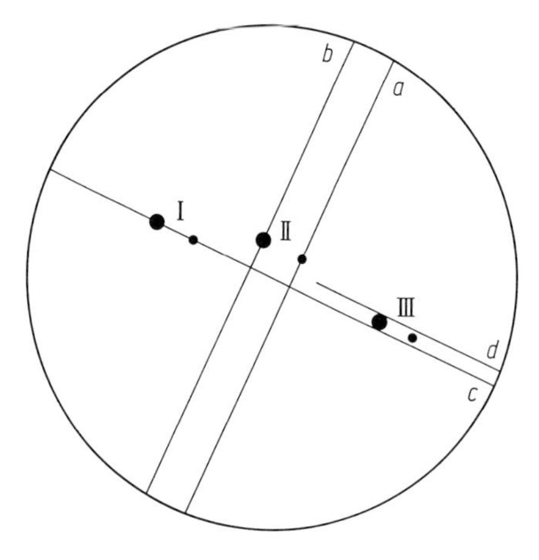

Fig. 26.3. Measuring with the filar micrometer.

such fine quality that the astronomical user should not have to worry about major screw errors; a test table supplied by the manufacturer should indicate where and to what extent progressive pitch errors occur. Divided degree circles obtainable are also sufficiently precise and reasonably affordable.

The best-known type of micrometer is the *filar micrometer*, which consists of a frame located in the focal plane of a telescope, at right angles to the optical axis and rotatable around the axis, and a slide moved by a long precision screw along the frame. The slide and frame each carry a thin wire at right angles to the screw (Fig. 26.3 *a,b*); spares and cross wires may be added as desired. The wires a and b should be closely adjacent in depth, but should not touch when passing; the short-focus eyepiece is focused on the middle distance so that both wires look equally thick. In order to protect the wires, the instrument is mounted in a closed case in which the eyepiece is inserted—the latter preferably movable on a small, independent slide. The screw is connected with a divided drumhead to read parts (1/100) of a revolution, and a scale to count full revolutions, so that the displacement of the movable wire can be precisely found. The position circle to read the orientation of the micrometer is divided into degrees (counterclockwise) and is firmly attached to the telescope tube. A circle of about 20 cm diameter with sharp divisions suffices to read tenths of degrees. The rotating parts of the device (case, frame, and index on the circle) should be rigidly connected, and the connection screw/slide should be free of backlash. Also needed is a faint, adjustable source of illumination. Some observers prefer directly illuminated, bright wires, while others choose a faint (preferably red) field illumination in which the wires appear black. To attach and tighten the wires is a game of skill, but a well-prepared set of wires may last many years. Spider web has now been replaced by thin, vinyl-type threads which are more durable (against humidity) and also superior with regard to homogeneity and sharpness of light reflection.

The degree of difficulty with which a particular double star can be observed with a given telescope depends (apart from atmospheric conditions and eye sensitivity) on three stellar properties: the separation ρ, the magnitude difference Δm, and the combined magnitude. It must be immediately emphasized that the often quoted rules about limiting magnitudes, seeing disks, magnifications, etc. do not apply at all to the present task.

The well-known Dawes formula states that doubles (of adequate brightness and with equal components) can be barely separated at angles of 12″ divided by the telescope aperture in cm. In closer pairs the stellar disks touch and merge into an oblong shape. The Dawes limit agrees closely with the half-width of diffraction images predicted by theory, but for larger apertures (over 0.5 to 1.0 m, depending on conditions) the gain in resolution is less than calculated, as more turbulence cells simultaneously in the beam affect the image. The magnification used is normally the highest that atmospheric conditions will permit, except for faint components. For modest apertures (less than the 0.5 or 1.0 m quoted), experienced observers prefer to use "overmagnification." (Since the eyepiece frame is close to the plane of the wires, beware of wire damage when changing eyepieces!) Stepping down the aperture is useful when the gain of better image definition and reduced glare outweigh the loss in resolution, that is, for bright stars observed with large-aperture instruments.

Difficulties and systematic errors occur in close pairs when the images are adjacent, and when the space between the wires is less than the wire thickness and therefore hard to estimate. Less experienced observers should avoid separations below twice the Dawes limit. In the other direction, toward wide pairs, the precision decreases when high magnification limits the convenient, simultaneous viewing of the stars—this occurs at about 10^4 arcsec divided by the magnification. Systematic errors are also much larger for extended objects such as planetary disks and features.

The resolution is fully exploited only for near-equal components. Even $\Delta m = 1$ mag is much more difficult and also substantially increases the risk of systematic errors. The impediment by Δm can be included in the separation by an empirical "scale of difficulty"

$$C = 0.22\ \Delta m - \log \rho. \tag{26.5}$$

When a 40-cm instrument, for instance, reaches $\rho = 0\farcs3$ at the Dawes limit, then $C = 0.5$, and the corresponding limit at $\Delta m = 3$ will be around $\rho = 1\farcs4$.

The seeing scale is very different from that in other observations, since the visual observer concentrates on the bright nucleus of the turbulent disk and not on the total spread of light. The author has often estimated the seeing as $0\farcs2$, while the photometrist at the telescope next door recorded 1″ or $1\farcs5$.

Observations are also impeded when the stars are too faint for convenient direct (foveal) viewing. Under good conditions, an aperture of 60 cm may just reach components of 13th magnitude, but the "convenient" limit (beyond which the uncertainty of measuring increases) is at magnitude +11. For an aperture of 25 cm these numbers become 2 magnitudes lower. The usual guidelines on how faint a star a given telescope will show are irrelevant here, as there is an enormous difference between a barely visible point of light and one whose position can be measured.

The measuring procedure is illustrated in Fig. 26.3: a fixed wire (c) is rotated until positioned so as to join the centers of the stellar disks (position I), and the pointer at the position circle read; the setting is repeated a few times. Separation is measured with a pair of wires perpendicular to the line joining the stars; the fixed wire (a) is placed over star 1, and the movable wire (b) then brought to bisect star 2 (position II), whereafter the wire positions are exchanged. The difference of the two drum readings corresponds to twice the separation. This also is repeated. Settings of angle may also be done with a close pair of wires (c–d in position III) bracketing the double star in the middle, or the crosswire (c) can be dispensed with by using (a) with a subsequent rotation by 90°. The trick of the trade is to make the settings quickly (to avoid eye fatiguing), yet precise, and to catch the fleeting moments of steady images. The line joining the eyes is parallel (or perpendicular) to that joining the stars; observations made with a slanted orientation of the eyes risks large errors in the position angles.

A pair is measured on two or more nights of the same year, so that comparison of the results shows the amount of random scatter. The observing record includes seeing, magnification (when using more than one eyepiece), and the quadrant of the secondary star according to Fig. 26.2 so that 180° confusion is avoided. A separation estimate may be added to safeguard against gross reading errors. Some experienced observers also record magnitude estimates; catalogue data on brightness are often unreliable for faint objects.

The measurements are simply reduced by the following procedure:

1. Calculate average of angle readings minus zero point of circle = position angle;
2. Find average of double-separation readings multiplied by half the screw value = separation in arcseconds;
3. Observing time is converted into fractions of the year; cf. Appendix Table B.9. (0.5 yr is added to the numbers in the *Astronomical Almanac*, Section B, in order to transfer from the middle to the beginning of the year.)

The zero point of the circle is the reading when the wire used for position angles points exactly north. It is found from the east–west direction, by placing the measuring wire in repeated attempts so that an equatorial star trails exactly (to $\pm 0\overset{\circ}{.}1$) along the wire across the field. These readings are performed at the beginning and end of every observing session; experience will show how far the constancy of the zero point can be relied upon. Systematic position angle errors will be caused by defective grading or centering of the circle; they can be checked by reading differences between auxiliary pointers at various angular positions. The usefulness of a field-reversing prism in front of the eyepiece (to find subjective errors by comparing readings with and without prism) is doubtful.

The screw value (the angular equivalent of the screw pitch in the focal plane) must be determined accurately, but it is then known once and for all. The interval of sidereal time required to travel a certain number of screw revolutions is noted and then converted into arcseconds by the factor $15 \cos \delta$. The author recommends star transits around $\delta = \pm 75°$; closer to the pole the motion is so slow that the scintillation is much more annoying. The screw value may also be found by measuring distances (total, or only in declination) between very widely separated stars with well-known positions (e.g., the Pleiades), allowing for differential refraction. For comparison, the screw value is computed:

$$\text{screw value} = 206265'' \times \frac{\text{screw pitch (mm)}}{\text{effective focal length (mm)}}.$$

Double-image micrometers are designed to measure stellar images relative to one another instead of against wires. Birefraction is employed in the Muller prism [26.2];

it consists of two parts of a birefringent prism glued together at $\pm 45^\circ$ tilt against the optical axis, and movable across part of the field by the screw. The device is less sensitive to imperfect telescope tracking and probably also to subjective measuring errors. The position angle and separation are measured by forming an easily judged configuration, say, an exact square, from the four points of light seen. The addition of a rotating polaroid plate with scale permits one to vary the relative brightness of the two images, making the instrument also a comparison photometer for closer double stars. The image-splitting process combined with some absorption in the prisms results in a total light loss of about 1 magnitude, which is partly compensated by the absence of illumination in the field; thus, there is some justification for the more laborious and expensive construction of the device.

The double star could also be compared with an artificial pair of stars whose angle, separation, and brightness is varied by screw and polaroid. But the contrast between the appearance of real, scintillating stars and that of pinhole light points has proven troublesome and has led to the abandonment of this idea.

26.4 Speckle and Photographic Observations

The quest for higher resolving power has brought interferometry into the technical arsenal used to study binaries. Monochromatic light of wavelength λ falls through two slits and is subsequently recombined to form a pattern of interference fringes which, in the case of double stars, disappears (or reaches minimum visibility) in a specific position: the slits are lined up with the position angle and are $\lambda/2\rho$ apart, which gives the angular separation ρ. The enormous light loss experienced by interferometry has—at least until the recent arrival of high-speed detectors—severely limited its astronomical applications, and the shortage of the requisite large-telescope time has also contributed to the low productivity of interferometric investigations. Most data came from Finsen [26.3], who skillfully constructed and operated an eyepiece interferometer, reaching reliable measures down to 1/2 of the Dawes separation limit of his modest refractor.

The small turbulence elements of the air, which are 10 to 20 cm large and have a timescale around 1 to 0.1 s, dissolve the image into rapidly changing spots, or *speckles*. The trained eye subconsciously perceives an instantaneous image in the spots, but this ability is hampered, as noted above, for large apertures—and thus for close pairs—by the increased number of simultaneous air cells in the image. The so-called speckle methods [26.4] purport to undercut the detrimental frequencies of scintillation by still shorter recording (exposure or integration) times of 1/30 to 1/100 s. The focal image is magnified by microscopic optics, recorded by high-speed charge-coupled (photon-counting) devices (CCDs)—which have replaced the previously employed image intensifiers and cameras—and the data fed online into a computer, which makes the image of the binary re-emerge from the morass of hundreds of speckles. The closest known pairs at separations down to about $0\overset{''}{.}03$ have been measured in this way; in particular, many objects known only as spectroscopic pairs became resolved and were subsequently processed to obtain masses. (This combination of data can, as stated earlier, be substituted for the parallax measurement needed otherwise.) Stellar diameters and brightness distributions on stellar disks have also been measured, starspot regions were identified, and the latest and fastest CCDs can reach to very faint stars within admissible integration times. Several (somewhat different) devices of this type have been successfully applied during the last decade, and are expected to soon become the dominant measuring technique. These devices are still very expensive prototypes and are not yet commercially available; their operation requires a technical support staff, and thus their use is feasible only for large observatories. In addition, the correct link of the composite speckle image with the binary star position requires a series of careful calibration measurements, including a safeguard against spurious "companion" speckles.

Compared with visual measurements, photographic observations of double stars admit of higher precision, as one plate accommodates multiple exposures of an object which can then be measured carefully and repeatedly. However, few objects exist which are photographically well separated and yet worthy of regular observation, and so the technique is little practiced. Exceptions are some nearby stars covered by parallax programs, and the common-proper-motion pairs and streams for studies of moving groups. Owing to the low quantum efficiency of photographic emulsions, more scintillation accumulates during the requisite exposure time; the seeing disk—much larger than in visual work—limits the resolution much more than does the plate graininess. The author much prefers fast fine-grain emulsions over slower, hyperfine grains.

The angular separation in radians is given by the linear separation distance in the focal image divided by the effective focal length. Converted into arcseconds: 1 mm on the plate = $206\overset{''}{.}265$ divided by the focal length in meters. Well-defined images, correctly exposed and made in good seeing, may be measurable down to a separation of 0.07 mm, but troublesome adjacency effects arising from both development and measurement of the images occur below about 0.15 mm. Thus, even for long-focus telescopes, good photographic resolution is above ten times the visual Dawes limit. Magnifying Barlow-type optics have been employed to lift the image size over the photographic grain and adjacency limitations, but they help only for small apertures when the seeing disk may not be the critical limiting factor. The tedious part of photographic astrometry is to measure the plates (films are, of course, not used), and this requires a good measuring machine. Elaborate automatic machines have improved and much expedited the production of basic astrometric data, but they are of only limited usefulness for double stars since the scanner cannot separate the close double image within the diaphragmed plate region.

Owing to the very specialized and long-term nature of astrometric observations, they are not treated in a special chapter in this book; in the present context, a few explanations will do. Proper motions of stars and binaries (and sometimes mass ratios of the latter) are found from plate measurements relative to unrelated comparison stars in the field whose positions and proper motions are already known from transit-circle catalogues; this requires long-focus coverage over considerable time spans, usually decades. Images of larger reference-star fields rely on precise orthogonality of the plate and the optical axis. (A 20-cm plate covering an angle of 1°, but slanted by 0.5 mm on one side, will displace the optical center relative to the geometric center by about 1 μm, and hence the central star will be shifted relative to comparison stars near the edges at the level of measuring precision.)

Measurements of wide star pairs will have to be corrected for differential refraction; this also applies when such measurements are used to determine the image scale or the visual micrometer screw value. The atmosphere lifts the lower star more strongly; per arcminute, the altitude difference is thus diminished by about $0''.04$ at 40° altitude, and by $0''.07$ at 30°. Reduction of reference-star fields by the formulae contained in Sect. 3.4 in Vol. 1 removes the linear part of differential refraction; the nonlinear remainder is negligible except at unusually large zenith distances or hour angles. The atmospheric spectrum and part of chromatic aberration in refractors are eliminated by suitable color filters, a double star of substantial Δm can be exposed through gratings so that side images of the overexposed primary star appear. If hypersensitization is desired (in order to reduce exposure times and diminish the influence of tracking errors), the preferred methods are *nitrogen baking* (done shortly before exposure) and *pre-flashing*; the efficiency of each varies much with the type of emulsion used.

Small changes in the optical adjustment over time may mimic irregularities in the motions observed, and the presence of unseen companions may then be suspected from what actually are varying aberrations in the imaging of stars of differing brightness and color. This effect has been especially problematic in the study of presumably low-mass objects; most of the low-amplitude effects and, in particular, the "planet" discoveries occasionally headlined in the news media were all subsequently found to have resulted from uncorrected optical errors. The criteria for real changes are that they either are clearly above the error level or are distinctly periodic. (A similar stability problem occurs in spectrographs; hundreds of apparently variable radial velocities were disallowed upon reinvestigation with higher accuracy.) In principle, small masses can be found by astrometry, but the optical effects must then be modeled and removed. The linear reduction (Chap. 3) is then replaced to include nonlinear terms, and the usual number of reference stars (about five) does not suffice; the plates must show ten to twenty reference stars. The existence of unseen stellar components has been proven astrometrically in some double or multiple systems and in some apparently single stars. Three masses seem to be substellar at the time of this writing (the visible components of Wolf 424 and the invisible companion of DT Vir), and one (Hei 299) is a suspect.

26.5 Orbital Elements and Ephemerides

Seven quantities, called *orbital elements*, determine the orbit of a visual binary star. They are defined in analogy to the elements of planetary orbits:

a = semimajor axis of the true orbital ellipse in arcseconds, and
e = numerical eccentricity, which fix the size and shape of the orbit.

The relative motion is expressed by

P = period of revolution in years, and
T = time of passage through periastron (the point of smallest true distance between the components).

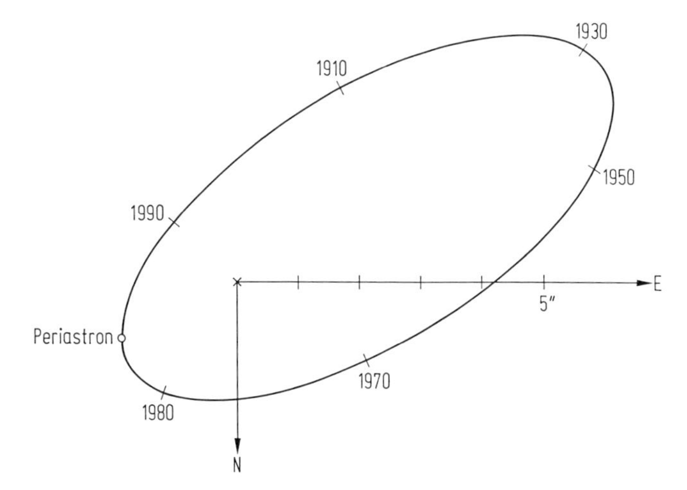

Fig. 26.4. The orbit of 70 Ophiuchi with $P = 88.3$ years.

The position of the companion at any other time is then determined by Kepler's laws I (law of elliptical motion) and II (law of equal areas). Depending upon the orientation of the orbit with respect to the line of sight, the observed, projected ellipse differs from the true orbit by projection, which is defined by three angles:

i = inclination, the angle between the orbital plane and the tangential plane at the "sphere," counted from 0° to 180°. The projection factor $\cos i$ is taken positive ($i < 90°$) when the motion is toward increasing position angles (direct, or counterclockwise); $i = 90°$ corresponds to edge-on viewing.

Ω = node, the position angle of intersection between the true and tangential planes. Usually $\Omega < 180°$ is stated; the other node is 180° opposite. Precession changes all position angles, and also the node, by the amount $+0\overset{\circ}{.}557 \sin\alpha \sec\delta$ per century. Measurements from different times are reduced accordingly in order to refer to a common equator.

ω = angular distance from node to periastron in the true orbit, counted in the direction of motion.

The *Thiele–Innes elements* serve to describe the orbit in rectangular coordinates. They correspond to the group (a, ω, Ω, i), and are composed thus:

$$\begin{aligned} A &= a(\ \ \cos\omega\cos\Omega - \sin\omega\sin\Omega\cos i), \\ B &= a(\ \ \cos\omega\sin\Omega + \sin\omega\cos\Omega\cos i), \\ F &= a(-\sin\omega\cos\Omega - \cos\omega\sin\Omega\cos i), \\ G &= a(-\sin\omega\sin\Omega + \cos\omega\cos\Omega\cos i). \end{aligned} \tag{26.6}$$

From the elements can be computed an ephemeris which gives the positions of the binary at any time t (Fig. 26.4). To optionally determine the polar coordinates p and ρ or the rectangular coordinates x and y, the following sequences of easily programmable formulae are employed:

$$\frac{360^\circ}{P}(t - T) = M = E - e \sin E; \tag{26.7}$$

$$\tan v/2 = \sqrt{(1+e)/(1-e)}\ \tan E/2, \qquad X = \cos E - e,$$
$$r = a(1 - e^2)/(1 + e \cos v), \qquad Y = \sqrt{1 - e^2}\ \sin E; \tag{26.8}$$

$$\tan(p - \Omega) = \tan(v + \omega)\ \cos i, \qquad x = AX + FY,$$
$$\rho = r \cos(v + \omega)/\cos(p - \Omega), \qquad y = BX + GY. \tag{26.9}$$

The conversion proceeds via an auxiliary angle E to coordinates in the true orbit (v and r, or X and Y, respectively) by Eqs. (26.8). Projection into the corresponding coordinates of the apparent orbit is afforded by Eqs. (26.9), which contain the direction cosines. Equation (26.7) is called the *Kepler equation*, expressing the law of areas, and is solved iteratively (see [26.1], p. 35). With degrees as units, the term $e \sin E$ is multiplied by 57.296 in order to obtain the same unit.

New observations sometimes suggest that a new orbit or a substantial improvement over a previous result may be obtainable, and that is usually done by the observer. The latest orbit catalogue in print [26.5] lists 850 pairs with orbital data as of 1982; the number has now (1990) increased to 1000, with some of the data having been revised. These are mostly close pairs; separations in the range $0\overset{\prime\prime}{.}1$ to $1''$ dominate, periods are mostly between 10 and 400 years, and distances from the Sun rarely exceed 200 pc. Many orbits are known with high precision, while others are still quite tentative—especially the long-period ones which may need another century or so to affirm by complete coverage. Differential corrections to initial elements are readily computed from a good and complete body of observations.

26.6 Photometric Binaries

The components of a binary star system may be so close together as to pass in front of one another, as seen from Earth, during their mutual revolution; the periodic brightness variation due to the mutual eclipses then reveals its binary nature. As is the case for genuine variable stars, the discovery and study of these binaries lies in the realm of photometry; however, the often used term "eclipsing variable" is not to be taken to mean *physical* variability of most of these components.

Eclipses are unlikely when the distance between the stars is large compared with their diameters. Thus, photometric binaries tend to be mostly short-period systems, and the components preferentially the larger stars (subgiants and giants) of advanced evolutionary status. Periods are typically a few days; they range from an extreme 27 years for the supergiant ε Aurigae down to nearly 1^{h} for the closest pairs of faint dwarfs. The rapid orbital motion is also seen spectroscopically by substantial variations in the Doppler shift. Owing to the importance of obtaining the parameters of evolving stars and for their comparison with theory, several hundreds of eclipsing binaries have been carefully studied by two-color and multi-color light curves, spectra, and computer modeling.

The light curve of an eclipsing binary graphs brightness (maximum intensity normalized to equal 1) against time in the cycle of period. The features relating to the abscissa (time) are essentially determined by the geometry of the eclipses. An interval of near-constant light around mid-eclipse shows a central (total or annular) eclipse; it is absent in partial eclipses. In the simplest case, when eccentricity $e = 0$ and inclination $i = 90°$ exactly, the stellar radii r_1 and r_2 can be found directly (as fractions of the orbital radius a) from the period P, the overall duration D of an eclipse (first to last contacts) and the duration d of totality:

$$r_{1,2} = \pi a(D \pm d)/2P. \tag{26.10}$$

This relation assumes that the distance between the components is large enough that the orbit curvature over the time D and distortions of the stars can be neglected. The orbital radius a (from which the masses follow) is inferred from the radial velocities. A component of a binary has the radial velocity

$$V = K[e\cos\omega + \cos(v + \omega)] \tag{26.11}$$

relative to the mass center, with the elements e and ω and the variable v as defined above. The semi-amplitude K is composed of a, e, i, and P (with i and P well defined by the light curve); the radial velocity of the mass center itself enters merely as an additive constant. As mentioned before, mass determination requires knowledge of the K values of both components.

The intensities as light-curve ordinates are complicated functions of the orbital parameters; even assuming analytical expressions for limb darkening and ellipticity of the stellar disks, they will be represented by elliptic and hyperelliptic integrals. Their derivatives with respect to the orbital elements are much more straightforward to evaluate (see [26.1], pp 98). The preferred synthesis thus starts with an approximate set of, for instance, six parameters (radii and limb-darkening factors of the two components, luminosity ratio, and inclination); differential corrections are computed and terms allowing for irradiation, ellipticity, etc. added as feasible. The spectral data help by supplying the mass ratio (and hence the shape of the equipotential surfaces) and the stellar surface temperatures, in order to test if the input model is realistic (detached or other configuration).

The demands on observation and computation (usually all parts of the light curve are relevant and have to be covered) give large telescopes and their favorable location in more consistently good weather a definite edge; the present outline is meant to provide a background for instructional rather than research projects. A significant contribution can still be made with modest instrumentation and without coverage of the complete light curve: the system is observed over widely spaced intervals to obtain the precise timing of the light minima. Mass flow in close binaries changes the periods and can thus be identified. Although the gradual changes are only by small fractions of seconds, they add up strongly to shift the instants of minima. At normal transfer rates, say $10^{-7}\,\mathfrak{M}_\odot$ per year, light phases may be displaced by minutes and hours when compared with ephemerides based on measurements taken a few decades ago. (Times of minima are reduced to the position of the Sun, removing the amount of light-travel time due to the Earth's orbital motion.) More detailed studies may then show if a period has changed gradually or in a discontinuous fashion (from sudden events in the flow pattern), or if a nonconservative model (with loss of total mass and angular momentum to the external environment of the system) is indicated. Sometimes a rotation of the entire orbit (apsidal motion) is found or suspected (similar to the motions of Earth-orbiting satellites, and due to gravity deviating from that of a point-mass field), or a light-time effect from the presence of a more distant third star which causes the distance of the eclipsing pair from the Sun to vary with a long period. It is not surprising that thousands of observing hours and printed pages have been devoted to obstinate cases like β Lyrae, and yet such studies are expected to yield more information on the patterns of stellar evolution.

References

26.1 Heintz, W.D.: *Double Stars*, Reidel, Dordrecht 1978.
26.2 Muller, P.: *Astronomical Techniques* (ed. W.A. Hiltner), Chap. 19, U. of Chicago Press, Chicago 1962.
26.3 Finsen, W.S.: *Astronomical Journal* **69**, 319 (1964).
26.4 McAlister, H.A.: *Ann. Rev. Astr. Astrophys.* **23**, 59 (1985).
26.5 Worley, C.E., Heintz, W.D.: *Publ. U.S. Naval Obs.* **24**, Part VII, 1983.

27 The Milky Way Galaxy and the Objects Composing It

T. Neckel

27.1 The Visual Appearance of the Milky Way

The band of the Milky Way spans the sky as a great circle. This is best seen when the north or south pole of the Milky Way (north galactic pole at $\alpha = 12^h49^m$, $\delta = 27\overset{\circ}{.}4$ (1950); south galactic pole at $\alpha = 0^h49^m$, $\delta = -27\overset{\circ}{.}4$ (1950)) is at the zenith and the Milky Way can be seen girdling the entire horizon. For this impressive view, it is of course necessary to have a sky which is haze-free down to the horizon, in addition to the optimum location (geographic latitude $\varphi = +27°$ or $\varphi = -27°$) and sidereal time ($\tau = 13^h$ or 1^h).

Galilei recognized that the light of the Milky Way is generated by innumerable faint stars, too weak to be distinguished individually by the naked eye. His just-invented, but still very primitive, telescope sufficed for this important discovery. The explanation for the apparent shape of the Milky Way is well known to modern astronomers: the Sun is a member of a strongly flattened, disk-shaped system of stars whose projection onto the sky appears as a luminous band encircling the sky. When viewed from Earth along the plane of the disk, the majority of stars will be so distant that they cannot be seen as individual points by the eye; their combined light therefore manifests itself as the glowing band of the Milky Way.

While the fainter of the naked-eye stars, especially those of the fourth and fifth magnitudes, are distinctly concentrated toward the plane of the Milky Way, this does not hold true for the much brighter first and second magnitude stars. In fact, young stars (i.e., types O and B) in particular show a distinct concentration to another great circle in the sky, and one which distinctly differs from the Milky Way. This circle is named *Gould's Belt*, and it is inclined to the plane of the Milky Way by about 20°. This deviation from the plane of the Milky Way is most pronounced in Orion and in Scorpius, where many young stars are found at unusually high galactic latitudes.

The Milky Way exhibits quite an irregular structure, with dark and nearly "starless" areas sometimes found between the bright star clouds. It was originally believed that these are directions in which there are no stars, but this impression is deceptive. In reality, this effect is caused by dense dust clouds which to varying degrees reduce the light from the stars behind them. Therefore, without additional information, one cannot tell whether a particular bright cloud in the Milky Way is a genuine accumulation of stars or a region of relatively low dust concentration. Sect. 27.14.5 will show that both possibilities occur.

The most pronounced region in the Milky Way is the one which contains the famous "rift" which begins in Cygnus and extends into Sagittarius. The best-known isolated dark cloud is the Coal Sack, which lies immediately next to the Southern Cross in the southern Milky Way.

Superposed upon the irregular brightness distribution in the Milky Way is a systematic increase in brightness in the direction toward Sagittarius. Here also, the Large Sagittarius Cloud, the brightest cloud in the entire Milky Way, is found. Figure 27.1 shows a picture of the Milky Way from Aquila to Centaurus, composed of two wide-angle exposures taken by H. Vehrenberg. The Large Sagittarius Cloud is close to the center of the picture. This picture illustrates the distinct similarity of the Milky Way with other spiral galaxies seen edge-on. Apparently, the Large Sagittarius Cloud is nothing more than the central region of the Milky Way system. The galactic center, as will be seen in Sect. 27.14.1, lies about 8.5 kpc from the Sun. This suggests that the major part of the light of the Large Sagittarius Cloud comes from stars in the denser region around the galactic center, and thus also lies about 8.5 kpc (28 000 light years) distant.

Infrared and radio observations show that the true center of the Milky Way system is not exactly within the Sagittarius Cloud, but instead lies a few degrees away in the direction of a rather dark portion of the Milky Way. As will be explained in Sect. 27.2, the light-attenuating effect of the interstellar dust is responsible for the fact that the true center of the Milky Way cannot be seen at all, but only outer portions of the central region which are less affected by the influence of the interstellar dust.

Radio observations have determined the location of the galactic center quite accurately. It is at the coordinates

$$\alpha(1950.0) = 17^{\mathrm{h}}42\overset{\mathrm{m}}{.}4; \quad \delta(1950.0) = -28\overset{\circ}{.}92.$$

Also, the central plane of the Galaxy is very well defined through observations of the 21-cm line of neutral hydrogen, since the hydrogen is very strongly concentrated toward it. Thus it is possible to define a natural *galactic coordinate system* whose coordinates are called *galactic longitude* l and *galactic latitude* b. The zero-point $(l, b) = (0^\circ, 0^\circ)$ defines the direction to the Galactic Center, the equator $b = 0^\circ$ the projection of the galactic plane into the sky, and galactic longitudes are counted eastward. In order to visualize galactic longitudes, the reader may find it helpful to memorize a few directions: Aquila 60°, Cygnus 80°, Cassiopeia 120°, Carina 290°, Centaurus 305°. The mathematical transformation from equatorial to galactic coordinates is given by

$$\begin{aligned} \cos b \cos(l - 33^\circ) &= \cos\delta \cos(\alpha - 282\overset{\circ}{.}25), \\ \sin b &= \sin\delta \cos 62\overset{\circ}{.}6 - \cos\delta \sin 62\overset{\circ}{.}6 \sin(\alpha - 282\overset{\circ}{.}25), \quad (27.1) \\ \cos b \sin(l - 33^\circ) &= \sin\delta \sin 62\overset{\circ}{.}6 - \cos\delta \cos 62\overset{\circ}{.}6 \sin(\alpha - 282\overset{\circ}{.}25). \end{aligned}$$

After the Sagittarius Cloud, the brightest clouds in the Milky Way are those in Norma, Scutum, Centaurus, and Carina (at l = 330°, 22°, 307°, and 290°, respectively). Of these, only the Scutum Cloud is visible at northern mid-latitudes, but it, like the Sagittarius Cloud, is never very high above the southern horizon. The full splendor of the Milky Way can be realized only when it is observed from the south-

Fig. 27.1. The Milky Way in the vicinity of the galactic center. At the center is the Large Sagittarius Cloud with the star Antares to the right. Photographs by H. Vehrenberg, taken from the *Atlas Galaktischer Nebel*, Vol. II.

Fig. 27.2. Part of the southern Milky Way, with the Coal Sack and Southern Cross. Photograph by B. Koch.

ern hemisphere, and the view is further enhanced at a remote observing site in, say, Namibia or Chile, which provides the additional substantial advantage of minimal light pollution of the night sky.

The finest and most familiar section of the southern Milky Way with the Coal Sack and the Southern Cross is shown in Fig. 27.2.

27.2 Interstellar Dust and Its Effects upon Distance Determinations Within the Galaxy

One component of the Milky Way's structure influences and impedes the visibility of all other galactic objects. That component is the *interstellar dust*. It weakens the light of all objects, so that they can be observed optically only with large instruments, if at all. This applies not only to the dust in the conspicuous dark clouds, such as the Coal Sack, but also to such dust as is present almost everywhere near the galactic plane, and therefore escapes direct observation entirely. The history of the discovery of its existence is here recounted.

In 1930, R.J. Trumpler completed an extensive study on the spatial distribution of open star clusters. He had investigated the member stars of many such clusters spectroscopically in order to assign absolute magnitudes M corresponding to their spectral types. Photographs of the clusters served to determine apparent magnitudes m so that the distances of the stars (in parsecs) could be derived from the relation

$$m - M = 5 \log r - 5, \tag{27.2}$$

and hence, by averaging, the cluster distance. From the apparent diameters of the clusters, their true linear diameters then followed directly. This led to a surprising result: the cluster diameters systematically increased with the distance, independent of the direction. If this were true, then it would be equivalent to stating that the Sun occupies a unique position within the Milky Way, an assertion which has been held to be suspect by nearly every astronomer since Copernicus. This strange systematic effect in cluster diameters could be avoided by assuming the existence of an absorbing medium which was homogeneously distributed throughout the Milky Way. Then Eq. (27.2) is augmented by a term $A(r)$, which indicates that the extinction of light steadily increases with distance:

$$m - M = 5 \log r - 5 + A(r). \tag{27.3}$$

This equation gives smaller distances than Eq. (27.2), and also, in the case of clusters, smaller diameters. The larger the distance and thus the amount of extinction $A(r)$, the more the respective distances derived from Eqs. (27.2) and (27.3) differ. Trumpler found that a constant amount of 0.79 mag kpc^{-1} is needed to obtain cluster diameters independent of distance. In the currently used UBV system, this corresponds to a visual extinction coefficient $a_v = 0.65$ mag kpc^{-1}.

A wholly different indicator for the presence of light-absorbing material near the galactic plane is afforded by the distribution of extragalactic objects in the sky. Their density (number per square degree) is largest at high galactic latitudes, and steadily decreases when approaching the plane of the Milky Way. A belt 15°–20° wide along the galactic equator shows almost no galaxies (see Fig. 27.3). This region is known as the *zone of avoidance*, and its discovery led astronomers to the premature conclusion that the galaxies had to be part of the Milky Way system. In 1934, E. Hubble recognized that the zone of avoidance was to be interpreted as indicating a layer of dust strongly concentrated toward the central plane of the Milky Way. From the count of nebulae as a function of galactic latitude, the extinction in the directions of the

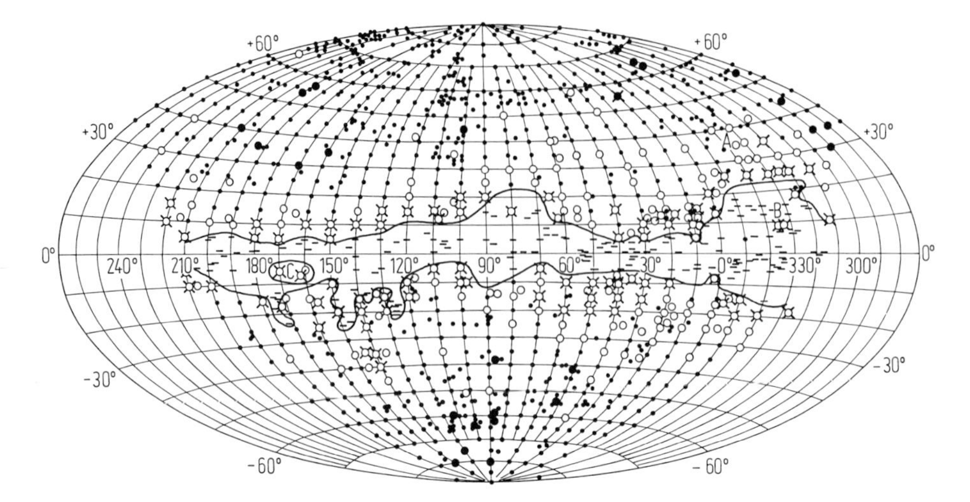

Fig. 27.3. The *zone of avoidance* (free of nebulae) around the galactic equator. The plot shows the apparent distribution of extragalactic objects on the celestial sphere, after Hubble. *Horizontal central line*: the galactic equator; *small dots*: normal density of nebulae; *large dots*: excess density; *circles*: deficit; *circles with bars*: large deficit; *dashes*: no nebulae (i.e., the zone of avoidance). After *Sterne und Sternsysteme* by W. Becker.

galactic poles could be determined as 0.25 mag, assuming that the absorbing matter is distributed homogeneously and in plane-parallel strata to the galactic plane.

One other observation revealed not only the existence of interstellar extinction but also one of its primary features: that the color indices of stars of equal spectral types increase on the average with larger distance; i.e., the stars become redder. This fact was also first discovered by Trumpler, and this allows of a simple explanation, assuming that interstellar extinction is caused by small dust particles whose absorbing effect continuously increases towards shorter wavelengths. Even when the blue light from a star has been appreciably dimmed, a relatively large fraction of its red light will reach the observer. Thus, the star will appear "reddened," just as is the case for the setting Sun or the rising Moon. This feature finally opened a way to determine, for any particular object in the Milky Way, what percentage of its light was blocked by the dust during its journey to Earth. The amount by which the color index of a star is reddened is easily measured. In terms of the currently most widely used *UBV* system, $(B-V)_0$ is the natural (unreddened) color of a star whose light has not been affected by extinction. It is closely related to the spectral type, and can be found in Table 8.3 of Vol 1. From the measured color indices $(B-V)$, the *color excess* is first found from

$$E_{B-V} = (B-V) - (B-V)_0. \tag{27.4}$$

It is just the difference between the extinctions in blue (B) and in visual (V) light, denoted by A_B and A_V respectively. This is immediately seen upon comparison with the magnitudes B_0 and V_0 which a star would have in the absence of extinction, and hence $B = B_0 + A_B$, $V = V_0 + A_V$, and $(B-V)_0 = B_0 - V_0$.

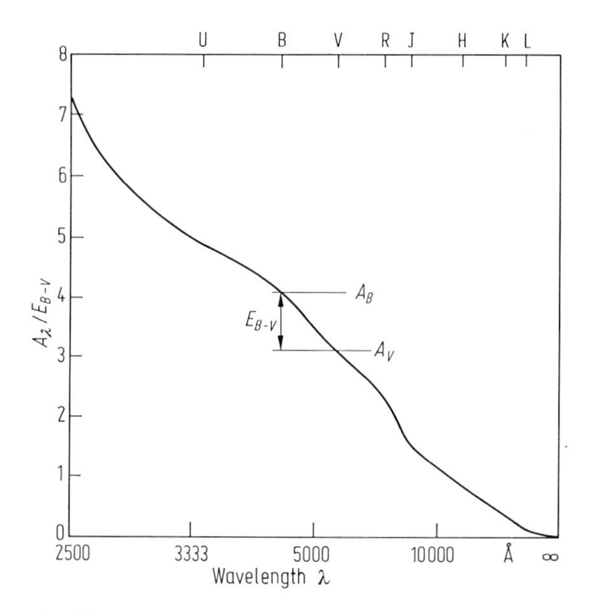

Fig. 27.4. The mean interstellar extinction curve, normalized such that it gives the extinction at the wavelength λ corresponding to the color excess $E_{B-V} = 1.0$.

The greater the influence of interstellar extinction, the larger will be the color excess E_{B-V}. In order to use the color excess to find the amount A_V of the extinction itself, the ratio

$$\mathfrak{R} = \frac{A_V}{E_{B-V}} \tag{27.5}$$

must be known. (The gothic symbol $\mathfrak{R}$ is used for this ratio to avoid confusion with the red magnitude R.) This enters into all distance determinations in the Milky Way system, and is therefore of paramount importance for all questions relating to galactic structure. Fortunately for astronomers, the function

$$f_\lambda = A_\lambda / E_{B-V}, \tag{27.6}$$

which describes the wavelength dependence of the extinction, is nearly constant throughout the Galaxy. This guarantees that $\mathfrak{R}$ is also constant. The value which is currently considered to be best is $\mathfrak{R} = 3.1$, with an error of at most ± 0.1. Departures from the standard curve f_λ occur in very young star-forming regions and are caused by the presence of large dust particles. $\mathfrak{R}$ values of up to about 5 have been found in such regions.

The shape of the mean interstellar extinction curve as depending on wavelength is shown in Fig. 27.4. The letters U, B, V, R, I, J, H, K, L, and M denote the centers of the photometric bandpasses used. Thus, to determine the distance to a star, the following separate steps are required:

1. Determination of the unreddened color, perhaps by means of the spectral type, with the aid of Appendix Table B.28;
2. Measurement of the color index $(B - V)$;
3. Calculation of color excess E_{B-V} using Eq. (27.4);
4. Calculation of visual extinction using Eq. (27.5);
5. Determination of the absolute magnitude M_V, also from the spectral type, with the aid of Appendix Table B.27;
6. Calculation of the distance using Eq. (27.3).

These steps may be modified in special cases. For instance, for δ Cephei stars the period may serve as the criterion for determining the unreddened color as well as the absolute magnitude. Other combinations, such as R and I, can be used in place of V and B magnitudes. This is advisable especially when the B magnitudes are too faint and unmeasurable owing to very high extinction.

When applying the method described to determine the extinction and distance for the members of an open cluster or for the exciting stars of a large HII region, one should not be surprised if quite discordant distances result; they may differ by fully a factor of two. The cause is that the absolute magnitude of a star can be derived from the spectral type to an accuracy of at best 0.5 mag. Also, the use of other photometric or spectroscopic methods to determine absolute magnitudes does not give more precise results. The best distance value of, for instance, an HII region is obtained by determining the individual distances of as many stars as possible which are embedded in it, and then assigning an average of these values.

27.3 The Galactic Objects: A Survey

Stars are the most numerous species among the galactic objects, apart from suspected but as yet undetectable planets. With the naked eye about 6000 stars can be seen, but even a small telescope shows several times this number. The bigger telescopes became, the greater the number of stars which came within reach. Some plates of the Palomar Sky Survey taken of rich Milky Way fields show several millions of stars. The total number of stars in the galactic system is estimated to be several 10^{11}! The basic properties of stars as individual objects are presented in Chap. 24, and their collective contribution to the structure of the Milky Way will follow in Sect. 27.14.

If one scans the Milky Way with a large-aperture telescope at low magnification, he gradually notices the large number of star clusters and of nebulous objects that come into the field of view. It also becomes readily apparent that there are two entirely different kinds of clusters. One type contains usually 100 or fewer stars, and the brightest members of the cluster are noticeably red. Such clusters are found throughout the Milky Way zone, but only seldom outside this zone. Owing to their loose structure, they are termed *open clusters*; see Sect. 27.5. Completely different in structure are the *globular clusters*; see Sect. 27.6. With a small telescope, say a 3-inch, a globular cluster will be seen only as a diffuse, round spot without any indication that it is composed of stars. For the smallest objects of this kind, the same is true even for a 50-cm telescope. When, however, one of the larger (and hence nearer) globular clusters is observed with a powerful telescope, it is recognized as consisting of thousands of stars strongly concentrated toward its center.

The finest and most diverse galactic objects are the large galactic emission nebulae such as the Orion, the η Carinae, the Lagoon, or the Trifid Nebulae, to name just a few. Unfortunately, many of these nebulae are too faint in their outer parts for their full, rich shapes to be revealed to visual observation; some of them have been discovered on photographs. It is thus seen that in the present age of high-speed photographic emulsions, these objects can be most rewarding for amateur astronomers devoted to celestial photography.

Five kinds of diffuse galactic nebulae are distinguished: *HII regions*, *reflection nebulae*, *supernova remnants*, *planetary nebulae*, and *Herbig–Haro objects*. Lists of clusters and galactic nebulae of all kinds are given in Appendix Tables B.33–37).

HII regions (see Sect. 27.7) shine in the light of numerous emission lines, most of which are generated through recombination of atoms which previously had been ionized by the ultraviolet radiation emitted by one or more hot stars. Only the highest-temperature stars, namely those with spectral types from O to B1, produce enough energy-rich UV photons to create an HII region. Since O- to B1-type stars are all very young, this also holds true for the HII regions.

Stars of later spectral types also may render the surrounding interstellar matter luminescent, but merely by illuminating the dust, which then reflects the stellar light. Such *reflection nebulae* (Sect. 27.8) are often seen in or at the edges of dark clouds. The best-known reflection nebulae are those surrounding the Pleiades, and the ones illuminated by Antares and ρ Ophiuchi. In some instances, a reflection nebula may be illuminated by an older star whose space motion happened to carry it through a cloud of interstellar gas and dust; usually, however, the stars illuminating reflection nebulae have only recently been formed in the clouds of which the nebula is a part.

In contrast to these two types of objects, remnants of supernova events (see Fig. 27.10), as well as planetary nebulae (Sect. 27.11), are objects associated with the end points of the evolution of a star. The best-known supernova remnant, which is also observable in visible light, is the Cirrus Nebula in Cygnus, and the best-known planetary nebula is the Ring Nebula in Lyra.

The most massive individual objects in the Galaxy are the *giant molecular clouds*. The dust they contain often makes them opaque to visible light, and so they appear as dark clouds. Comparison of a red and a blue photograph of the same galactic region usually shows many much-reddened stars at the edges of such dark clouds. This observation also unambiguously shows that most apparently star-poor regions in the Milky Way are indeed an artifact caused by the dimming and reddening of starlight by interstellar dust. *Bok globules* are small, roundish, isolated dark clouds named after astronomer B.J. Bok, who contributed substantially to their investigation.

The dark molecular clouds contain the necessary raw material from which new stars are born. Almost all HII regions can be directly linked with molecular clouds. In a very early phase of star formation, the nebulae associated with younger stellar objects undergoing high velocity mass outflow can be recognized by their characteristic shapes: bipolar or comet-shaped. Closely related to these are very tiny, almost starlike nebulae known under the name *Herbig–Haro objects* (Sect. 27.13).

27.4 Catalogues and Atlases of Galactic Nebulae

The first known list to contain some nebulae and clusters is the famous one compiled by C. Messier from 1784. It contains 39 extragalactic nebulae, 29 each of open and globular clusters, 4 planetary nebulae, and only 7 diffuse galactic nebulae. Object No. 42 (designated simply as M42) on the list is the best-known representative of its kind, the Great Nebula in Orion. The Messier list is reprinted in Appendix Table B.33.

Much more comprehensive is the *New General Catalogue* (NGC) published by J.L.E. Dreyer in 1888, and which contains 7840 objects. Most of the observations in the NGC are by Sir John Herschel, who in 1864 had already compiled a *General Catalogue* with 5079 entries. Supplements to the NGC are the *Index Catalogue* (1895) and the *Second Index Catalogue* (1908), both published by Dreyer. A revised version of the NGC by J.W. Sulentic and W.G. Tifft appeared in 1973 as the *Revised New General Catalogue*.

Users of the NGC and the IC should keep in mind that these catalogues are based almost entirely on visual observations. There is currently widespread opinion that faint nebulae can be reached only by long-exposure photographs. In view of the numerous faint nebulae in the NGC, the amateur observer with a large-aperture telescope should endeavor also to target some of the inconspicuous nebulae visually.

A survey of the POSS (Palomar Observatory Sky Survey) for diffuse galactic nebulae led to the *Catalogue of Bright Nebulae* by B. Lynds. The large number of entries in this catalogue does not necessarily imply that all of them are individual nebulae; often various parts of one nebula have been entered separately.

The best-known catalogue of HII regions was compiled by Sharpless, and it is also based primarily on the POSS. It is supplemented for the southern sky by the *Catalogue of Hα Emission Regions in the Southern Milky Way*, published by Rogers, Campbell, and Whiteoak in 1960.

Catalogues of northern reflection nebulae have been prepared by Dorschner and Gürtler (1964), and by van den Bergh (1966). The southern sky is covered by a list of reflection nebulae compiled by van den Bergh and Herbst (1975). The *Catalogue of Galactic Nebulae* by Perek and Kohoutek contains many data on most of the presently known planetary nebulae, as well as their photographs.

Important data and literature references on over one thousand star clusters and associations were compiled by Alter, Balazs, and Ruprecht. Data on 434 open clusters determined by *UBV* or *RGU* 3-color photometries have been edited and published by Janes and Adler (1982).

An even approximately complete compilation of photographs of galactic nebulae (except for planetaries) did not exist until recently. This gap has now been closed for the northern sky with the appearance of the first two volumes of *Atlas of Galactic Nebulae* by Neckel and Vehrenberg. After completion of the third volume, this atlas will contain all diffuse galactic nebulae seen in the POSS and in its southern extension, the *ESO/SRC Atlas*, except for planetary nebulae lying north of $\delta = -33°$. While the first two parts of the *Atlas of Galactic Nebulae* contain all objects north of $\delta = -33°$, the southern limit of the POSS, Part III contains the southern objects, primarily from the *ESO/SRC Atlas*. The total number of nebulae included is about 1500. Scales and areas of the photographs are chosen so that all essential details seen in the POSS at magnification 10× also appear in the atlas.

27.5 Open Clusters

Open clusters of stars are among the most attractive objects for visual observations, even when using just a small telescope. Some of these clusters, and even the brighter stars in them, can be seen with the naked eye. Thus, most observers can easily see the six bright stars in the Pleiades, which, however, are called the "Seven Sisters." Photographs with large telescopes show the Pleiades to be embedded in an extended complex of a reflection nebula (see Fig. 27.5) which is characterized by its unique, cirrus-like structure.

Perhaps the most important open cluster is the Hyades, which is so nearby that its distance (about 45 pc) has been measured by trigonometric methods (see Chap. 24). The absolute magnitudes of the Hyades stars are thus well known (see Sect. 27.5.2). The brightest star Aldebaran (α Tauri), despite its position at the cluster center, is not physically associated with the Hyades. An especially impressive object for visual

Fig. 27.5. The Pleiades. This loose open cluster illuminates a group of reflection nebulae. Photograph by Birkle at the Schmidt telescope in Calar Alto.

Fig. 27.6. The two adjacent open clusters h and χ Persei. Both are quite young and still contain many high-luminosity stars. Photograph by T. Neckel at the Schmidt telescope in Calar Alto.

observations is the double cluster h and χ Persei (Fig. 27.6). These two clusters, which lie at the same distance from Earth, are characterized by an exceptional richness of stars.

27.5.1 Classification of Open Clusters

Open clusters are substantially different with respect to their central concentrations and the number of member stars. The latter, of course, depends upon the limiting magnitude, and is, for that reason, ambiguous. Using the degree of central concentration as the primary criterion, Trumpler classified the open clusters into the following four groups:

I Strongly concentrated clusters, clearly standing out against the background;
II Weakly concentrated clusters, but still standing out clearly against the background;
III Clusters without noticeable concentration, but still standing out;
IV Clusters resembling merely random accumulations of background stars.

The coding p = "poor," m = "moderate," r = "rich" specifies the number of cluster members to be less than 50, between 50 to 100, or over 100, respectively. The Hyades are coded as IIm, and Praesepe is Ir.

27.5.2 Color-Magnitude Diagrams and Two-Color Diagrams

The stars which comprise a cluster, be it open or globular, obviously have a common evolutionary history. Star clusters provide a natural setting for showing how stars with a broad range of masses evolve by different amounts within equal time intervals. The study of clusters is thus of fundamental importance for developing criteria for theories of the structure and evolution of stars.

The observing technique which is ideally suited for studies of clusters is three-color photometry. Some clusters have been observed in the Johnson *UBV* system, others in the *RGU* system of Becker. These two color systems are largely similar, and data in one can be reduced to the other without much difficulty. Of the about 1200 known open clusters, 434 had been photometrically observed by 1982. The results have been edited by Janes and Adler and homogenized in the *UBV* system. Their conclusions regarding the spatial distribution of open clusters will be returned to in Sect. 27.14.

From the observed data on U, B, and V colors, a color–magnitude diagram, which in principle is nothing more than a Hertzsprung–Russell diagram, is constructed. However, the colors and magnitudes will be affected by the extinction A_V and the reddening E_{B-V}, both caused by interstellar dust. The amount of extinction A_V can be found without requiring any knowledge of the spectral types of the individual stars in the clusters, from which the *individual* color excesses could be derived. It may be assumed that the wavelength dependence f_λ of interstellar extinction (Sect. 27.2) is universally valid, and thus the ratio

$$\frac{E_{U-B}}{E_{B-V}} = \frac{A_U - A_B}{A_B - A_V} \tag{27.7}$$

has a nearly constant value. Based on results of a large number of photometric measurements of early-type stars with different reddening factors, it has been found that

$$\frac{E_{U-B}}{E_{B-V}} = 0.72 + 0.05\ E_{B-V}. \tag{27.8}$$

The weak dependence of this ratio on E_{B-V} is due to the fact that the central wavelengths of the U, B, and V photometric bands shift toward the red with increasing extinction. Since these bands are several tenths of nanometers wide, the extinction at the short-wavelength end is noticeably larger than at the long-wavelength end. The numbers in Eq. (27.8) are also slightly dependent upon the spectral type of the star, but this fact can be neglected here.

The two-color diagram which graphs $U-B$ against $B-V$ shows how the extinction shifts a star in the direction of the "reddening path" defined by Eq. (27.8), as shown in Fig. 27.7. Stars in a cluster are all subject to the same extinction, and will be displaced by a constant amount along this path.

As was seen in Chap. 24, the unreddened main-sequence stars are ordered along a characteristic curve in the two-color diagram (Fig. 27.4). When the two-color diagram of the cluster shows a similar pattern, but displaced relative to Fig. 27.4, the data points can be shifted back onto the reddening path to coincide with the main sequence. This is illustrated in Fig. 27.8 for the stars in h Persei. The amount of displacement in the direction of $B - V$ directly gives the color excess E_{B-V}, and hence from Eq. (27.5) the amount of visual extinction A_V.

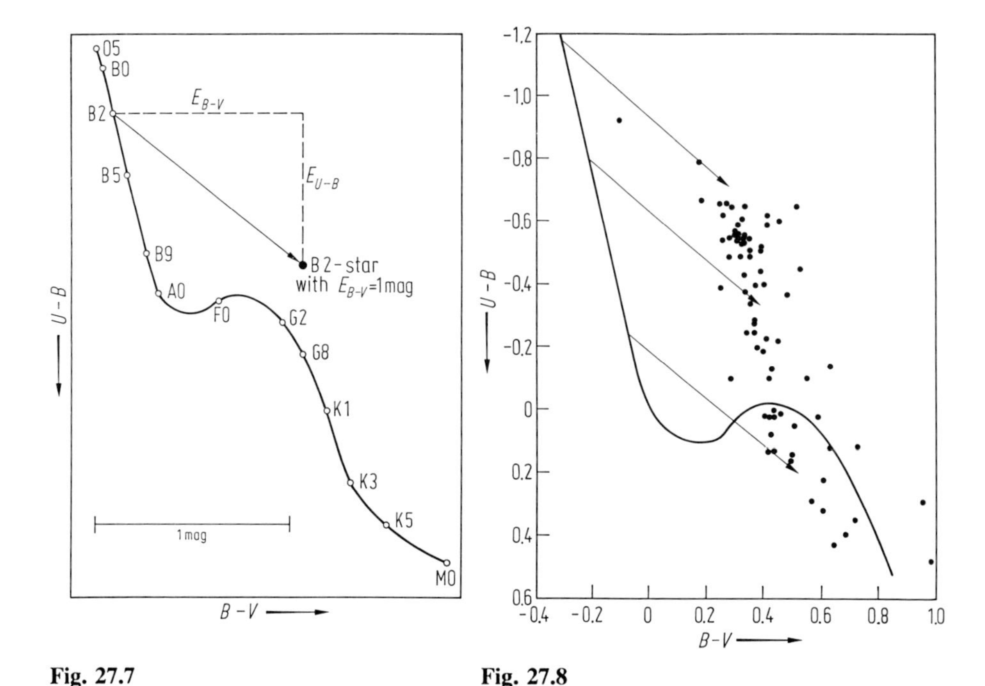

Fig. 27.7 **Fig. 27.8**

Fig. 27.7. The interstellar extinction shifts the position of a star in the two-color diagram along the reddening vector whose components are linked by Eq. (27.8). The *solid curve* shows the locus of unreddened main-sequence stars.

Fig. 27.8. The two-color diagram for *h* Persei demonstrates that all cluster members are shifted off the main sequence along the reddening vector by about the same amount. The solid curve is the unreddened main sequence.

Having determined the true colors $(B - V)_0$ by subtracting E_{B-V} from each of the measurements, a color–magnitude diagram with $(B - V)_0$ as abscissa and $m_V - A_V$ as ordinate can be drawn. Equation (27.3) shows that the ordinate differs from the absolute magnitude M_V by a constant, which is just the true distance modulus $5 \log r - 5$ of the cluster.

Figures 27.9 a and b show the color–magnitude diagrams of respectively the Hyades and Pleiades obtained in this fashion. Comparison with the HR diagram shows that in both clusters the majority of stars are placed along the main sequence, and only a few of them above it. The earliest stars of spectral type O are absent in both clusters. In the Pleiades, the main sequence is populated beginning with classes A1, in the Hyades beginning with F8. This pattern in color–magnitude diagrams is typical for all open clusters: the lower part of the main sequence is occupied up to a certain spectral type. The HR diagram of any cluster can be shifted in the ordinate so that the lower sections of the main sequence coincide.

The observed feature described above is exactly what would be expected from the evolution of an ensemble of stars with different masses: every star spends the greater

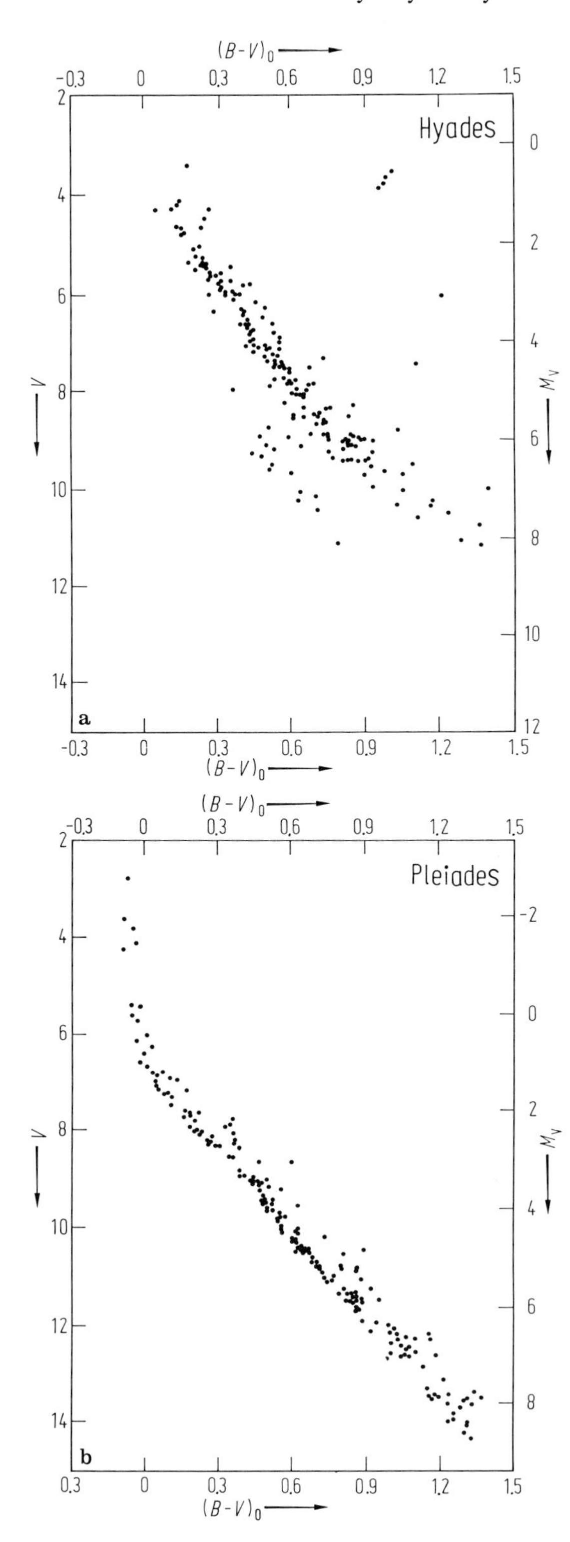

Fig. 27.9. a The color–magnitude diagram of the Hyades after G. Hagen. Some red giants above the main sequence and several subdwarfs lying below it are shown. **b** The color–magnitude diagram of the Pleiades contains no giants. The stars lying above the main sequence are likely unresolved binaries.

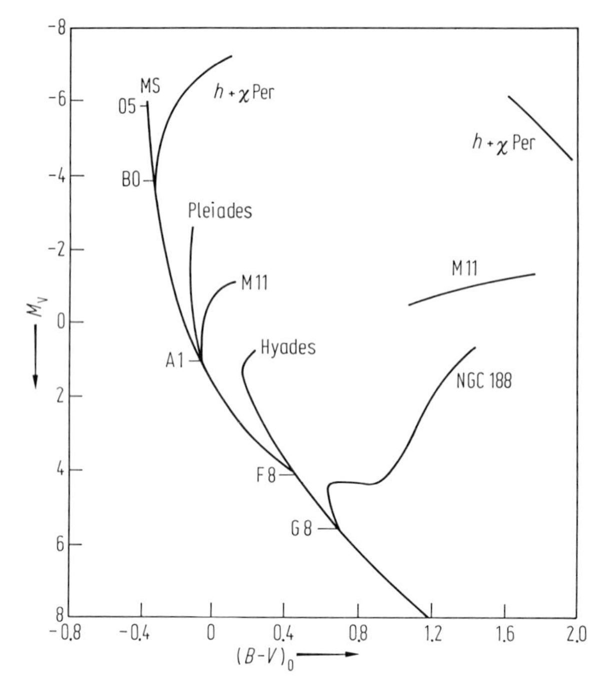

Fig. 27.10. Composite schematic color–absolute-magnitude diagram of five open clusters of different ages. With increasing age, the turnoff point of the main sequence (MS) shifts to later-type stars.

part of its life on the main sequence, dwelling there the longer the lower its mass. In a group of stars with equal ages, as occurs in an open cluster, all those stars whose main-sequence lifetime is less than the age of the cluster must have departed toward the red-giant branch. The earliest spectral type among the main-sequence stars thus indicates the age of the cluster.

If the color–magnitude diagrams of some well-observed clusters are superimposed so as to make the lower parts of the main sequences coincide, the combined diagram (see Fig. 27.10) shows how the pattern of the color–magnitude diagram of open clusters changes with age. Very young clusters contain a few red supergiants separated by a large gap from the main sequence (e.g., h and χ Persei). Old clusters whose main-sequence turnoff point is found around class F possess a giant branch which smoothly departs from the main sequence (e.g., NGC 188). The color–magnitude diagrams here approach similarity with those of globular clusters.

The well-defined lower envelope of the composite HR diagram is the locus of unevolved main-sequence stars, called the *zero-age main sequence*, or ZAMS. Its zero-point is fixed by the Hyades, which has a known distance modulus. The evolutionary timescales for G-, K-, and M-type stars are so long that these stars are still on the ZAMS. Stars of earlier types will already have reached the end of their main-sequence lives. Normal O- to F-type main-sequence stars not belonging to an open cluster will thus be a mixture of unevolved stars and of slightly evolved stars, but not so far

Table 27.1. The zero-age main sequence.

Spectral Type	Absolute Magnitude M_V	Color Index $(B-V)_0$	Color Index $(U-B)_0$
O4	−5.2	−0.33	−1.20
B0	−3.25	−0.30	−1.08
B1.5	−2.1	−0.25	−0.90
B3	−1.1	−0.20	−0.69
B6	−0.2	−0.15	−0.50
A0	+1.3	−0.02	−0.05
A4	+1.9	+0.10	+0.08
A7	+2.4	+0.20	+0.10
F0	+2.8	+0.30	+0.03
F4	+3.4	+0.40	−0.01
F8	+4.1	+0.50	+0.00
G0	+4.7	+0.60	+0.08
G6	+5.2	+0.70	+0.23
K0	+5.8	+0.80	+0.42
K5	+7.3	+1.15	+1.08
M0	+8.8	+1.40	+1.22
M2	+10.3	+1.50	+1.17
M5	+12.6	+1.65	+1.26

advanced that changes could be noted in the spectrum. Over the range of O- to F-type stars, the ZAMS is thus located at slightly fainter absolute magnitudes than the observed main sequence.

The construction of the ZAMS serves as the basis of luminosity calibration for stars as a function of spectral type. While the known Hyades distance permits the derivation of absolute magnitudes of F V and later main-sequence stars (as exist in the Hyades), the calibration of absolute magnitudes may be extended to B- and A-type stars on the assumption that the F- to M-type stars in the Pleiades have the same absolute magnitudes as in the Hyades. Clusters with O-type main-sequence stars or with giant or supergiant members thus also allow calibration of these MK types. With the interstellar extinction determined from color excesses by Eqs. (27.4, 5, 7), the ZAMS becomes the basis for all distance determinations within the Milky Way system, outside the small range of the solar neighborhood for which trigonometric parallaxes can be measured.

The data on the ZAMS relevant for practical application are compiled in Table 27.1. To determine the distance of a star cluster, it is advisable to graph a curve M_V as a function of $(B-V)_0$ and superpose it upon the $(m_v - A_V)$ versus $(B-V)_0$ diagram of the cluster, so that it runs symmetrically through the measured points and the two abscissa scales coincide. The distance modulus of the cluster can then be read directly as the difference of the ordinate scales.

In conclusion, it should be mentioned that the period–luminosity relation for Cepheids, which is of paramount importance for measuring extragalactic distances, was calibrated using several galactic star clusters containing a few δ Cephei-type stars.

27.5.3 Membership of Stars in Open Clusters

Almost all open clusters lie within the band of the Milky Way. Therefore, among the stars within the cluster, there will be many nonmembers. In numerous cases, the two-color or color–magnitude diagrams of the cluster may indicate whether a star is or is not a cluster member. If, for instance, the cluster has a distinct turnoff point, for example at A5, an O star in the cluster must certainly be a field star unrelated to the cluster but seen projected onto it.

An objective method for deciding on the membership of stars is based on the study of proper motions within the cluster. If the proper motions differ systematically from those of the surrounding field stars, it can be fairly reliably decided whether or not a particular star is a true cluster member. Finding the proper motions of numerous stars is a demanding project, and the number of clusters thus investigated is still small. Of course, the radial velocities, like proper motions, can also be used to decide upon the cluster membership of stars, but this requires even more observational effort.

27.6 The Globular Clusters

A globular cluster may contain up to 10^7 stars, far higher than in an open cluster. The density of stars within globular clusters increases strongly toward the center, where it sometimes becomes so high as to prevent resolution into individual stars. In some globular clusters, the lines of equal star density appear circular, while other clusters are distinctly flattened. For instance, the cluster M19 (NGC 6273) has an apparent axial ratio of about 6:10.

Following Shapley, globular clusters are classed into 12 groups, designated I, II, ..., XII, depending on the degree of central concentration, the most strongly concentrated clusters being those at the beginning of the scale. Figure 27.11 shows the cluster M13, the largest globular cluster seen from the northern hemisphere. The largest of all the 131 globular clusters known up to 1979 is ω Centauri, which is visible, however, only from the southern hemisphere.

27.6.1 Color-Magnitude Diagrams of Globular Clusters

Globular clusters show color–magnitude diagrams, for instance that of M3 in Fig. 27.12, which are substantially different from those of open clusters: the main sequence is populated by stars with spectral types from F5 and later. This feature is common to all globular cluster diagrams. Therefore, the brightest stars of even the nearest globular clusters have apparent magnitudes only in the range +19 to +21, so that the observable segment of the main sequence in a globular cluster is at most 2 to 3 magnitudes wide. As in old open clusters, this main-sequence segment bends upward into a sequence of evolved stars which is, however, much more diversified than in open clusters. First, a subgiant branch extends from the turnoff point to a color–magnitude position characterized by $M_V \approx 0$ and $(B-V)_0 \approx 0.8$. From this point, the horizontal branch departs to the left (i.e., toward bluer colors). The hori-

Fig. 27.11. The globular cluster M13. Photograph taken by Birkle with the 1.2-m telescope, Calar Alto.

zontal branch in all globular clusters has a characteristic gap between $(B-V)_0 = 0.2$ and $(B-V)_0 = 0.4$. The giant branch extends from the turnoff point mentioned to the upper right.

Less well defined is the asymptotic giant branch, which lies somewhat above and runs approximately parallel to the giant branch. The precise location of the giant and subgiant branches varies somewhat from cluster to cluster. For instance, the horizontal branch in the color–magnitude diagram of M3 is actually horizontal, but in M13 it is distinctly inclined, with the blue end at appreciably fainter absolute magnitudes than the red end.

The pattern of stars in cluster diagrams can be understood from the evolutionary paths of stars of different masses subsequent to the main-sequence state (see Chap. 24). Giant and subgiant branches are composed of stars just evolving off the main sequence, that is, on the way to the red-giant region in the upper right. The horizontal-branch stars, however, have already passed through this stage, and in the color–magnitude diagram are on the way back from the red-giant branch across the main sequence toward white dwarfs. This is illustrated in Fig. 27.13 for a star with a mass of 1 $\mathfrak{M}_\odot$.

The dissimilarities in appearance of color–magnitude diagrams for various globular clusters result from differences in their chemical compositions. Most globular clusters are substantially poorer in heavy elements than are normal stars in the solar vicinity or stars in open clusters. Their metal abundances range mostly between 10% and 0.5% of

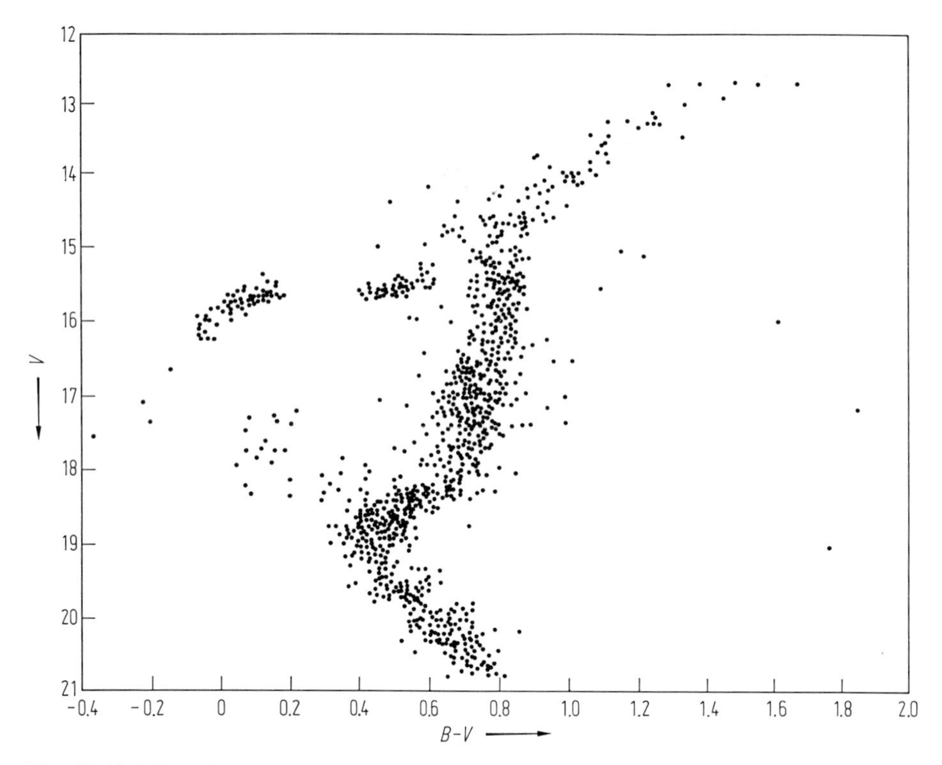

Fig. 27.12. The color–magnitude diagram for the globular cluster M3. Note that only those stars with spectral types F5 and later are still on the main sequence, as all stars with earlier types have evolved into red giants. The variability gap in the range $0.2 < B - V < 0.4$ is quite distinct.

that for the Sun. Evolutionary model calculations indicate, from the color–magnitude diagrams of globular clusters and in particular their turnoff points, ages in the range $9–12 \times 10^9$ years. Globular clusters thus were formed at about the same time as the Milky Way system itself, when heavy elements were still scarce.

27.6.2 Determination of Distances to Globular Clusters

If a globular cluster is near enough that its brightest main-sequence stars are within photometric reach, then, of course, its distance can be determined by comparing the observed main sequence with the ZAMS, as for open clusters. There also exists an expedient method applicable to more distant clusters: the typical gap found in the horizontal branch in the color–magnitude diagram of all globular clusters is occupied by RR Lyrae-type variables. The mean absolute magnitude of RR Lyrae stars is well known to be +0.5. As these stars are easily identified via photometry, the cluster distance is readily deduced. If, however, the RR Lyrae stars in the cluster are too faint to be found and photometrically measured, then other, less reliable methods can

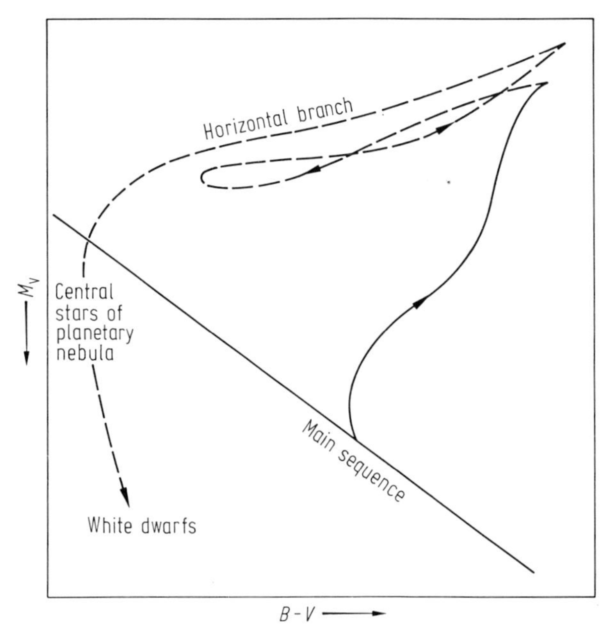

Fig. 27.13. Evolutionary track of a 1 $\mathfrak{M}_\odot$ star in the color–magnitude diagram. After R.J. Taylor.

be employed. One such method is based on the assumption that the mean absolute magnitude of the 25 brightest stars in the globular cluster is a characteristic quantity, and has about the same value for all clusters. After this average has been calibrated from clusters whose distances are known by other methods, a distance—albeit a much more uncertain one—can be estimated for even those faint globular clusters.

27.7 HII Regions

27.7.1 The Exciting Stars in HII Regions

The hydrogen in an HII region is ionized by the short-wavelength radiation from a star with spectral type in the range O3 to B1. (In an HI region the hydrogen is neutral and atomic, while in a molecular cloud it occurs in molecular form H_2.) The energy of 13.60 eV required to ionize a hydrogen atom can be supplied by a UV photon at a wavelength below the Lyman limit at $\lambda = 91.2$ nm. Stars with spectral types of B2 and later do not generate enough of such *Lyman continuum photons* to create a visible HII region. Therefore, any HII region must have at least one star in the O3–B1 range inside it or in its immediate vicinity.

The exciting star of an HII region can ionize a certain amount of hydrogen. If it is more or less homogeneously distributed, the ionized hydrogen will fill a spherical volume called a *Strömgren sphere*, with the exciting star at its center. The radius r_S

Table 27.2. Production rates N_L of Lyman-continuum photons and radii r_S of Strömgren spheres for main-sequence stars of different spectral types. The latter is given for $n_H = 1\ cm^{-3}$. For other densities, the radii r_S should be multiplied by the factor $n^{-2/3}$.

Spectral Type	r_S (pc)	N_L (10^{48} photons s^{-1})
O5	108	47
O6	74	17
O7	56	6.9
O8	51	4.0
O9	34	1.7
B0	23	0.47
B0.5	12	0.068

of the Strömgren sphere depends on the spectral type of the star as well as on the density n_H of hydrogen. The values of r_S listed in Table 27.2 for different stellar spectral types hold for the density $n_H = 1\ cm^{-3}$. Corresponding Strömgren radii r_S for other densities are obtained by dividing the numbers in Table 27.3 by $n_H^{2/3}$.

Round, symmetrically shaped Strömgren spheres with a single exciting star at the center are rare. Figure 27.14 depicts two such regions, S31 and S32, which are close to the ideal case mentioned.

The search for and subsequent photometry and spectroscopy of the exciting stars of HII regions is of great importance for a variety of problems. The only direct method for finding distances to HII regions is through the distances of the exciting stars. It is to be noted that the distance of one star, obtained using photometric data in conjunction with the absolute magnitude and unreddened color derived from the MK type, carries an uncertainty of at least ±25%. Of course, the distance to an HII region will be more reliably determined if found by averaging the distances of several exciting stars.

27.7.2 The Spectra of HII Regions

The spectrum of an HII region consists primarily of emission lines set against a usually weak continuum. The hotter the exciting star, the more diversified are the emission lines. The hydrogen lines dominate in every HII region, and in the red segment of the spectrum Hα, the Balmer line with the longest wavelength, is far more intense than all other lines. Thus, most HII regions appear reddish in color, a fact which is particularly pronounced on red-sensitive photographs.

Many spectral emission lines of HII regions come from the recombination of ions H^+, He^+, N^+, O^+, S^+, and others with free electrons. The preceding ionization of H, He, N, and other atoms is via absorption of UV photons in the radiation flux of the exciting star. When a positive ion captures a free electron, usually in an excited energy level, it simultaneously emits a continuum photon. Thereafter it gradually falls stepwise to lower energy levels and generates the recombination line photons. The

Fig. 27.14. The HII regions S31 and S32 come close to the ideal case of a round, homogeneous Strömgren sphere. From the *Palomar Observatory Sky Survey* (1960), National Geographic Society, by kind permission of the California Institute of Technology.

simplest energy diagram is that of the hydrogen atom (see Fig. 27.15). It consists of discrete energy levels characterized by the quantum number n. The lines Hα, Hβ, Hγ, and so on of the Balmer series occur via transitions from a level $n > 2$ to the level $n = 2$, while a photon emitted during the capture of a free electron directly into the second energy level is a called a Balmer continuum photon. Similarly, $n = 1$ generates the lines of the Lyman series and Lyman continuum, and $n = 3$ the Paschen series and continuum.

Forbidden lines are characterized by an extremely low probability for a spontaneous transition from an upper to a lower level. This translates to a long mean lifetime in the upper level, where the electron can stay in the excited state. The typical lifetimes

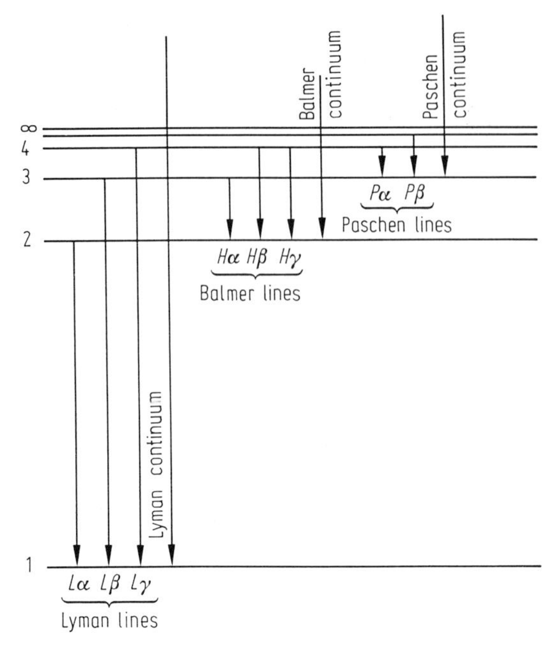

Fig. 27.15. The energy-level diagram for the hydrogen atom.

of upper levels of "permitted" recombination lines are only about 10^{-8} s, but may be as long as several seconds or minutes for forbidden lines. When during that lifetime the excited atom collides with a free electron, the energy may be transferred to the electron without the emission of a photon in the forbidden line. Only if the electron density is low enough that no substantial collisional de-excitation can occur will the forbidden lines be formed. Electron densities in HII regions are in almost all cases between 10 to 10^3 cm^{-3}, i.e. several orders of magnitude less than the critical densities for de-excitation. The latter differ from one ion to another. For [NII] and [SII], they are 8×10^4 cm^{-3} and 3×10^3 cm^{-3}, but for [OI], 10^7 cm^{-3}. Therefore, at densities in the range between 10^5 and 10^6 cm^{-3}, the forbidden [OI] lines 630.0 and 630.3 nm can appear but not the [NII] and [SII] lines. The most prominent forbidden lines in HII regions are the [NII] lines at 654.8 nm and 658.4 nm, the [SII] lines at 671.7 nm and 673.0 nm, and the [OIII] lines at 495.9 nm and 500.7 nm. The generating mechanism of the [OIII] lines was first discovered by I. Bowen in 1927. Up to that time, their emission had been ascribed to a then-unknown element called "nebulium."

The ionization of OII to form OIII requires an energy of 35.12 eV, three times the amount for hydrogen ionization (13.60 eV). The OIII lines indicate an exciting star with exceptionally energy-rich radiation. Should these two lines exceed Hβ ($\lambda = 486.1$ nm) in intensity, then it is strongly indicative of there being an O-type star in the vicinity.

A list of the primary emission lines is given in Appendix Table B.38. Also given are the relative intensities as observed in the Orion Nebula.

The continuum in the spectrum of an HII region originates, as was mentioned, primarily from scattering by dust particles of the radiation from the exciting star. This was unambiguously shown in the case of the Orion Nebula, as the absorption line of He II at 468.6 nm was found in the continuum and could have originated only by the radiation from the brightest Trapezium star θ^1 Ori C.

27.7.3 The Radio Continuum Radiation from HII Regions

HII regions are composed primarily of hydrogen ions (protons) and free electrons. An electron which passes close to an ion is deflected and emits a photon of radiation whose wavelength lies in the decimeter or centimeter range. Such radiation can be received by radio telescopes. The intensity S_ν of the radio continuum is proportional to $\nu^{-0.1}$ for $\lambda < 10$ cm (where ν is the frequency). A measurement of S_r makes possible the determination of important properties of the HII region. In particular, S_ν is a measure of the number $N_{c'}$ of Lyman continuum photons to be generated per second by the exciting stars of the HII region:

$$N'_c = \text{const.} \times \nu^{0.1}\, T_e^{-0.45} S_\nu r^2. \qquad (27.9)$$

Here, T_e (in kelvin) is the electron temperature of the HII region, and r its distance from the Sun in kpc. S_ν is given in janskys (1 Jy $= 10^{-26}$ W m^{-2} Hz^{-1}). From the production rate of Lyman-continuum photons (Table 27.2), one can estimate whether the known young stars of an HII region can produce the requisite continuum photons to maintain the observed radio continuum flux. If this is not the case, then it is concluded that the HII ionization must be due to as yet unidentified stars. The search for such stars often succeeds in such places distinguished by particularly high concentrations of dust. The exciting stars of some HII regions have not been found; however, from Table 27.2 and Eq. (27.9), even in those cases some sound statements regarding the exciting stars can be made.

While the intensity S_ν of the radio continuum is proportional to $\nu^{-0.1}$ for $\lambda < 10$ cm, at wavelengths longer than 1 m, $S_\nu \sim \nu^2$. (In other words, the HII region is optically thin at $\lambda < 10$ cm, but optically thick at $\lambda > 1$ m. In between is a range of transition which depends somewhat on the electron temperature of the gas in the HII region.

27.7.4 The Radio Recombination Lines of HII Regions

For increasing quantum numbers n (see Fig. 27.15), successive energy levels of the hydrogen atom become closer together, and thus the energies freed in a transition from one level to a neighboring one are correspondingly decreased. From about $n = 60$ and upward, the wavelengths of the recombination lines fall into the range of detection of radio telescopes.

The line made by the transition from $n' = 110$ to $n = 109$ is called H109α, while the transition from $n' = 111$ to $n = 109$ is termed H109β, and so on. The frequencies of these lines are computed from the simple relation

$$\nu = 2RcZ_{\mathrm{eff}}\left(\frac{1}{n^2} - \frac{1}{n'^2}\right), \tag{27.10}$$

where R is the Rydberg constant, c the speed of light, and Z_{eff} the effective nuclear charge. Observation of radio recombination lines permits the measurement of radial velocities even in those HII regions which are not optically observable owing to excessive extinction.

27.7.5 The Infrared Radiation from HII Regions

The dust in an HII region is heated via the absorption of starlight. Depending on the temperature reached, radiation is emitted which can be approximated by a Planck curve. The temperature of the dust may reach up to 1300 K, but the grains evaporate at higher temperatures. Within HII regions, temperatures of several hundred K are most common.

Very young stars in HII regions are often surrounded by dust clouds so thick that the bulk of their radiation is absorbed by the dust. Essentially all of the energy of these hot stars is then converted into thermal energy of the dust and subsequently, after reaching an equilibrium, re-emitted by the dust in the infrared. Thus, by integrating the infrared radiation over all wavelengths, the bolometric magnitude (see Chap. 24) of the stars embedded in the dust can be found. This furnishes a second method for determining the integral properties of optically invisible stars.

Figure 27.16 shows the spectrum of a typical infrared source in an HII region in the wavelength range 1 μm to 3 m. From 1 μm to 1 mm, the thermal radiation of the dust dominates, but in the centimeter range the radio continuum of ionized hydrogen prevails.

27.7.6 Typical Structures in HII Regions

The number of recombinations per second inside an HII region is proportional to both the density of ions and that of free electrons. Since hydrogen in an HII region is completely ionized out to nearly its periphery, and an HII region consists predominantly of hydrogen, both densities are in first approximation identical with the density n_{H}. The number of recombinations is thus proportional to n_{H}^2, and this determines the density of recombination lines. Hence, the brightness of the various parts of an HII region is proportional to the square of its density.

The densest parts of an HII region often are not ionized; they appear as inclusions of dark clouds. Large proportions of line radiation are emitted in the immediate vicinity of such neutral inclusions, which have substantially increased density but still contain ionized hydrogen. The edges of these neutral enclosures thus often appear quite bright, and are called *bright rims*.

Assuming that an HII region is excited by a single star, then its UV radiation evidently does not reach the side averted from the star of such a neutral enclosure. It thus remains dark in contrast to the opposite side of the dark cloud. The most

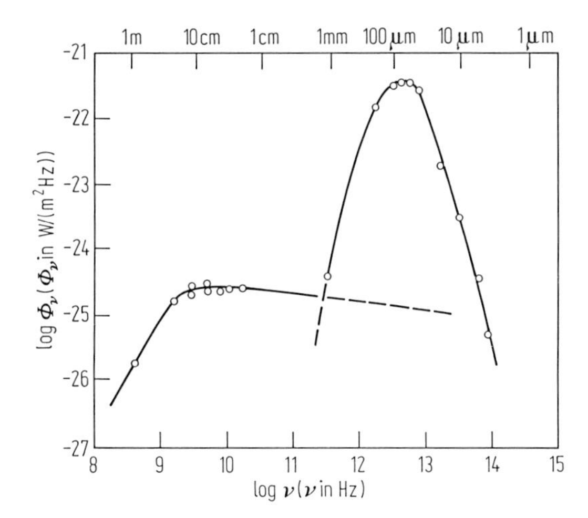

Fig. 27.16. Spectrum of the infrared source IRS1 in the HII region W3A. At wavelengths below 3 mm, the radiation from the hot dust dominates, while above this value the radio continuum of the ionized hydrogen is strongest.

luminous part of the bright rim should thus be turned toward the exciting star. This fact can sometimes be used to decide whether or not a particular star contributes to the excitation of an HII region.

27.7.7 Physical Parameters of HII Regions

Table 27.3 compiles distances, densities, masses and data on the exciting stars for several HII regions. The wide variety of properties of HII regions reflects the diverse properties of the exciting stars, as well as the time development of the HII regions, which range from the small and compact to the very extended, low-density nebulae.

27.7.8 The Best-Known HII Regions

There are several large and unusually bright HII regions which are famous. They are unquestionably among the most beautiful of celestial objects, not just for photographic studies, but also for direct visual telescopic observation.

The best-known galactic nebula is the Great Orion Nebula M42/M43, which can be found pictured in virtually any book on astronomy; it is therefore not reproduced here. This nebula is excited by the Trapezium θ_1 Orionis, which consists of four early-type stars as listed in Table 27.3.

Several factors contribute to making the Orion Nebula so marvelous an object. First, the two O-type stars provide a high flux of Lyman-continuum photons. Second, the interstellar extinction in front of the Orion Nebula is negligibly small. Third, the molecular cloud associated with the Orion Nebula lies behind it, and thus does not

Table 27.3. Physical parameters of some HII regions. Here, r = distance, d = diameter, N_e = electron density, $\mathfrak{M}_{HII}$ = mass of ionized hydrogen, N'_c = required number of Lyman-continuum photons per second from the exciting stars.

Object	S 171	Orion Nebula	Lagoon Nebula	Omega Nebula
Angular Size	$20' \times 4'$	$90' \times 60'$	$45' \times 30'$	$20' \times 15'$
r (kpc)	0.9	0.5	1.4	2.2
d (pc)	30	0.6	3.5	5
N_e (cm^{-3})	< 20	5×10^3	6×10^2	5×10^2
$\mathfrak{M}_{HII}$ ($\mathfrak{M}_\odot$)	$> 1.6 \times 10^3$	~ 10	2×10^2	6×10^2
N'_c (photons s^{-1})	1.2×10^{49}	5×10^{48}	2×10^{49}	2×10^{50}
Exciting Stars	BD+66°1661; O9 V	θ_1 Ori A; O7	HD164794; O5.5	Chini 1; O4
	BD+66°1675; O7 V	θ_1 Ori B; B0 V	HD164816; O9.5 IV	Chini 2; O5
	BD+66°1673; O5	θ_1 Ori C; O6	HD165052; O6 V	Chini 3; O6–O8 V
	BD+67°1598; O9	θ_1 Ori D; B0.5 V	HD164816; O9.5 IV	Chini 18; O9 V
			Herschel 36; O7	

cause an extinction of nebular light. Finally, the Orion Nebula is relatively close to the Sun, at only about 500 pc.

Infrared photographs of the Trapezium region reveal a strongly reddened star cluster which, however, lies *behind* the Trapezium in the associated molecular cloud. The cloud also harbors the well-known *Becklin–Neugebauer object*. This and other infrared sources are nascent stars still deeply embedded in dust and therefore optically invisible.

The North American Nebula and the neighboring Pelican Nebula form a connected complex, as radio-astronomical observations have shown. These two nebulae together coincide with the radio-continuum source W80. The gas density of 20 cm^{-3} in the North American Nebula is unusually low, indicating an advanced evolutionary state. The North American Nebula is still associated with very dense, dark clouds, which also make it appear separated from the Pelican Nebula. It is not yet entirely clear which star is the source of excitation in these nebulae, and thus its distance is not reliably known. Pictures of the North American and Pelican Nebulae may be found in the *Atlas Galaktischer Nebel*, Vol. II.

The Trifid Nebula (M20) is characterized by dense filaments of dust which divide it, as the name implies, into three parts (Fig. 27.17). The UV photons are provided by an O7 main-sequence star at its center. The nebular matter extends much beyond the range of ionization visible on the red photograph; pictures taken in blue light reveal fainter outer parts of the nebula shining by reflected starlight.

The Lagoon Nebula (M8) contains three O-type stars and several early B-type stars. In its brightest part is located the Hourglass Nebula, whose shape is reminiscent of a bipolar nebula (see Sect. 27.13). Immediately adjacent to it is the O7 star Herschel 36. M8 is particularly rich in distinct bright rims and in small dark clouds projected onto the brightest parts of the nebula.

The Omega Nebula (M17) is located at the edge of a giant molecular cloud. Most of its UV photons come from a binary system consisting of two O4 stars. They and other O- and early B-type stars lie in the prominent dark cloud at the western edge of M17. The brighter part of the nebula forms a bar aligned roughly in the

Fig. 27.17. The Trifid Nebula is one of the brightest and most beautiful HII regions in the Milky Way. Photograph by T. Neckel using the 2.2-m telescope of the Max-Planck Institute für Astronomie, La Silla, Chile.

direction southeast to northwest. Toward east, some faint, extended nebular filaments are attached. Remarkably, no bright rims and no enclosed dark clouds are found in M17, as are common in other HII regions of comparable size.

The Eagle Nebula (M16) surrounds the young star cluster NGC 6611 and is rich in prominent bright rims and in small, completely opaque dark cloudlets.

All of these HII regions are observable from the northern hemisphere, and are very rewarding objects to look at even in small telescopes. To these must be added the most beautiful nebula in the southern Milky Way, the η Carinae Nebula. Photographic reproductions of all nebulae mentioned here can be found in the aforementioned *Atlas Galaktischer Nebel.*

27.8 Reflection Nebulae

Reflection nebulae are illuminated by stars with spectral types B2 and later. These do not emit sufficient UV photons to substantially ionize the hydrogen which surrounds them. Nebulae around B1.5-type stars are exactly on the dividing line between HII regions and reflection nebulae. The light of nebulae ionized and illuminated by such stars consists of about equal parts emission lines and scattered continuum.

Logically, the brightest stars will create the brightest reflection nebulae, and indeed the illuminating stars in many such nebulae are of B-type main-sequence or late-type giants such as Antares. Since almost all stars illuminating reflection nebulae are substantially lower in luminosity than the stars which excite HII regions, most reflection nebulae are much less conspicuous than typical HII regions. The few exceptions include the nebulosities around Antares and ρ Ophiuchi (see Fig. 27.18), and those around the Pleiades. In the case of Antares, two favorable circumstances combine to provide the extraordinary brightness of its reflection nebula: Antares is a luminous supergiant, and it lies only about 160 pc from the Sun.

Compared with HII regions, nearly all reflection nebulae, such as the nebula VdB125 (see Fig. 27.19), are quite dull and structureless. This lack of structure is partly caused by the intensity of the scattered light being proportional to the density of the scattering medium, whereas the line intensity in HII regions is proportional to the *square* of the gas density.

The stars illuminating reflection nebulae are usually easily identified. In almost all cases, a particular bright star is so obviously associated with the nebula that no other object in the vicinity need even be considered as a source of illumination. Such an assumption can easily be tested: if the spectrum of nebula and star are identical, then the star is without doubt the source. Another indication is the color of the nebula, which is always a bit bluer than the illuminating star. For instance, Antares is easily identified as the light source of the nebula which surrounds it, as both are very red, but the nebula a bit less so than the star itself.

A more demanding method of observation, which also serves to identify the illuminating stars of reflection nebulae, will here be mentioned. Light scattered by dust is in general very strongly linearly polarized, with the vibrational direction of the electrical vector anywhere in the nebula perpendicular to the illuminating star. Therefore, reflection nebulae show a typical *centrosymmetric polarization pattern*, which makes identification of the illuminating star quite easy. This method is of value particularly when the illuminating star is not optically visible, as often occurs in bipolar or cometary reflection nebulae (see Sect. 27.13). The characteristic structural element of these types of nebulae is an equatorial disk of dust, which completely absorbs the starlight emitted in the equatorial plane, but is rather transparent toward the poles. If the line of sight lies in the plane of such a dust disk, then the central star will be obscured, but not the parts of the nebulae which lie in the direction of the polar axis.

The apparent photographic magnitude m_{pg} of an illuminating star in most reflection nebulae is related quite simply to the apparent nebular diameter a:

$$m_{pg} = 11 - 4.9 \log a, \tag{27.11}$$

where a is in arcseconds. This relation can also be used to identify illuminating stars. It fails, of course, when the star suffers heavy extinction in the direction of the observer, as is the case for most bipolar nebulae.

Fig. 27.18. Reflection nebulae and dark clouds around Antares, including the globular clusters M4 = NGC 6121 (west of Antares) and NGC 6144 (northwest of Antares). From the Palomar Observatory Sky Survey, ©1960, National Geographic Society, by kind permission of the California Institute of Technology.

Most HII regions as well as reflection nebulae are associated with regions in the Galaxy where new stars have recently formed. Since, however, they can be observed only at comparatively small distances from the Sun, they are of limited significance for studies relating to galactic structure.

Fig. 27.19. The reflection nebula VdB125 (GN 19.23.7). In the center of this structureless nebula is its illuminating F0 star, HD182830. From the Palomar Observatory Sky Survey, ©1960, National Geographic Society, reprinted by kind permission of the California Institute of Technology.

27.9 Galactic Cirrus or Hagen Clouds

At galactic latitudes $|b| > 20^\circ$, wide areas of the sky are covered by faint nebulosity showing perceptible shapes only in a few places. An example of such a nebular complex at high galactic latitudes is pictured in Fig. 27.20. Called *Hagen clouds*, these nebulae were known to exist even before the invention of photography. As W. Becker has stated in the book *Sterne und Sternsysteme*:

"... in Hagen clouds, even simple observations encounter severe obstacles. Because these objects are so faint, they are accessible to only the most sensitive method, namely the visual, and even then are close to the threshold value."

At first glance, therefore, it seems hardly likely that these faintest of all galactic nebulae should actually be visually observable.

The Hagen clouds appear equally bright on the red and blue prints of the POSS. They are redder than typical reflection nebulae, and thus cannot be HII regions, since HII regions so faint would be completely invisible in the blue. In contrast with normal HII regions or reflection nebulae, the Hagen clouds do not seem to be associated with any unambiguously identifiable stars as the exciting or illuminating sources. All aspects of observations are satisfied by assuming that the Hagen clouds reflect the integral light of the Milky Way. The clouds are redder than normal reflection nebulae, evidently because the latter are usually illuminated by blue stars of type B. The integrated color of the Milky Way, however, best matches that of a K-type star. Those rare reflection nebulae illuminated by individual K-type stars show brightness differences between red and blue prints of the POSS similar to those of Hagen clouds.

Fig. 27.20. Faint reflection nebulae at high galactic latitudes. Such nebulae are not dominated by individual stars but rather by the Milky Way. The nebula no. 471 from the *Catalogue of Bright Nebulae* by B. Lynds is a typical representative of this kind. From the Palomar Observatory Sky Survey, ©1960, National Geographic Society, by kind permission of the California Institute of Technology.

In spite of the large apparent extension of many Hagen clouds, most galaxies shine through them without suffering substantial reddening. This implies that the density of matter in such clouds is extraordinarily low.

Hagen clouds did not attract much attention until recently, when they were rediscovered by the *IRAS* satellite. They now are known by another name: *galactic cirrus*. It is truly noteworthy that the cirrus structures just "discovered" in the Milky Way with the most modern technical equipment were already known to visual observers prior to the astronomically useful photographic plate.

27.10 Supernova Remnants

The evolution of a massive star is often terminated by a supernova explosion. Most of its matter is then ejected and subsequently forms a continuously expanding shell known as a *supernova remnant* (SNR). Remaining inside the supernova remnant is a neutron star, which may be detected as a *pulsar*. The pulsar PN 0532 at the center of the Crab Nebula emits about 30 pulses each second, and is observable optically as well as in the X-ray and radio ranges.

The number of optically observable supernova remnants is small. 29 nebulae have thus far been identified as SNRs, most of them by S. van den Bergh. Many of them are inconspicuous, and can only be found by taking long-exposure photographs in the light of Hα. The best-known SNRs are the Crab Nebula (M1 = NGC1952) in Taurus, the Cirrus Nebula in Cygnus, IC443, the Vela SNR, and Shajn 147.

The Crab Nebula (Fig. 27.21) consists of an amorphous core surrounded by numerous bizarre filaments. The light of the core is of synchrotron origin, emitted by relativistic electrons traveling in a magnetic field. The short-wavelength portion (with wavelengths below the Lyman continuum limit) is presumably the source of ionization of the outer filaments.

In all other known SNRs, the observed line radiation also originates by recombination or collisional excitation of forbidden lines, but their energy source is the energy liberated in the collision of the expanding remnant shell with the surrounding interstellar matter. The Vela SNR with the Vela pulsar at its heart consists of numerous peculiar filaments, as does Shajn 147 (Fig. 27.22).

By the time the endpoint of evolution of a massive star is reached, when the supernova occurs, lighter elements, such as H, He, and so on, have already been largely converted to heavier elements via thermonuclear processes. The supernova explosion itself generates still heavier elements. Since the bulk of the stellar matter is dispersed as a supernova remnant into the surrounding interstellar medium, the latter is continually enriched with heavy elements as a consequence of supernova events. The reader may be interested in knowing that the gold in his/her tooth fillings originated inside a supernova.

27.11 Planetary Nebulae

When viewed through a small telescope, some of the brighter planetary nebulae display greenish disks which are similar to those of the planets Uranus and Neptune, hence their name. Other planetary nebulae are ring-shaped, such as the Ring Nebula in Lyra

Fig. 27.21. The Crab Nebula (M1) is the best-known supernova remnant. Photograph taken by Thiele with the 2.2-m telescope in Calar Alto, Spain.

(M57 = NGC 6720), which can easily be found between the stars β and γ Lyrae with just a small-aperture telescope.

Some planetary nebulae have distinct bipolar symmetry, a typical example being NGC 6302 (see Fig. 27.23). It bears an amazing similarity to S106 (see Fig. 27.35), the prototype of visible bipolar nebulae which, however, characterizes an *early* stage of stellar evolution.

The smallest planetary nebulae cannot be distinguished from stars on ordinary photographs of the sky. They are identified on objective-prism photographs. Most of the planetary nebulae so far known (about 1200) are listed in the *Catalogue of Galactic Nebulae* by Perek and Kohoutek, which contains 1034 entries.

The spectra of planetary nebulae resemble those of HII regions, but are usually distinguished from the latter by higher excitation levels. That is to say, they contain strong lines of ions whose ionization requires a more energetic radiation than is available in an ordinary HII region. Intense lines of HeII, or singly ionized helium, are commonly found in the spectra of planetary nebulae. The best-known HeII line is the one at 468.6 nm, which is completely absent in the spectra of most HII regions. Also characteristic is the high intensity of the [OIII] lines at 500.7 and 495.9 nm, the famous "nebulium" lines, which are responsible for the greenish tint of most planetary nebulae. They do occur in HII regions, but at low intensity. The ratio of the intensity of the two [OIII] lines to that of the Hβ line varies in HII-region spectra from 0.0 to about 5.0, while in the spectra of nearly all planetary nebulae this ratio is larger than 7.0. There exist a few low-excitation planetary nebulae which cannot be spectroscopically distinguished from normal HII regions.

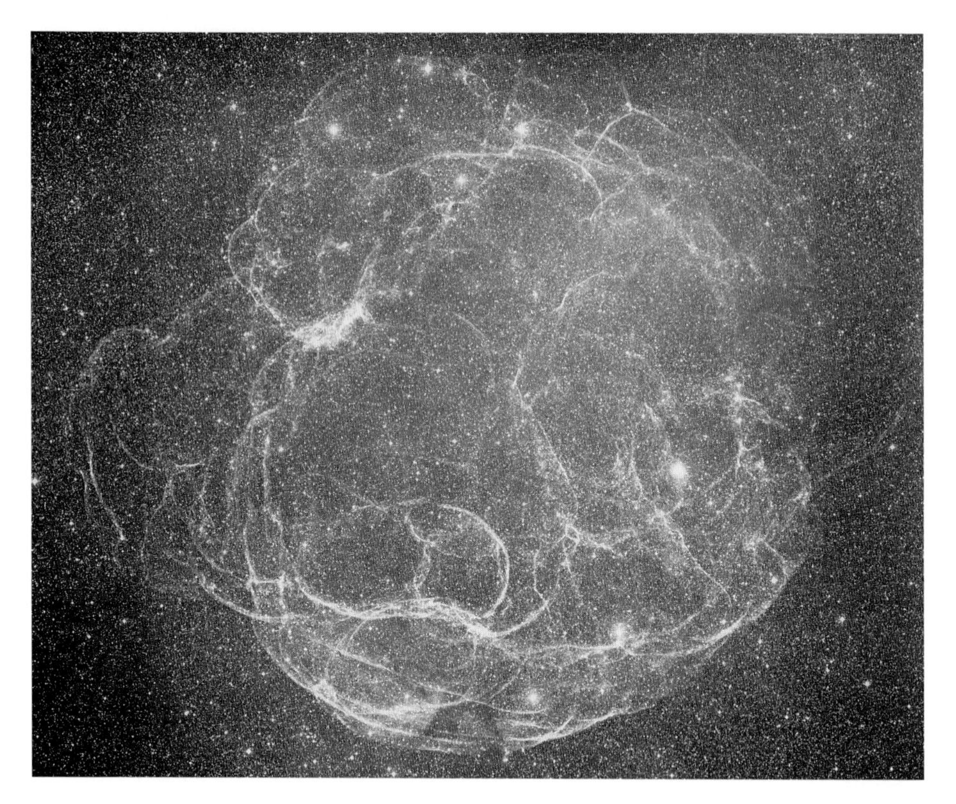

Fig. 27.22. The supernova remnant Shajn 147, which was discovered in 1952 by Shajn and Hase at the Crimean Observatory. Red photograph taken with the Schmidt telescope at Mt. Palomar.

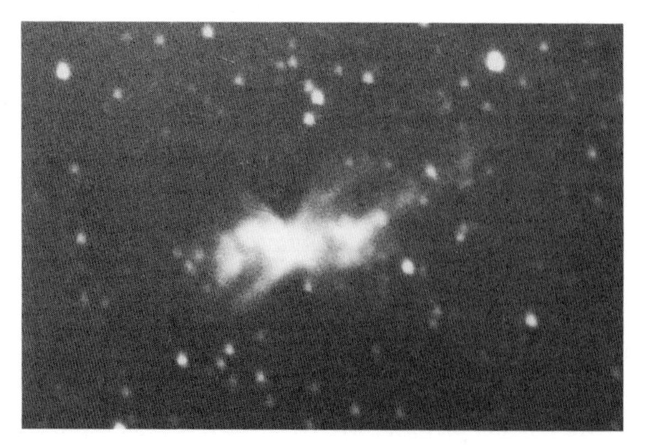

Fig. 27.23. The bipolar nebula NGC 6302 is generally considered to be a planetary nebula. Photograph by Birkle with the 1.2-m telescope at Calar Alto, Spain.

The spectral lines of planetary nebulae are often split into a redshifted and a blueshifted component, this being due simply to expansion of the optically thin nebula: as one side of the nebula approaches Earth, the simultaneously visible opposite side recedes.

At the heart of a planetary nebula is found a hot, bluish star, called the *central star* or *planetary nebula nucleus* (PNN). It almost always has an O-type spectrum, but is distinguished from normal galactic O-type stars by an abnormally low luminosity. PNNs have absolute magnitudes usually between 0 and −3, and thus are fainter than normal O-type stars. On the other hand, PNNs are hotter, reaching temperatures of up to 200 000 K, while a normal O5 star gets up to only 50 000 K. This extreme temperature is the reason for the high degree of excitation found in planetary nebulae, as almost all of the radiation of the central stars lies at wavelengths below the limit of the Lyman continuum ($\lambda = 91.2$ nm).

By its expansion, a planetary nebula indicates that it is a shell which has been ejected by its central star. It is estimated that within only 10 000 years the density of a planetary nebulae, which lies between 10^2 cm^{-3} and 10^4 cm^{-3}, will have dropped so much that it will no longer be seen. In any event, a planetary nebula is an object which originated only recently, but not so its central star. The latter is an old star well past the evolution in the red-giant region and tending toward its final state as a white dwarf.

The rather high ages of PNNs is also shown by their spatial distribution: their strong concentration toward the galactic center indicates their membership in the Old Disk Population, whose members are in the age range 2–10 × 10^9 years.

27.12 Molecular and Dark Clouds

27.12.1 The Spatial Distribution of Interstellar Extinction

Returning to the discussion of interstellar dust, whose light-dimming and reddening effects where the subject of Sect. 27.2, the question of its spatial distribution is now to be explored. Any star for which the distance r and the extinction A_V of its light are known supplies information on the distribution of the dust: if $a(r, l, b)$ is the extinction per unit distance, which depends, in addition to r, also on the direction characterized by the galactic coordinates l and b, then

$$A_V = \int_0^V a(r, l, b)\; dr.$$

Two stars with extinction values $A_{V,1}$ and $A_{V,2}$ at distances r_1 and r_2 in the same direction (l, b) permit the identification of an extinction of amount $A_{V2} - A_{V1}$ to occur in the distance range $r_1 < r < r_2$. A practical method to investigate the distribution of the dust is thus to graph A_V against the distance r as abscissa for a particular celestial region. In such a diagram, it is often possible to draw a curve through the points to document the monotonic increase of extinction with distance. The slope of this curve,

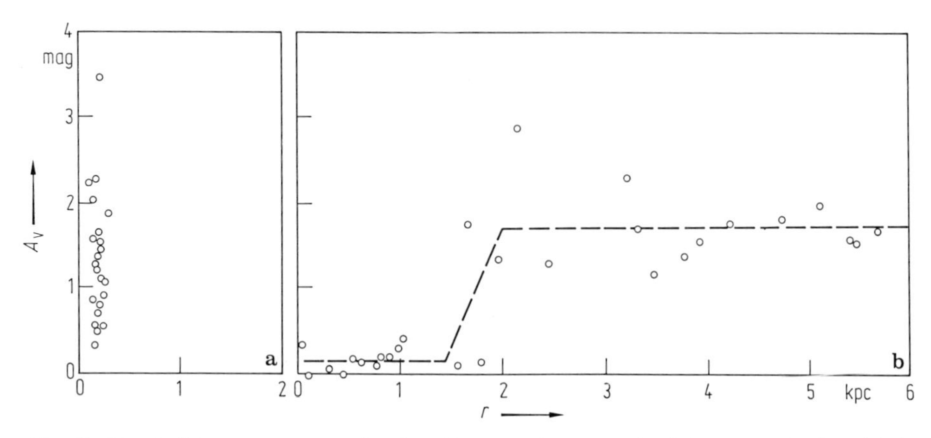

Fig. 27.24a, b. Distance-versus-extinction diagrams for two celestial areas. **a** In the vicinity of ρ Ophiuchi, A_V suddenly increases at $r \approx 200$ pc by more than 2 magnitudes. Data are not available on more distant stars. **b** A field in Puppis at $l = 245°$, $b = 0°$ has a very small extinction to $r \approx 1.5$ kpc. Then, a cloud between $r = 1.5$ kpc and $r = 2.0$ kpc raises A_V to about 1.7 magnitudes.

i.e., the derivative dA/dr, gives the extinction per unit distance interval as exists at various distances.

Useful $A_V(r)$ diagrams can be constructed for many different regions in the sky. The method is limited by errors of distance determination ($\Delta r \sim 0.25r$) and the strong irregularity of the extinction distribution: A_V changes not only with distance r, but also with direction (l, b). To allow for the latter dependence, it is necessary to draw $A(r)$ versus r diagrams for as small areas of the sky as possible, so that the directional dependence of the extinction can be neglected in approximation. This, of course, is limited by the small number of stars with known distances and extinctions, which is currently about 14 000 over the entire sky.

Figure 27.24 illustrates two examples of extinction–distance diagrams. It is seen that wide regions are free of any extinction, while in other places steep increases occur. Evidently, the $A_V(r)$ diagrams have detected the presence of unseen dust clouds.

The dust distribution in the galactic plane has been studied using a compilation of a large number of these diagrams. The results are illustrated in Fig. 27.25 at maximum distance of 3 kpc. It shows that the dust forms large cloud complexes, with diameters of up to 500 pc. In between, there are regions of the same size which are nearly free of dust. These cloud complexes are by no means dark clouds of the Coal Sack type, which can be partially seen with the naked eye. In the region of such a dense, dark cloud, an $A_V(r)$ diagram shows at best a short, steep increase in extinction, but has no measured points thereafter since the more distant stars are too faint to be observed at this large extinction. In Fig. 27.24 a, for instance, the increased extinction in the nearer parts of the dark cloud in Ophiuchus is distinctly seen at $r = 160$ pc. No conclusion can be reached from Fig. 27.24 a on the radial extinction of the dark Ophiuchus cloud nor or the extinction of a star behind it. The stars of known distance

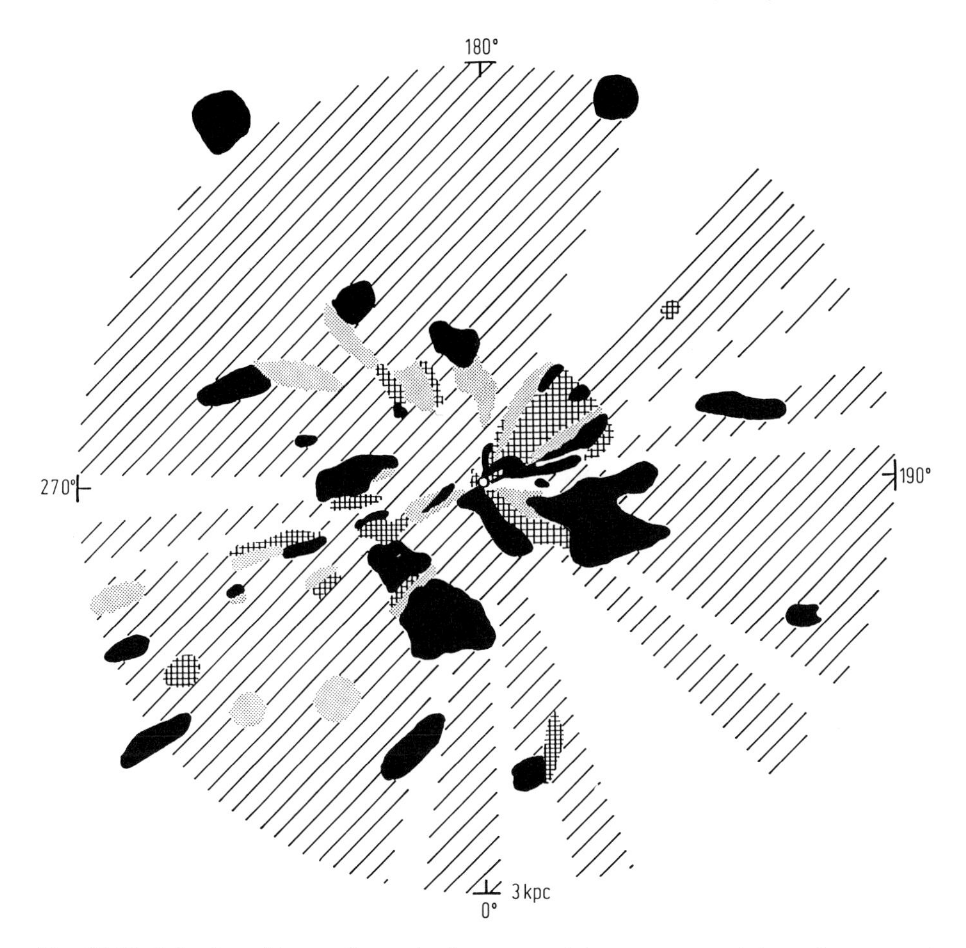

Fig. 27.25. Behavior of interstellar extinction to $r = 3$ kpc, constructed from distances and extinction values of individual stars. The blank areas lack sufficient data. The extinction is less than 1 mag kpc^{-1} in the *hatched regions*, between 1 and 2 mag kpc^{-1} in the *dotted areas*, between 2 and 3 mag kpc^{-1} in the *double-hatched areas*, and more than 3 mag kpc^{-1} in the *black areas*. The numbers at the edge of the figure are galactic longitudes, the Sun is at the center. After Neckel and Klare.

and extinction thus serve only to explore the dust component of large cloud complexes of low density.

The gas component of clouds is explored by measuring equivalent widths of the Lyman α absorption line ($\lambda = 121.6$ nm) in the spectra of stars with distances up to 1.0 kpc. These measurements were obtained with the *Copernicus* satellite, and it was found that the equivalent width of Lα is approximately proportional to the amount of extinction in the stars observed. This proportionality, however, is noted only in regions of low extinction. Thus, the extinction regions shown in Fig. 27.25 contain hydrogen in its neutral atomic form in their less dense portions.

Analyses of color excesses give fairly complete data up to approximately $r = 1$ kpc. For $r > 1$ kpc, the stars with the largest A_V are missing, which seriously hampers the statistics. For $r < 1$ kpc, the extinction is on the average 1.8 magnitudes per kpc. The extension of dust clouds perpendicular to the galactic plane is remarkably small; at a distance of only 40 pc from the galactic plane the dust density has dropped to one-half of that in the plane. The mean gas density out to $r = 1$ kpc, the limiting distance of the *Copernicus* satellite, reaches 1 H atom per cm^3. Extinction (or color excess E_{B-V}) and the hydrogen column density are, as stated, nearly proportional to each other. On the average,

$$N_H/A_V = 2 \times 10^{21} \text{ H atoms cm}^{-2} \text{ mag}^{-1}. \tag{27.12}$$

At larger distances from the Sun, neutral hydrogen can be found only by radio-astronomical observations of the 21-cm line. The resulting distribution of neutral hydrogen in the Milky Way and its interpretation will be dealt with in Sect. 27.14.

27.12.2 The Distances and Total Extinction of Dark Clouds

The distance of a dark cloud can be deduced indirectly from the number of foreground stars in its direction. For example, in the vicinity of a dark cloud seen in the *Palomar Atlas*, one can calculate how large the volume of a cone of aperture $5'$ should be to contain 50 stars to the limiting magnitude of the POSS at the same star density as in the solar neighborhood. This presupposes that stars to the distance of the dark cloud are evenly distributed, an assumption at least approximately permitted since all dark clouds recognizable as such by a diminished star density are at most 2.5 kpc distant. There are so many foreground stars in front of more distant dark clouds that a distinct reduction of the apparent star density is not produced.

Star counts in front of a dark cloud should of course exclude stars which actually lie behind the cloud. These are reddened and can be identified by comparing two photographs in two different colors. Figure 27.26 shows how the number of stars in front of a $10'$ section of a dark cloud increases with the distance for various limiting magnitudes.

The amount of extinction throughout a dark cloud may be deduced with the aid of a *Wolf diagram* (after Max Wolf who introduced it in 1923), which compares the function $N(m)$ in the dark-cloud region with that in a "dust-free" comparison field. $N(m)$ is the total number of stars per square degree brighter than apparent magnitude m. In the comparison field, $N(m)$ merely expresses the increase in number of stars with increasing distance (Fig. 27.27). In the obscured region, $N(m)$ runs the same for bright stars as in the comparison field, so long as all counted stars lie in front of the cloud. At an apparent magnitude m_1, where the most distant stars are just in or behind the dark cloud, $N(m)$ bends downward. When all stars with apparent magnitude m_2 lie behind the cloud, then for $m > m_2$, $N(m)$ runs parallel to the curve $N(m)$ of the comparison field, but is displaced. The horizontal amount of the displacement between the two $N(m)$ curves equals the total extinction by the dark cloud. This method is impressive because of its simplicity, but is only of limited use. In most cases, one cannot say for sure whether or not the comparison field is at least partly affected by

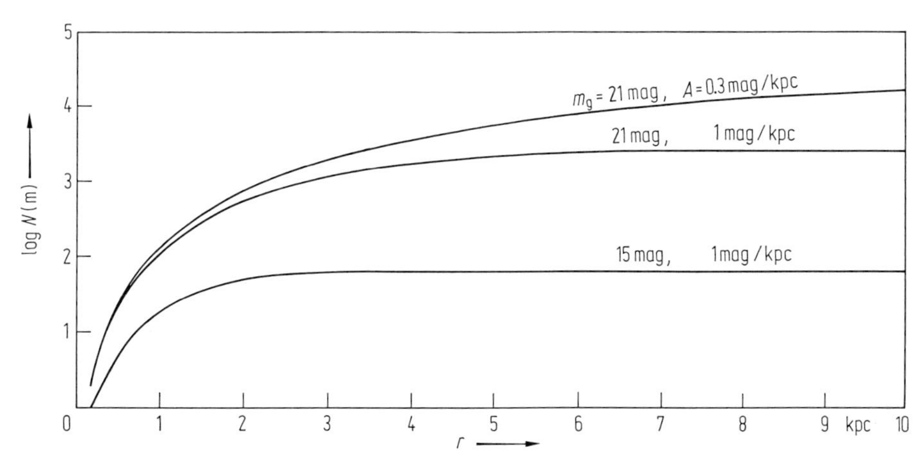

Fig. 27.26. Numbers of stars in front of a dark cloud to a limiting magnitude m_{lim} as a function of the distance r of the cloud for a field $10' \times 10'$, graphed for various limiting magnitudes and different extinctions in front of the dark cloud.

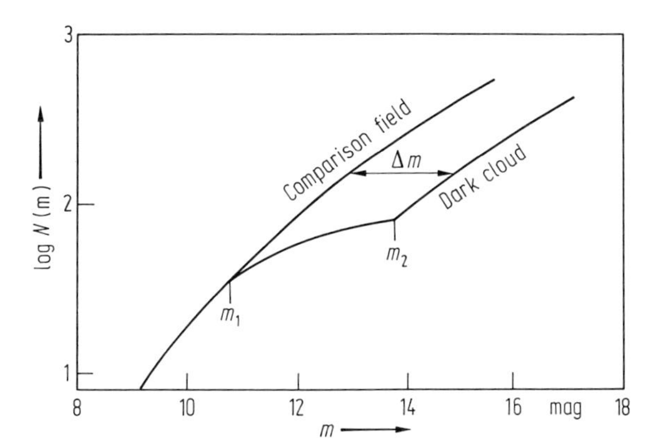

Fig. 27.27. Schematic Wolf Diagram to determine total extinction Δm by a dark cloud. It graphs the number of stars down to magnitude m in the obscured field and in an equally large comparison field next to it.

the dark cloud. This field should not be chosen too far off from the dark cloud or else there would be a risk of real differences in star densities which would vitiate the result of the counts. Also, owing to the large differences in the absolute magnitudes of stars, the bend at m_1, which characterizes the onset of the dark cloud, is poorly defined.

27.12.3 Molecules in Dark Clouds

Measurements of the Lyman α absorption as well as the 21-cm emission of neutral hydrogen show that atomic, neutral hydrogen is absent in dense clouds ($E_{B-V} > 0.3$).

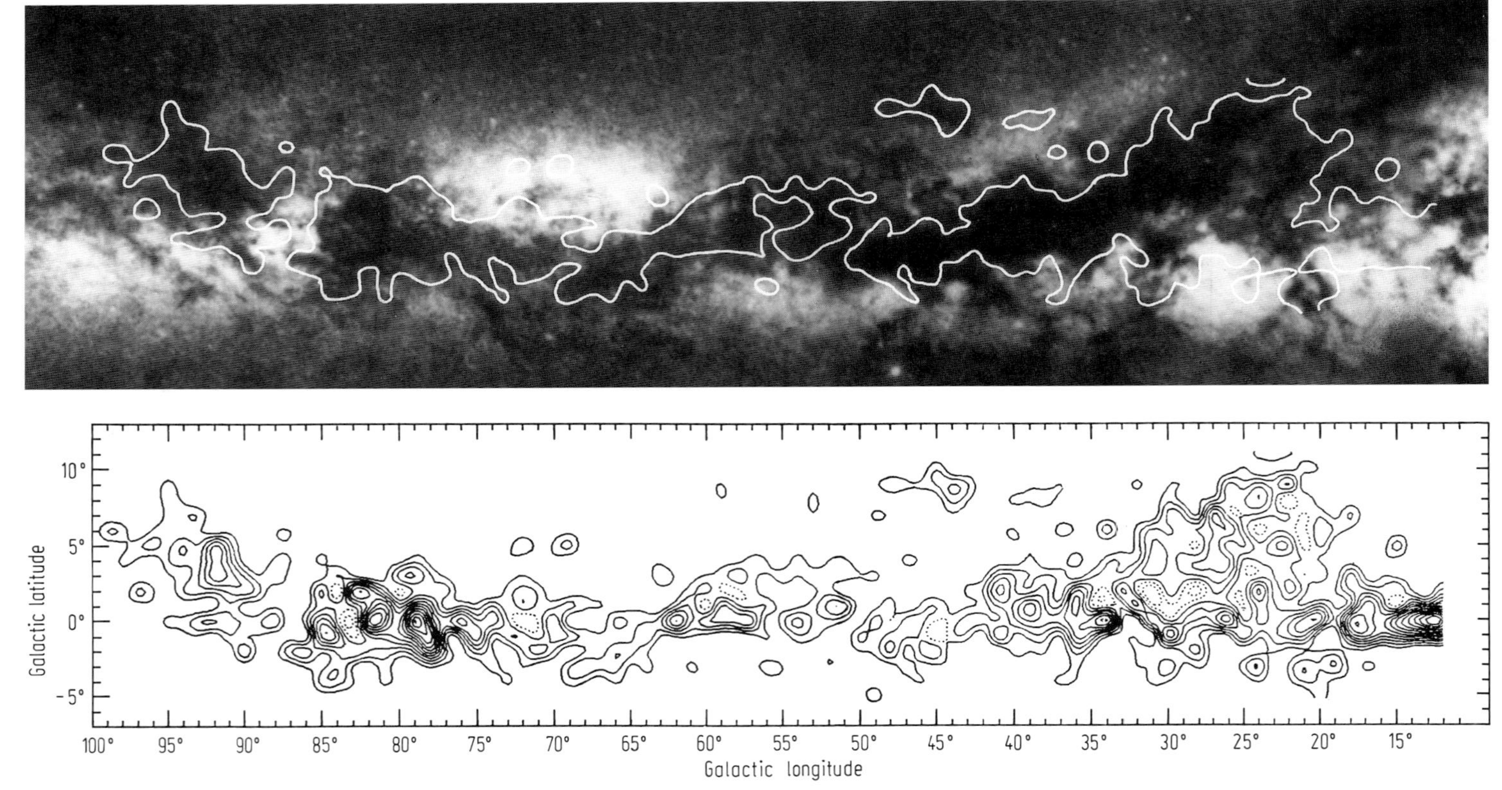

Fig. 27.28. Comparison of the distribution of dark clouds in the Milky Way, and the CO (carbon monoxide) emission. A mosaic of the Milky Way from Sagittarius to Cassiopeia is here compared with a map of CO emission drawn to the same scale, and integrated between -10 and $+34$ km s^{-1}. After T.M. Dame and P. Thaddeus.

These clouds screen off the short-wavelength radiation field of neighboring stars, so that they can no longer dissociate simple, long-lived molecules. Also, the dust in dense clouds directly promotes molecule formation as hydrogen atoms may be adsorbed onto the surface of dust grains, combine to produce H_2 molecules—with the release of the binding energy, and again separate from the dust particle, the latter being a catalyst for the production of H_2. This is the only mechanism known which can efficiently generate H_2 molecules; in a cloud of dustless gas, H_2 molecules could not form.

Most emission lines of molecules are from transitions between their various rotational states, the wavelength being in the millimeter or centimeter range. The strongest CO line is observed with radio telescopes at $\lambda = 2.6$ mm, that of NH_3 (ammonia) at 6 cm, and that of the OH radical at 18 cm. The CO molecule is excited through collisions with H_2 molecules.

Unfortunately, the H_2 molecule has no observable spectral lines. The column density $N(H_2)$, which has the dimension number per cm^2, can still be derived from the easily observed CO, and requires knowledge of the abundance ratio of CO and H_2 molecules. It has the value 2×10^{-5} for ^{12}CO formed with the carbon isotope ^{12}C. If the CO density becomes so high that the ^{12}CO radiation becomes optically thick, then no information can be gained on the CO density. However, the radiation of ^{13}CO formed with the forty-fold less abundant isotope ^{13}C is generally still not optically thick, and can be used in place of ^{12}CO. The CO molecule is therefore quite important, and can be found in all dark clouds. Of all the observed molecules, it has the longest lifetime against dissociation from the UV portion of the undiluted interstellar radiation field, and thus may exist even in dark clouds of low density. Most other molecules, on the other hand, can survive in the interstellar radiation field only when protected by many magnitudes of extinction. The similar distribution of CO and that of dark clouds is seen in Fig. 27.28. The arrangement of dark clouds on Milky Way photographs resembles the CO column density in nearly every detail.

Certain molecular lines are emitted only in very dense clouds. $N(H_3)$, for instance, is always an indicator of moderate to high densities (10^4 cm^{-3}); formation of CS lines requires even higher densities. The different parts of a dark cloud which varies in density can thus be analyzed through different molecules.

In recent years, new molecules have continually been found in interstellar space. Meanwhile, molecules composed of up to 13 atoms (cyanodecapentyne) have been found. Most molecules are typical organic molecules which were formerly assumed to originate exclusively in living organisms. The greatest variety of molecules was found in the Orion molecular cloud and in the immediate vicinity of the galactic center (Sgr B2).

27.12.4 Molecular Clouds

Dark clouds can generally be recognized to distances of only 2 kpc, or at most 3 kpc. The molecules contained in them, however, are observed without difficulty out to the most remote parts of the Milky Way. A survey of the entire Milky Way for CO emission thus could bring about a fairly complete picture of the distribution and physical properties of the interstellar medium (ISM).

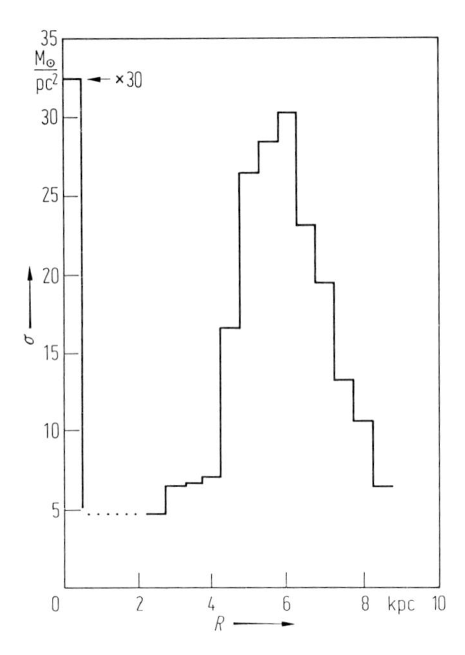

Fig. 27.29. Contribution of molecular clouds to the column density as dependent on the distance R from the galactic center. The value for $R = 0$ is actually 30 times higher than graphed. After P.M. Solomon, D.B. Sanders, and N.Z. Scoville.

One of the most important realizations was that the ISM is concentrated in *giant molecular clouds* (GMCs). A typical GMC has a diameter of 40 pc, though some clouds are much larger: the molecular cloud associated with the HII region M17 has a transverse extension of 170 pc. GMCs can contain up to about 10^6 $\mathfrak{M}_\odot$, although an average mass is 5×10^5 $\mathfrak{M}_\odot$. They are thus the most massive single objects in the Milky Way system, representing the reservoir of interstellar matter, and contribute about 1% of the total mass of the Galaxy.

The mean density of a GMC is $N(H_2) \approx 300$ cm^{-3}, which is 600 times larger than the mean density in the extended, diffuse clouds containing atomic, neutral hydrogen, and which are detectable in the analysis of color excesses. In some places inside a GMC, densities can reach several 10^4 cm^{-3}; these are places where star birth may occur (see Sect. 27.12.5). The overall extinction through a GMC may amount to several hundred magnitudes, so that its interior will always be hidden to optical radiation.

The temperature inside a GMC is in many cases only about 10 K. This low temperature arises because the interstellar radiation field is quickly absorbed in the outermost layers of the GMC. If temperatures only a few degrees higher are found in a molecular cloud, it is an indication of an internal heat source, namely young stars forming inside the cloud.

The frequency of GMCs as dependent on the distance from the galactic center has a pronounced maximum between $R = 4$ kpc and $R = 8$ kpc (Fig. 27.29). The CO layer is quite narrow in extent, similar to that for the dust based upon the analysis of color excesses: at a height of only $z = 60$ pc above the galactic plane, the density of CO has fallen to one-half of its value in the plane.

GMCs are gravitationally stable over a lifetime of at least 3×10^8 years, which is substantially longer than the time which a GMC would need to collapse under its own gravity (about 10^6 years). The complete gravitational collapse is thus somehow compensated by certain forces, the most significant contributions coming from the rotation of the cloud and turbulent motions. Also, the inhomogeneous density distribution in a GMC causes stars to form in the densest parts first. The stellar birthing process, however, returns large amounts of energy into the cloud in the form of stellar winds, ionizing radiation, and bipolar matter flows, which stop the collapse or may even completely dissipate the cloud.

27.12.5 Molecular Clouds and Star Formation

About 80% of all HII regions are associated with molecular clouds. The largest clouds occur in connection with HII regions in which exceptionally large numbers of high-mass O-type stars have been or are currently being born. Other young objects have also been found to be linked exclusively to dark clouds: reflection nebulae are nothing more than parts of dark clouds which are illuminated by young stars, and T Tauri stars—low-mass stars so young as to still be in the pre-main-sequence stage—are also found preferentially in close proximity to dark clouds.

In summary, molecular clouds are (and have been) the birthplaces of stars in the Galaxy. This conclusion may seem quite trivial, since what could stars be made of, if not of interstellar matter? Those parts of the ISM with the highest density are of course considered the most favorable locations for the formation of new stars.

The important aspects of stellar origin depend critically upon the sizes of molecular clouds. Only high-mass clouds are capable of forming high-mass O-type stars. The latter originate predominately, though not always, just below the surface of the molecular cloud. Thus, the Orion Nebula is situated on the Sun-facing side of the molecular cloud associated with it. This explains why the Trapezium stars in the Orion Nebula are nearly unreddened. In other cases, the projection is such that an HII region is just seen at the edge of the molecular cloud, a classic example of this kind being S 140 (Fig. 27.30). A similar configuration is seen in M17. In fact, the western edge of M17 marks the beginning of one of the largest molecular clouds in the Galaxy, 170 pc across! Its highest density occurs at the edge toward M17 (see Fig. 27.31). H_2O masers and infrared sources are observed in the transition zone between molecular clouds and HII regions, indicating the current birthing of new stars there.

As soon as an O-type star reaches the main sequence, it begins to ionize the surrounding part of the molecular cloud. The result is a "compact HII region," which will in some cases be observable only at radio or infrared wavelengths. The ionization continuously penetrates deeper into the cloud. As soon as the radiation reaches the surface of the molecular cloud, the ionizing gas expands with nearly triple the speed of sound (about 30 km s^{-1}) into the thin surrounding medium. This gives rise to a rapidly expanding, diffuse HII region which now becomes optically observable. The ionization fronts, which are the boundaries between the ionized and neutral parts, appear as bright rims. When inhomogeneities such as small lumps of dense, neutral matter impede the propagation of ionizing radiation, then as seen from the direction

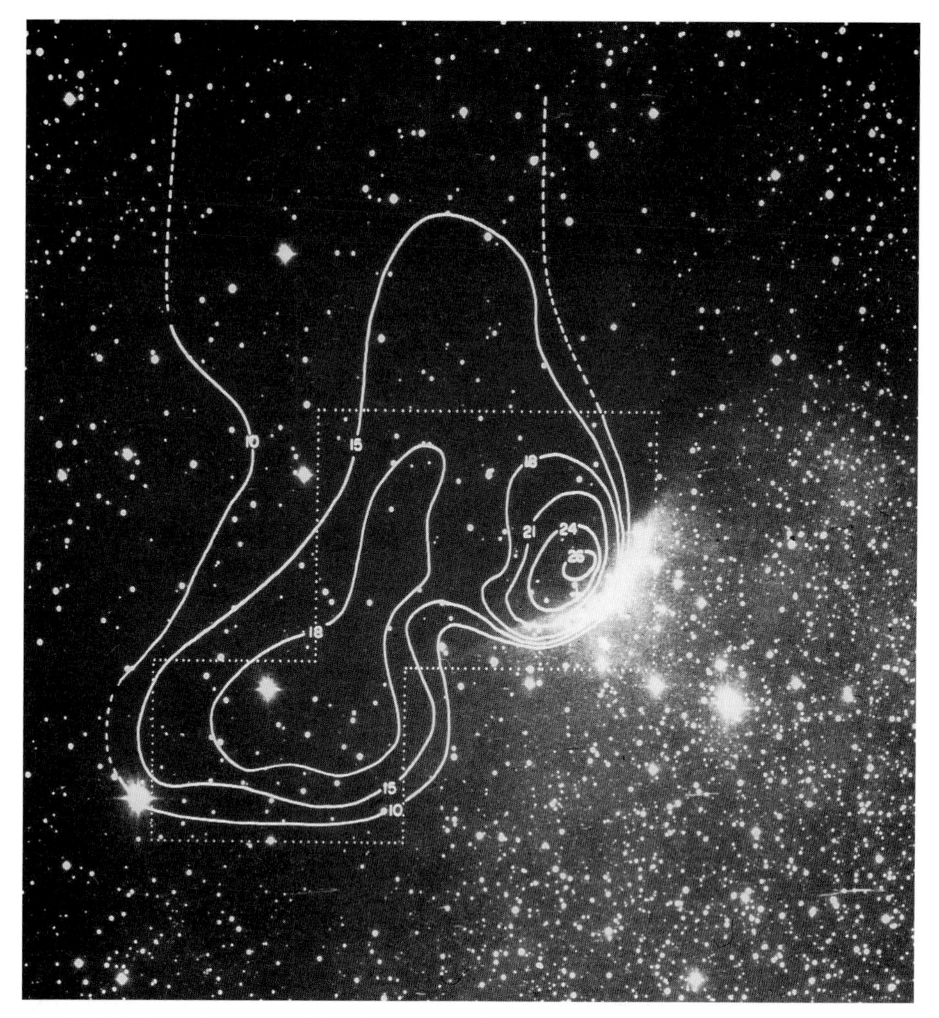

Fig. 27.30. The dark cloud LDN 1204 (LDN = *Catalogue of Dark Nebulae* by B. Lynds) with isophotes of CO emission. The HII nebula S 140 is seen in the southwest corner of the dark nebula. After G.N. Blair et al.

of the exciting star behind them, neutral channels, often called "elephant trunks," are produced.

In small molecular clouds, the stellar birthing proceeds at a much more leisurely pace. The lowest known mass for a cloud, one in which a just-forming star was identified, is only 20 $\mathfrak{M}_\odot$. Obviously, such clouds cannot make a star of 50 $\mathfrak{M}_\odot$. The most massive stars born of clouds with masses up to 1000 $\mathfrak{M}_\odot$ are late B-type stars, and stars with about 1 $\mathfrak{M}_\odot$ are most frequent.

In contrast to the GMCs, stars in small clouds are apparently formed near the cloud center. A typical example is the Bok globule LDN 810. A late B-type star of obviously

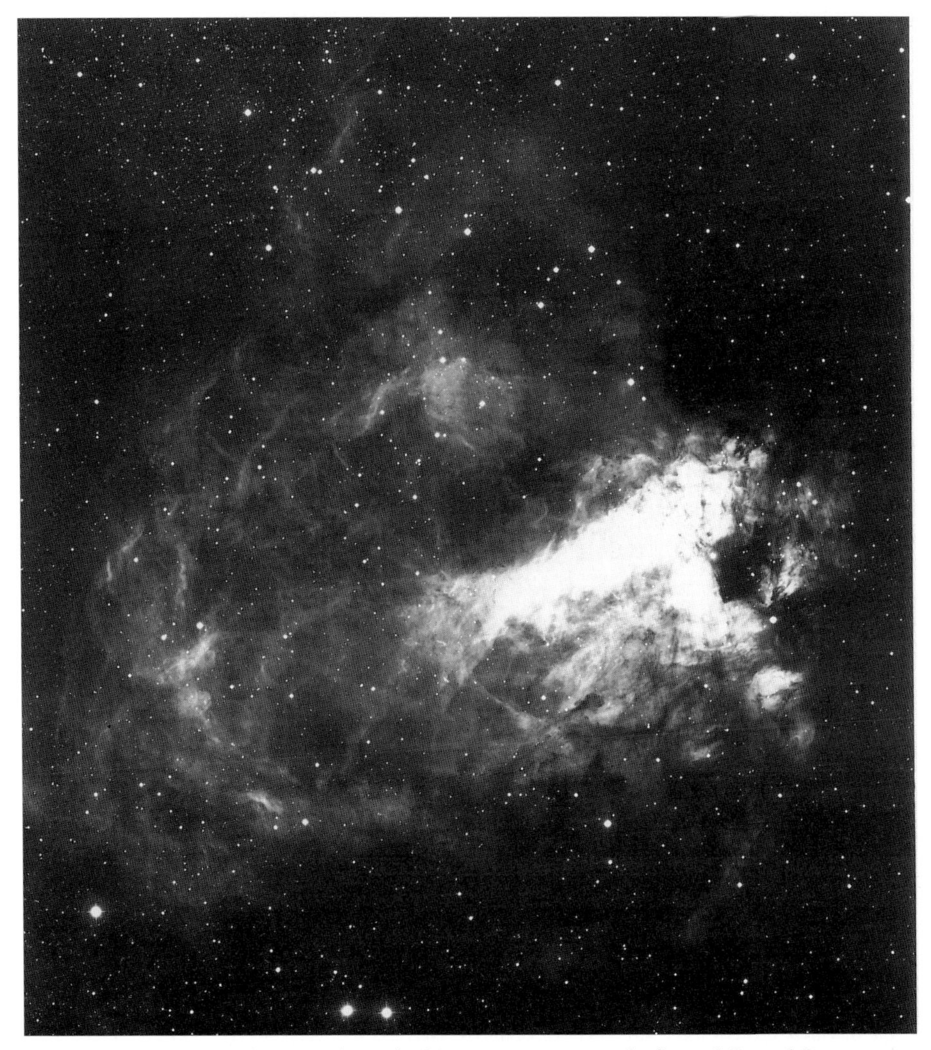

Fig. 27.31. The HII region M17 is excited by numerous recently formed O- and B-type stars; it has formed at the edge of one of the largest molecular clouds in the Galaxy. Photograph by T. Neckel using the 2.2-m telescope of the Max-Planck Institute for Astronomy, La Silla, Chile.

recent origin is located exactly at its center, which is marked by a maximum of NH_3 radiation as well as of temperature. Its extinction is about one-half of that of the total extinction in the cloud as derived from measurements of H_2CO (formaldehyde).

The best-studied example for stellar formation in a cloud of average mass is LDN 1551 (see Fig. 27.32). The best-known young star in it is IRS 5, which is currently a K-type star, still far above the main sequence and thus classified as a giant. When it reaches the main sequence, it will be a solar-type star, as its mass is in the range 1–2 $\mathfrak{M}_\odot$ (see Sect. 27.13). Besides IRS 5, other stars have formed in LDN 1551, including the T Tauri-type stars HL Tau and XZ Tau.

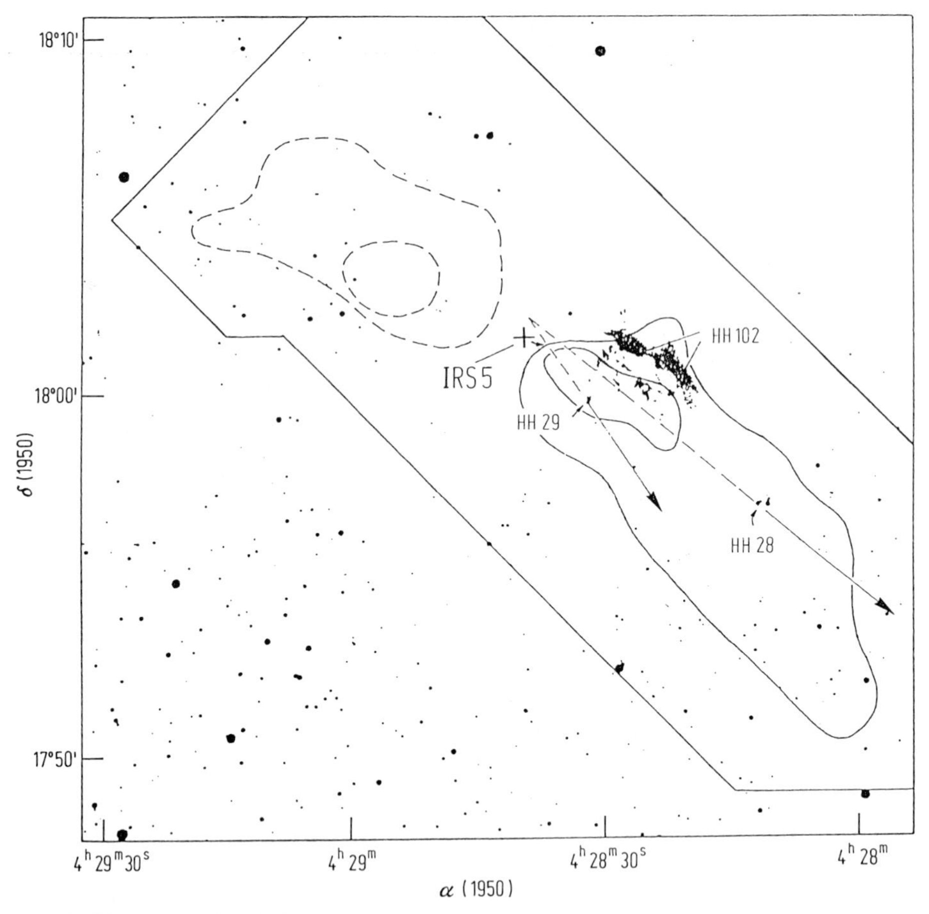

Fig. 27.32. In the dark nebula LDN 1551, two molecular clouds are observed on opposite sides of the pre-main-sequence star IRS5. The *solid lines* represent the CO flow toward the Sun, the *dashed lines* the receding flow. Owing to high extinction, IRS5 is observed only in the infrared. The Herbig–Haro objects HH28 and HH29 are also moving away from IRS5 with high velocity in the arrowed directions.

27.13 Herbig–Haro Objects, Bipolar and Cometary Nebulae

Many dark clouds contain small, almost starlike nebulosities which are found to shine in the light of emission lines, similar to HII regions. Their spectra are of lower excitation, as is typical for "shock-excited" gases. Shock excitation occurs through collision of gaseous masses with relative velocities of several hundred km s^{-1} through the ensuing heating. During the subsequent phase of cooling, characteristic lines are emitted; the forbidden [OI] line at 630.0 nm or the [SII] lines at 671.7 and 673.1 nm are, relative to Hα, appreciably stronger than in HII regions. These small nebulae are named *Herbig–Haro objects*, in honor of their discoverers.

A hallmark of nearly all Herbig–Haro (HH) objects are their large radial velocities, which in most cases are negative. Thus, they move predominantly toward the Sun, that is, out of the dark nebula onto which they are projected. An HH object of positive radial velocity would be located behind the dark nebulae associated with it and thus presumably be invisible.

Three HH objects were also found in the dark cloud LDN 1551 (Fig. 27.32). Two of them, HH28 and HH29, are particularly intriguing. Both have high negative radial velocities. They also have large proper motions, as was shown by Herbig and Cudworth, with their motion exactly directed away from the previously mentioned protostar IRS5. This fact will be shown to be of basic importance in the understanding of Herbig–Haro objects as well as of the entire stellar birthing process. This observation indicates that IRS5 has ejected matter which, after collision with surrounding matter at rest, becomes visible as an HH-type object.

CO observations show two streams of molecular gas originating at IRS5 and flowing into opposite directions; one of these streams moves in the same direction as HH28 and HH29 (see Fig. 27.32). The CO flow in which HH28 and HH29 are embedded has the same direction as the two HH objects.

On "deep" CCD images, Mundt and Fried in 1983 found a "jet" 20″ long originating at IRS5 and also flowing into the direction of propagation (see Fig. 27.33). This jet consists of lumps of matter with masses of about that of the Moon. Like the HH objects, they have shock-excited spectra with radial velocities of up to -200 km s^{-1}.

The IRS5 jet forms the axis of a reflection nebula whose edge is about of parabolic shape with IRS5 at the vertex. Such a nebulosity is called a *cometary nebula*. The prototype is NGC 2261, better known as *Hubble's variable nebula*, which is illuminated by the star R Monocerotis.

The density in the vicinity of the cometary nebula in LDN 1551 is at least one order of magnitude lower than in the surrounding molecular cloud. This nebula is actually a cavity inside the dark cloud and shines in the light from IRS5 scattered at the boundary surface between the cavity and the dark cloud. This cavity probably originated when matter was ejected from IRS5 not only in the jet direction but into the entire solid angle occupied by (as seen from IRS5) the cometary nebula. The outflowing matter has subsequently swept away the matter which was originally in the cavity and, in the process, has accelerated the surrounding molecular gas and generated the bipolar CO flow.

IRS5 itself suffers at least 20 magnitudes of extinction at visible wavelengths. Just a few arcseconds away at the commencement of the jet, the extinction has dropped to 2 mag, and at the end of the jet (as seen from IRS5) it is barely noticeable. Also between IRS5 and the cometary nebula it illuminates, there is no significant extinction, as can be inferred from its rather bluish color.

These different and, at first glance, conflicting, observations can be interpreted by the assumption that IRS5 is surrounded by a dense disk of gas and dust which, however, is largely transparent in the axial direction. The perspective under which LDN 1551 is viewed is such that the disk covers IRS5 but not the jet features 3″ away. The existence of such disk-shaped dust structures has been suggested in similar objects by numerous observations.

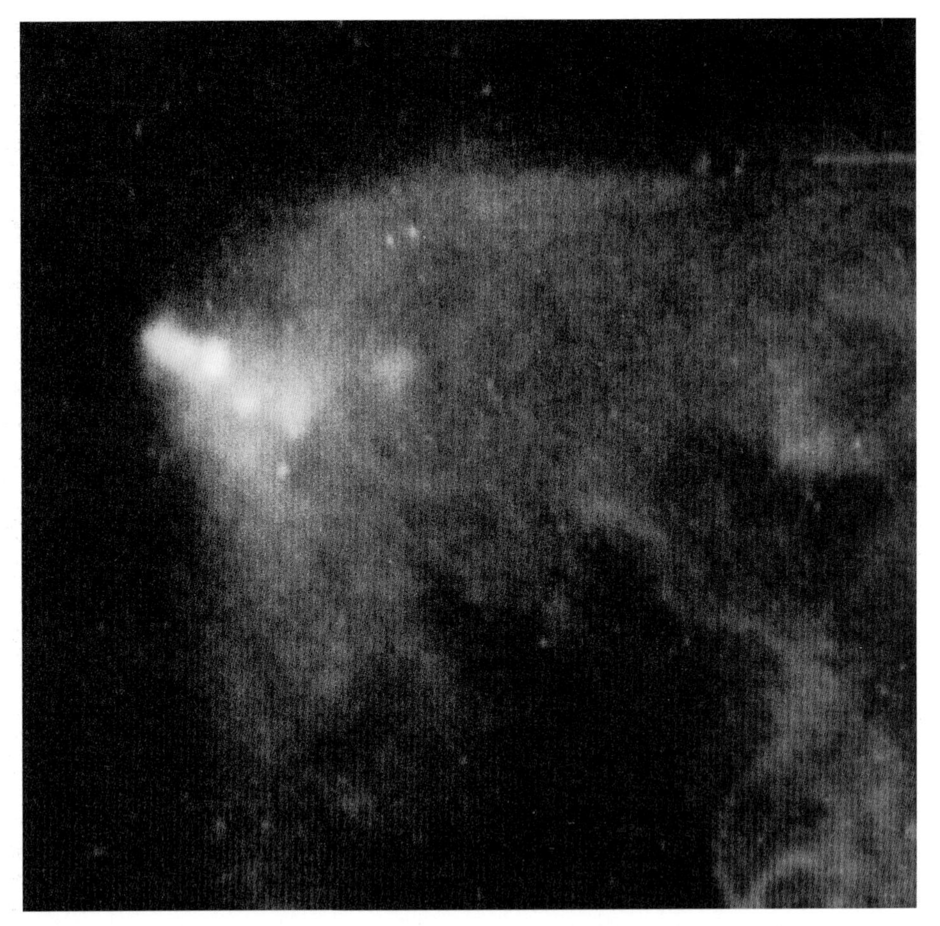

Fig. 27.33. A jet emanating from L 1551-IRS5 and into the surrounding cometary nebula. Photograph by Neckel with the 2.2-m Calar Alto telescope with CCD camera.

In the case of LDN 1551, the visible phenomena are noted only in the flow toward the Sun. Presumably the other flow toward the deep interior of LDN 1551 has resulted in the formation of similar objects, but which are concealed by the large extinction.

In two objects similar to LDN 1551, the cometary nebulae illuminated by R Monocerotis and by PV Cephei, deep CCD images found strongly reddened and dimmed counterparts to the visible cometary nebulae. These two objects are actually the two halves of a bipolar nebula, each of whose lobes has the shape of a cometary nebula.

Among the bipolar reflection nebulae with two lobes of about equal brightness are LkHα 208 and the "Boomerang Nebula" (see Fig. 27.34). Also, a Herbig–Haro object was found on the symmetry axis of the Boomerang, indicating its close kinship with the phenomenon in LDN 1551.

The best-known bipolar emission nebula is S106 (Fig. 27.35). Its central star of type O9 was found in the dark lane between its almost symmetrical lobes on an

Fig. 27.34. The bipolar "Boomerang Nebula."

infrared image-converter photograph taken with the 1.2-m telescope on Calar Alto. Visually, its light is dimmed by about 20 magnitudes. Also, S106 apparently contains an equatorial dust disk which is seen approximately edge-on. This explains the fact that the extinction is high in the direction of the central star, but almost nil toward the two ionized lobes. Moreover, the brighter of the two lobes is hollow, as is the case for the cometary nebula in LDN 1551. Matter flows away from the central O star at about 100 km s^{-1}. S106 also is embedded in a molecular cloud, which dims the lobes by from 3 to 15 magnitudes.

In previous years, so many bipolar CO fluxes were discovered that it seems justified to assume that every star during its formation goes through a phase in which substantial amounts of matter flow away from it in the form of bipolar molecular streams. The cause of this process is not yet clearly understood by theory. The most attractive theory ascribes the bipolar matter flows to a conversion of angular momentum of the collapsing cloud within which the central star has formed. The amount of energy returned by these matter flows to the collapsing nebula is substantial and may be comparable to the amount of energy emitted as stellar radiation. Certainly this is one of the processes which terminates the further contraction of the cloud. Thus, only a small percentage of the mass of the cloud (typically 7%) is converted into stars.

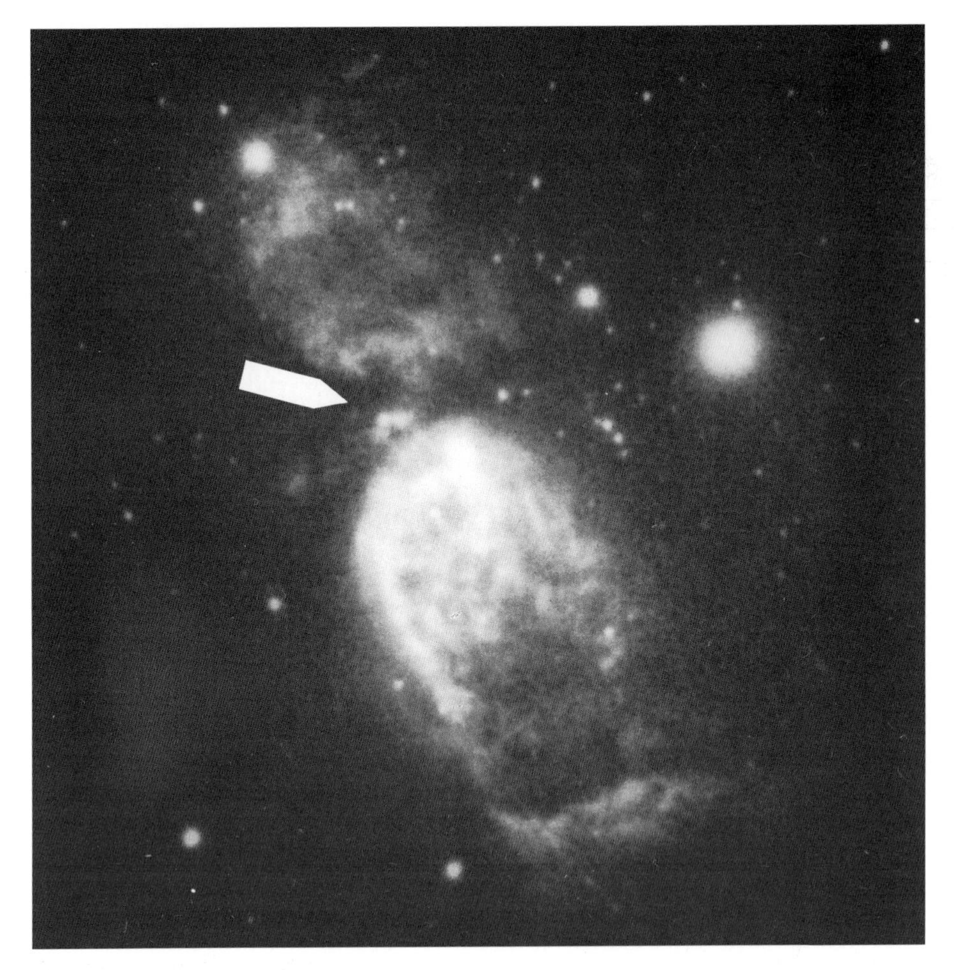

Fig. 27.35. The bipolar nebula S 106. Its exciting star of type O9 (indicated with an *arrow*) suffers 21 magnitudes of visual extinction, which originates in the dust disk between the two bright lobes. Photograph by Birkle with the 3.5-m telescope with infrared image-converter camera, Calar Alto.

27.14 The Stucture of the Milky Way System

27.14.1 The Galactic Center

The optical view of the center of the Milky Way is obstructed by dense clouds of interstellar dust. Over the whole length from the Earth to the galactic center, the dust effects an extinction $A_V \approx 27$ mag, which is a weakening by a factor 1.6×10^{-11}. This enormous extinction is, however, narrowly confined only to the galactic plane, as the layer of interstellar gas and dust is very thin. By the time galactic latitude 3° is reached, the line of sight runs largely outside the dust layer for distances over 2 kpc, and so the outer parts of the region near the galactic center are directly observed as the *Large Sagittarius Cloud.*

The distance to the galactic center can be determined only indirectly, by studying the spatial distribution of objects very strongly concentrated with respect to the center. Such objects are invariably members of the Halo Population II, which originated during a very early epoch when the Milky Way system was still approximately spherical, and the galactic disk had not yet taken shape. Globular star clusters and RR Lyrae stars are best suited for this task.

Globular clusters indeed concentrate distinctly toward the galactic center in the direction of the constellation Sagittarius. Their center of mass was first determined by H. Shapley early in the 20th century, after distances to globular clusters had been found with the aid of the RR Lyrae variable stars they contained. Shapley found that the center of the system of globular clusters lies at a distance of about 10 kpc.

A similar figure was reached by W. Baade from the frequency distribution of RR Lyrae stars in "Baade's window." Through this window, in the galactic dust clouds at $l = 0\overset{\circ}{.}9$, $b = -3\overset{\circ}{.}9$, the extinction is so low that Baade could observe field RR Lyrae stars far beyond the galactic center. The globular cluster NGC 6522 in the window has, in spite of its large distance of 6.5 kpc, an extinction of only $A_V = 1.5$ mag. Baade found that 76 RR Lyrae stars in the field had a distinct maximum of occurrence at a distance of about 9.1 kpc. Though the line of sight in that direction misses the center by 700 pc, it nevertheless makes sense to adopt this value as the distance to the galactic center. Other determinations of the distance R_0 have subsequently been performed, with results ranging from 8.2 kpc to 10 kpc. At present, the most reliable value, which has been adopted by the International Astronomical Union, is $R_0 = 8.5$ kpc.

In exactly the direction toward the galactic center, the dust is transparent only for radio and infrared wavelengths. In the radio continuum, the region around the galactic center is dominated by the bright source Sgr A, which VLA (Very Large Array) observations have shown to consist of four components: Sgr A West is a thermal source, as is any normal HII region. Sgr A East, a nonthermal source, is presumably an old supernova remnant. Both sources are embedded in a halo with a diameter of about 6′. Finally, a "point source" of diameter less than 10 AU embedded in Sgr A West stands out with high intensity. This point source presumably is the real center of the Milky Way system. Its position is at $(l, b) = (-3\overset{\prime}{.}34, -2\overset{\prime}{.}75)$. Perhaps it should be defined as exactly $(l, b) = (0°, 0°)$, but the difference is so small that introducing a new system of galactic coordinates is simply not justified.

This strong radio source is also a very conspicuous and bright object in the near- and far-infrared. The major part of the radiation is presumably emitted by warm dust which has been heated by the combined radiation from many luminous stars. The volume around the galactic center out to a distance of 150 pc, corresponding to an angular radius of 1°, generates a total luminosity of about $10^{10}\ L_\odot$. The total mass within this volume is estimated to $10^9\ \mathfrak{M}_\odot$. It has sometimes been speculated that a *black hole* may occupy the center of this region, but this matter is unclear.

The 21-cm radiation of neutral hydrogen shows that a rapidly rotating disk surrounds the galactic center; the radial velocities observed range from −220 to +200 km s^{-1}. Its radius is about 300 pc, its thickness 200 pc, and, strangely enough, it is tilted to the galactic plane by 22°.

Besides atomic hydrogen observed at 21 cm, the galactic center region is rich in molecular clouds whose structure strongly deviates from that of normal molecular clouds. First, the surface brightness of these clouds in CO light is unusually high. This fact is caused by the high density, averaging around 10^5 cm^{-3}. Also, the temperatures are considerably higher than in the usual molecular clouds, to over 100 K, as NH_3 observations have shown. Also, the velocity dispersion in these clouds is abnormally high. Both high temperature and high velocity dispersions presumably are due to the differential galactic rotation, which has a particularly pronounced effect near the center.

The volume to about $R = 500$ pc contains about $2 \times 10^8\ \mathfrak{M}_\odot$ in molecular hydrogen, or about 10% of its total amount in the entire Galaxy. Needless to say, all of these data should be viewed with great caution, because such calculations tacitly require certain conversions, such as that of CO column densities to H_2 densities. These correlations could be quite different near the galactic center than those in the solar neighborhood. In any event, it appears that the gas contribution near the center is only about 1% of the overall mass, while near the Sun a value of over 10% is assumed.

Low though the relative gas abundance near the galactic center may be when compared with the outer parts of the Milky Way, the absolute density of the gas is still intrinsically high. This explains why those molecular clouds are found with the greatest richness of complex molecules. The record is held by the thermal radio source Sgr B2 at $(l, b) = (0\overset{\circ}{.}67, -0\overset{\circ}{.}04)$, which in this respect surpasses Sgr A as well as the molecular cloud associated with the Orion Nebula.

27.14.2 Stellar Populations

The increase in star density toward the galactic equator implies that the Milky Way system is strongly flattened. Yet, different types of objects have differing degrees of concentration toward the galactic plane, and thus form differently flattened subsystems of the Galaxy which are characterized by their kinematic properties. The galactic concentration of a particular class of objects is said to be strong when the orbits of all members of that class lie in or near the galactic plane. The velocity components in the directions perpendicular to the plane are thus always quite small, and the orbits themselves are approximately circles centered on the the galactic center.

On the other hand, groups of objects which display little concentration toward the galactic plane move on elliptical orbits which are more or less inclined to the plane. Their velocity components at right angles to the plane are distinctly nonzero, and the velocity components in the direction $(l, b) = (90°, 0°)$, the direction toward which the Galaxy is rotating as seen from the Sun, are noticeably different from the circular velocity. Such objects are called *high-velocity stars*, since they move with substantial speeds relative to the nearly circular velocity of the Sun.

The degree of galactic concentration of various groups of objects is correlated with age. The oldest objects in the Galaxy are the globular clusters, which formed more than 10×10^9 years ago. Their distribution shows no significant concentration toward the

Table 27.4. The stellar populations. From Scheffler and Elsässer [27.1].

Population	Typical Members	$\overline{\lvert z \rvert}$ (pc)	Metal Content (%)	Age (10^9 yr)
Halo Population II	Globular clusters; RR Lyrae stars with periods $< 0\overset{\mathrm{d}}{.}4$	2000	0.01–0.1	10
Intermediate Population II	High-velocity stars; Long-period variables	700	0.01–0.1	10
Disk Population	F- to M-type stars	400	0.03–0.3	2–10
Old Population I	A-type stars; F- to K-type stars with strong metallic lines	150	0.3–1	0.5–5
Extreme Population I	O- and B-type stars; Open clusters; T-Tauri stars	70	1	< 0.5

galactic plane. At the other extreme are the O- and B-type stars, OB associations, and young open clusters, all of which have only recently formed out of interstellar matter. Almost all such objects are found in the immediate vicinity of the galactic plane. The substantial differences in space distribution of various types of stars has led to the notion of *stellar populations*. A population comprises all objects which—with respect to galactic concentration, to kinematic properties, and ultimately to age—are related to each other (the term "stellar generations" would perhaps be a more befitting term). Table 27.4 lists the characteristic trademarks of the five populations under which the various objects in the Galaxy may be subsumed. The various divisions reflect structural as well as chemical evolution of the Galaxy: at the beginning it was nearly spherically symmetric. At this time, the globular clusters along with long-period RR Lyrae stars were the most typical representatives of the *Halo Population II*. Gradually, the interstellar matter in the Galaxy flattened out and was continually enriched in heavy elements from, for instance, supernova explosions. Successively later generations of stars were thus formed at ever decreasing distances from the galactic plane, and with correspondingly lower velocity components perpendicular to the plane and increasing metal content. Stars with current ages from 5×10^8 to 5×10^9 years comprise the *Disk Population* of the Galaxy, its major constituent. The disk is densest in the central region and is also there most extended away from the galactic plane. This volume, which has a diameter of about 2 kpc, is called the *central bulge*.

The most luminous stars in the disk and bulge include red giants with strong mass loss. Such stars were identified by Habing among the point sources recorded by the *IRAS* satellite. Their spherical distribution (Fig. 27.36) shows the galactic disk and bulge as distinctly as if it were a photograph of an external galaxy seen edge-on.

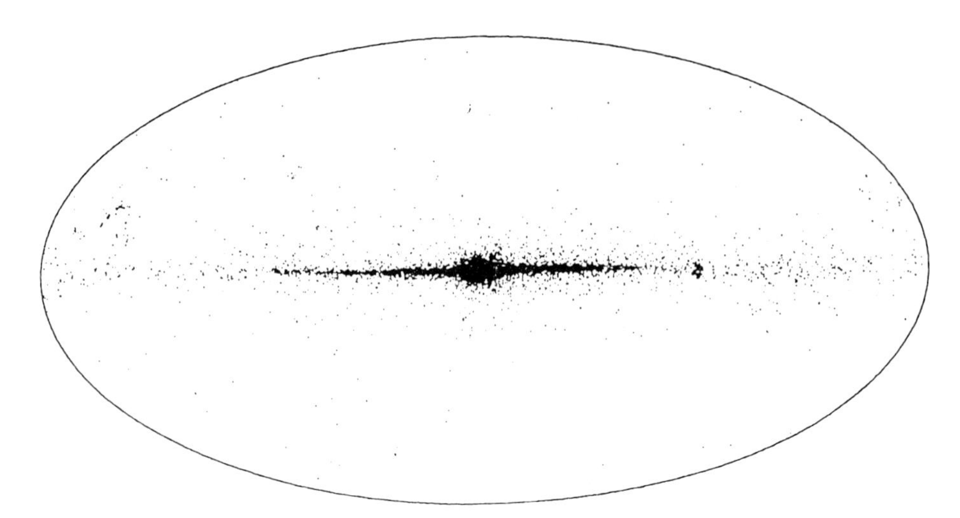

Fig. 27.36. The *IRAS* satellite has identified the brightest stars of the galactic disk and bulge out to large distances. Their spherical distribution matches the picture of a spiral galaxy seen edge-on (H.J. Habing 1987).

Currently, gas and dust remain in a layer only about 100 pc thick on both sides of the galactic plane. Here the members of the *extreme Population I* are still being formed (see Sect. 27.4).

By far most of the total mass in the solar system is concentrated in the Sun, and therefore the periods of planets are given by Kepler's third law solely from their mean distances from the Sun, their own masses being negligible. Hence, the orbital velocities of the planets are proportional to $a^{-1/2}$. This is not so in the Milky Way system: the periods of revolution and orbital velocities are given by that part of the galactic mass contained inside a sphere of radius R, where R is the distance of the star considered from the galactic center. The increase of the rotation curve at large R shows that the matter density there is still large enough to influence the kinematics of the outer parts of the Galaxy.

27.14.3 Differential Galactic Rotation—Kinematic Distances

The rotation of the Milky Way is determined by its rotation curve, which gives the rotational velocity $V(R)$ in km s^{-1} as a function of the distance R from the galactic center. As in all galaxies, $V(R)$ shows first a steep increase near the center, reaching a maximum of 280 km s^{-1} at $R = 1$ kpc. After declining to 220 km s^{-1} at $R = 3$ kpc, $V(R)$ begins to increase again, reaching about 250 km s^{-1} at $R = R_0$, the distance of the Sun from the galactic center. Recent measurements by R. Chini and coworkers have shown that $V(R)$ increases still further outward, to values over 300 km s^{-1} to at least $R = 20$ kpc.

The rotation of the Galaxy is thus not in analogy to planetary orbits, nor is it like a rigid body, which in the latter case $V(R) \sim R$ would hold true. Hence, the stars

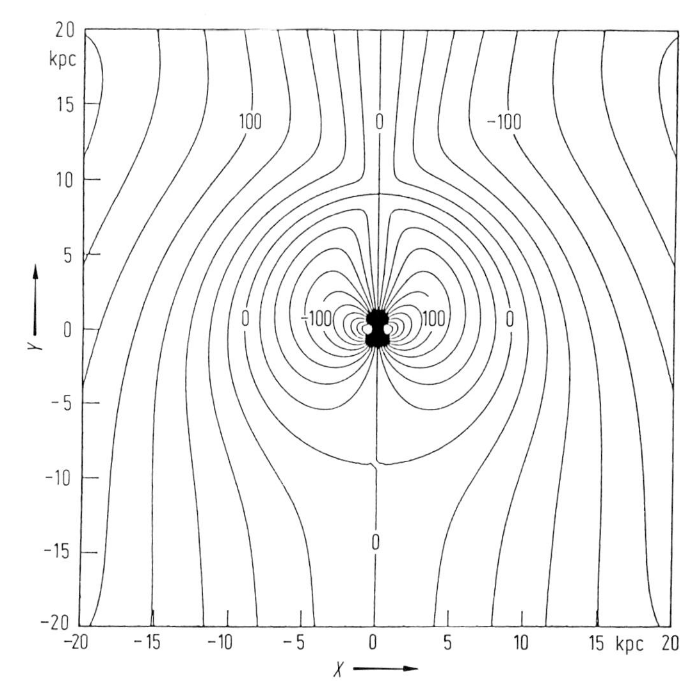

Fig. 27.37. Components of radial velocity caused by differential galactic rotation. The galactic center has the position $(X, Y) = (0, 0)$, the Sun is at $(X, Y) = (0, 8.5)$ kpc. After J.V. Feitzinger and J. Spicker.

shift relative to each other during the course of time and also relative to the Sun, thus giving rise to what is termed *differential rotation.* The effects of this shear are manifested as systematic components in the the radial velocities as well as in the proper motions relative to the Sun. For stars near the Sun, they are described by the relations

$$V(R) = Ar \sin 2l, \tag{27.13}$$

$$\mathrm{EB} = 0.21 \ (A \cos 2l + B). \tag{27.14}$$

Here A and B are the well-known Oort constants with values

$$A = 15 \text{ km s}^{-1} \text{ kpc}^{-1} \quad \text{and} \quad B = -10 \text{ km s}^{-1} \text{ kpc}^{-1}.$$

Equation (27.13) shows that the radial velocity of a star contains information about its distance, unless $\sin 2l$ is very small. Superimposed on the contribution provided by the differential galactic rotation are small, random velocities which may amount to up to 10 km s^{-1} for young objects. Only when the radial velocity of an object is distinctly larger than 10 km s^{-1} can it be used to infer a *kinematic distance.* Distances can be derived from radial velocities for objects outside the solar neighborhood using a diagram like the one in Fig. 27.37, where the connection between distance and $V(R)$ can be read.

The galactic rotational velocity at the position of the Sun is 250 km s^{-1}, which makes the time for one revolution of the Sun around the galactic center about 250×10^6 years. The motion of the Sun itself deviates from the purely circular velocity by 20 km s^{-1}, which is the *peculiar motion* of the Sun in the direction $(\alpha, \delta) = (18^h, +30°)$, the *apex* of solar motion.

Radial velocities are given relative to the Sun, but can be reduced to refer to a fictitious point at the position of the Sun and which moves in an exact circle around the galactic center. The coordinate system defined by this point, by the direction to the galactic center, and by the direction of galactic rotation, is called the *local standard of rest*. Radial velocities referred to this system are denoted by V_{LSR}.

27.14.4 The Spiral Arms of the Milky Way System

The optically visible spiral structure of spiral-type galaxies is delineated by luminous OB-type stars and by HII regions. In the Milky Way, the O associations give the first hint of the existence of spiral-arm filaments: their spatial distribution shows distinct aggregations in three oblong regions. More distinctly, these structures are shown in the pattern of young open clusters and of HII regions. The spatial distribution of such objects, which was first compiled by W. Becker and coworkers, is graphed in Fig. 27.38. The three filaments in which O associations, galactic clusters, and HII regions aggregate are considered segments of three different spiral arms known as the *Perseus arm*, the *Cygnus arm*, and the *Sagittarius arm*. The Cygnus arm is often called the "local arm," as the Sun is located at its edge. Attempts to substantially refine this picture by optical "spiral-arm indicators" have thus far been unsuccessful.

The difficulty in locating galactic spiral arms by optical methods has three causes. First, the Sun is very close to the central plane of the Galaxy, that is, in the plane of the arms. To find the configuration of the arms, observations of a number of objects which occur only in spiral arms are needed, their distances must be found, and then the spiral arms traced point-by-point. Were the Sun several hundred parsecs away from the galactic plane, the spiral arm structure would be seen at a single glance.

The second difficulty lies in the fact that the standard error in the distance moduli of individual stars is 0.5 mag, thus making the distances accurate to at best 25%. The same holds true even for open clusters. Upon comparing independent distance determinations for several clusters, Janes and Adler found that their distance moduli have standard deviations of 0.55 mag. In other words, they are by no means more accurate than the distance moduli of O- and B-type stars.

The third and most imposing obstacle to probing the galactic spiral structure is generated by interstellar extinction. The extinction out to only 3 kpc and beyond in most directions in the galactic plane is so high that more distant objects become too faint to be observed. A combination of distance errors and the apparent accumulation of stars or clusters in regions of low extinction creates the false impression of spokes aligned radially toward the Sun. These structures dominate many graphs purporting to show the spatial distribution of spiral-arm objects.

This effect is also seen in the distribution of young cluster as compiled by Janes and Adler (Fig. 27.39). The spiral-arm filaments deduced by W. Becker from a smaller

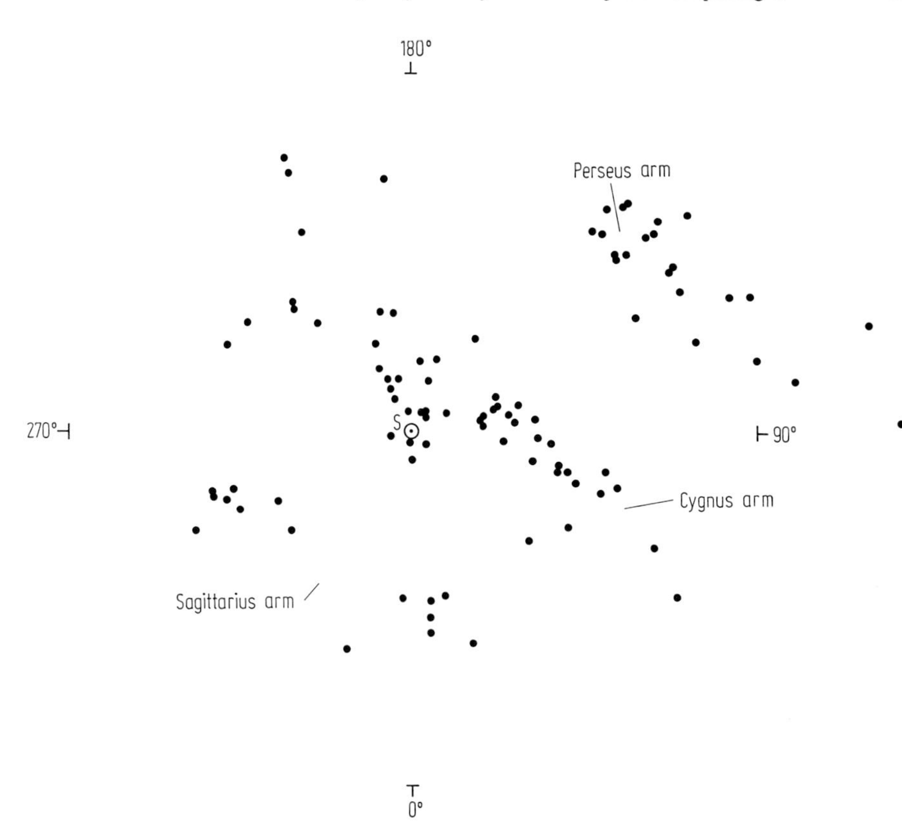

Fig. 27.38. The spatial distribution of HII regions and young open clusters shows accumulations in three filaments, interpreted as segments of galactic spiral arms. After W. Becker.

number of objects are seen only in traces. Janes and Adler come to the rather pessimistic conclusion that the inferred distribution of young clusters is influenced predominantly by the much-varying structure of extinction, and that optical studies generally are not suited for gaining insight into galactic spiral structure.

However, these attempts should not be devalued: using individual O- and B-type stars instead of open clusters, a much larger corpus of data becomes available, with the quality of distances being the same. Figure 27.40 shows the spatial distribution as deduced by Neckel and Klare from data on 6173 OB-type stars. Owing to the large number of objects, they are not graphed as individual points; instead their numbers in 250 pc × 250 pc squares are indicated by different hatching. The figure displays how, as a consequence of the unavoidable distance errors, the genuine maxima of star densities are stretched into long spoke-like formations aligned toward the Sun. Yet, the local and the Sagittarius arms clearly stand out from their surroundings as curved structures. The presence of the Perseus arm here is only marginal, as it is in Fig. 27.38. The limiting distance of all of these studies is between 3 and 4 kpc. Occasionally,

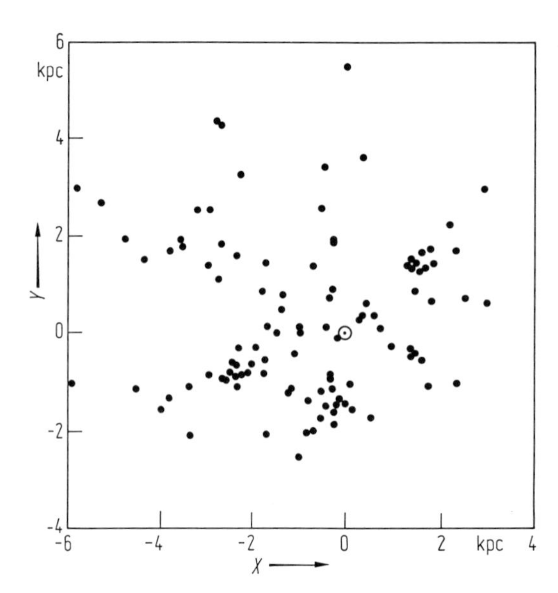

Fig. 27.39. Spatial distribution of young open clusters. After Janes and Adler.

rare, more distant objects are interpreted as indicators for farther-out spiral arms, but the reliability of such inferences is on weak footing.

For larger distances, only radio-astronomical observations of HII regions, neutral hydrogen, and molecular clouds, whose kinematic distances (Sect. 27.14.3) have been deduced from their radial velocities, can be used. In recent years the familiar picture of galactic spiral arms was constructed by Y.M. and Y.P. Georgelin, based on optical as well as radio-astronomical results (Fig. 27.41 a). A first glance appears to show that the majority of the objects are located close to the graphed spiral arms. But the reader should look at the same distribution of HII regions without the graphed spirals (Fig. 27.41 b) and try to construct, without bias, the spiral arms. (For this reason, Fig. 27.41 b is rotated and mirror-inverted with respect to 41 a.) One will find that there are simply too few objects for an unambiguous solution.

Observations of giant molecular clouds, in which prominent spiral-arm objects continuously form, should in principle provide the "ideal" method for locating galactic spiral arms. Indeed, it has repeatedly been claimed in the literature that distances of giant molecular clouds found from CO observations have finally solved the problem of mapping spiral structure. Nevertheless, there are other researchers who are just as convinced that the distribution of large molecular clouds shows no distinct spiral structure.

Similarly ambiguous is the interpretation of the density distribution of neutral hydrogen which follows from 21-cm observations (Fig. 27.42) based solely on kinematic distances. It does show approximately circular arcs around the galactic center, but these bear little resemblance to structures in typical spiral galaxies. Also, a comparison with the distribution of open clusters, OB-type stars (Fig. 27.40), or HII regions (Fig. 27.41 a) does not provide convincing agreement.

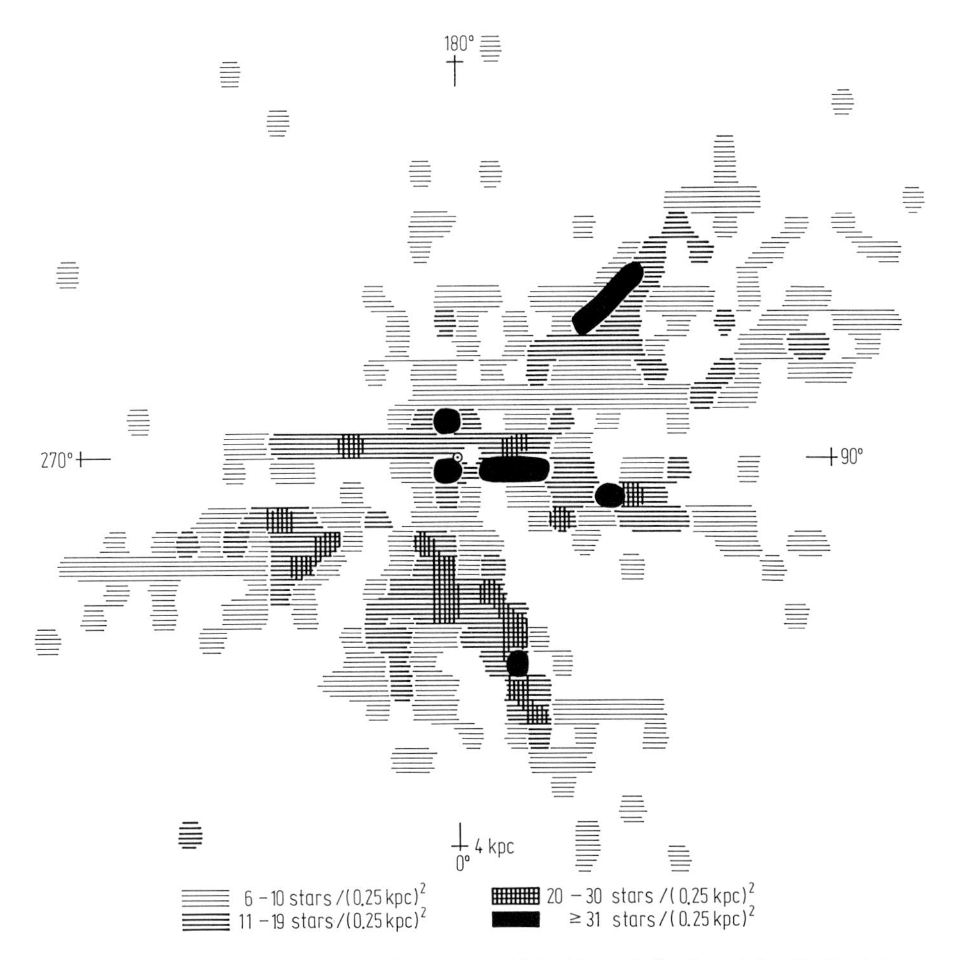

Fig. 27.40. The spatial distribution of OB stars within 4 kpc of the Sun. After T Neckel and G. Klare.

Presumably, neutral hydrogen is not associated with the star-forming regions of which the spiral arms are composed: the densities of the HI regions in Fig. 27.42 are at most 1.6 atoms cm^{-3}, and densities two to four orders of magnitude below those of HII regions or of GMCs. Another serious argument against the association of neutral hydrogen with spiral structures is its density distribution at right angles to the galactic plane. The HI density drops by $z = 220$ pc to one-half of its value in the plane, and therefore corresponds to typical values of Disk Population objects. GMCs are represented by a half-width of $z = 60$ pc, OB-stars and HII regions by $z = 50$ pc. A closer correlation of neutral hydrogen with the distribution of typical spiral-arm indicators is not to be expected.

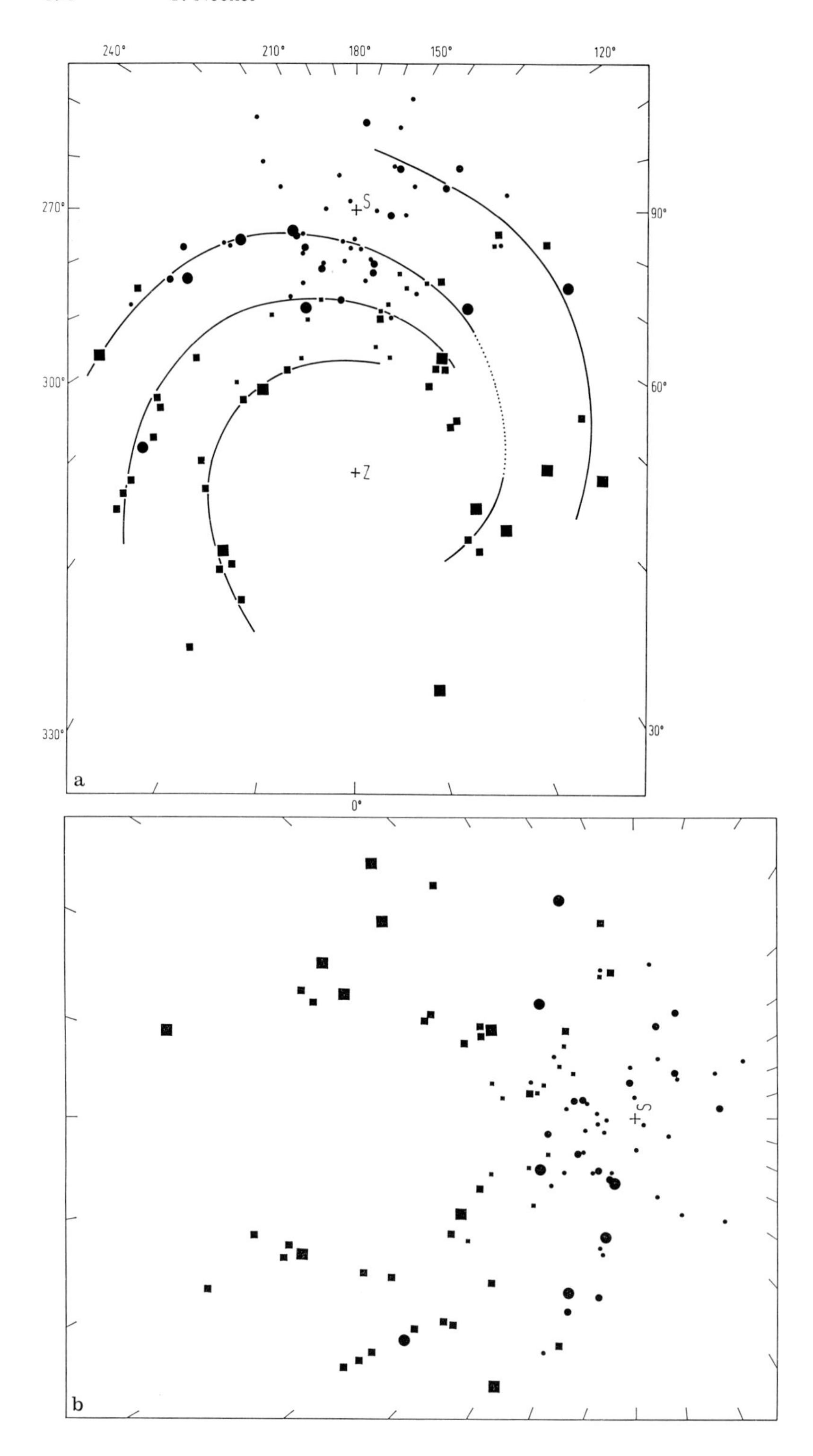
240°
210°
180°
150°
120°
270°
90°
300°
60°
330°
30°
S
Z
a
0°
S
b

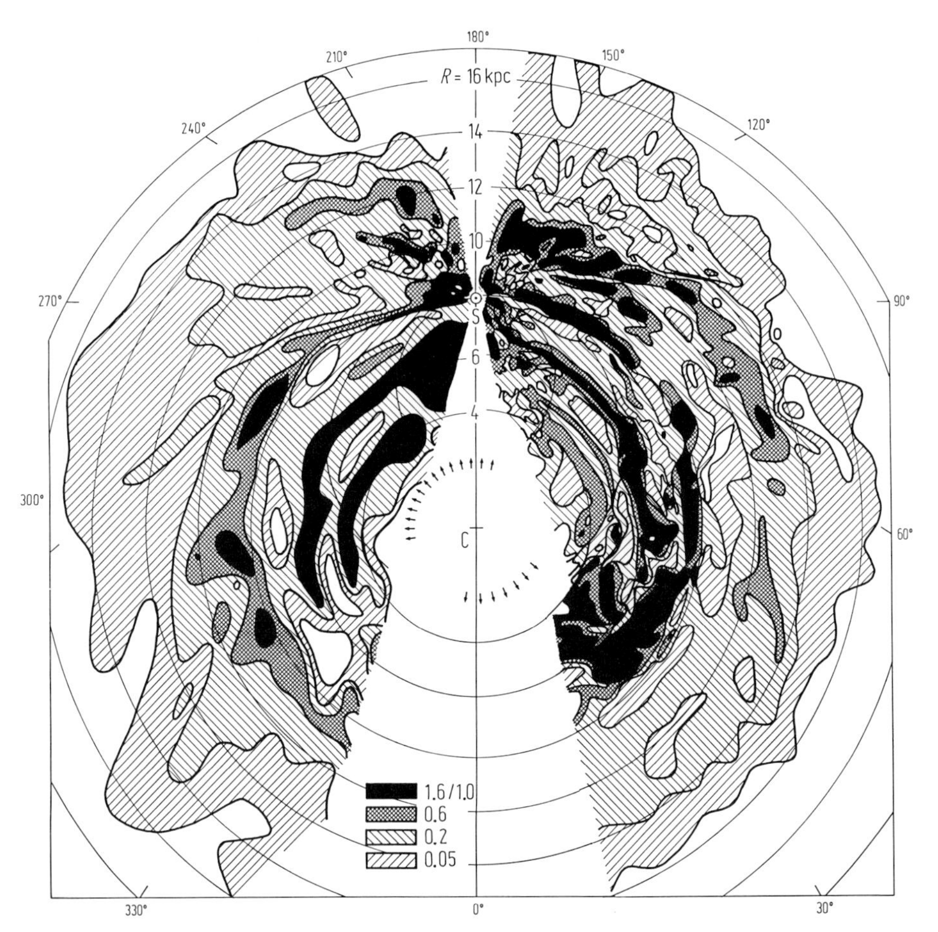

Fig. 27.42. Density distribution of neutral hydrogen in the galactic plane near the position of the Sun as deduced from 21-cm observations. No convincing similarities with the distribution of OB-type stars or other typical spiral arm indicators are noted. After J.H. Oort.

27.14.5 The Surface Brightness of the Milky Way

To describe the surface brightness of extended objects, the brightness of a unit area of the object is compared with that of a 10th magnitude star. The usual unit is 1 S_{10}, which is the brightness of a 10th magnitude star spread out over a square degree.

←

Fig. 27.41. a The large-scale distribution of HII regions in the Milky Way, as determined by Y.M. Georgelin and Y.P. Georgelin using the model of galactic spiral arms constructed from it. **b** Upon removing the drawn spiral arms, the arrangement of HII regions appears largely irregular. (This graph is rotated 90° and mirror-imaged with **a**.)

Typical values of surface brightness in the Milky Way are in the range 150 S_{10} to 500 S_{10}. The brightness of a unit area at position l, b is contributed to by all stars along the line of sight in this direction between the Sun and the edge of the Galaxy, with extinction $A(l, b, r)$ dimming the stars at distance r by the factor $10^{-0.4\,A(r)}$. If $\varepsilon(l, b, r)$ is the volume emission of all stars in the volume element at (l, b, r), then the surface brightness I is related to A and ε via the relation

$$I = 3.05 \times 10^3 \int_0^{\infty} \varepsilon(r)\ 10^{-0.4A(r)} dr. \tag{27.15}$$

Here, the unit for ε is chosen as one star of absolute magnitude $M = +5$ per cubic pc; the unit of the distance scale is kpc. The contribution of one star to the surface brightness is proportional to $1/r^2$, but the size of the volume element at distance r is proportional to r^2. Thus, the number of stars contributing to the surface brightness also increases with the square of the distance, and the influence of the two exactly cancels. Therefore, without extinction, the distant parts of the Galaxy would contribute the same surface brightness as the nearby ones do. This is approximately realized in some parts of the Milky Way where extinction occurs only at small distances from the Sun ($r < 500$ pc), as in many regions 5° or more away from the galactic plane.

Figure 27.43 illustrates the distribution of surface brightness in the Milky Way. For all directions l, b, where $A(l, b, r)$ is adequately known, the surface brightness $I(l, b)$ can be computed with the aid of a model $\varepsilon(l, b, r)$ of the galactic volume emission. The author has found good agreement between observed surface brightnesses and those computed using Eq. (27.15), with the following assumptions on the spatial distribution of volume emission: the emission of the galactic disk is represented in the galactic plane by a function $n_{\text{disk}}(R)$ which depends only on R; its half-value thickness $H_{\text{disk}}(R)$ is also a steady function only of R. The optimal functions $n_{\text{disk}}(R)$ and $H_{\text{disk}}(R)$ are graphed in Fig. 27.44. The Large Sagittarius Cloud, which lies about 5° away from the galactic plane, contributes most to the surface brightness of near-center regions of the Milky Way. In the plane itself, the view to the center is blocked by the dense dust clouds with a total visual extinction of more than 20 magnitudes.

The contribution of the spiral-arm population in the solar neighborhood must also be considered. Brightness maxima in Cygnus ($l = 80°$), in Vela ($l = 60°$), and in Carina ($l = 90°$) are caused by the line of sight in those directions being tangential to the local and the Sagittarius arms. The heights of these maxima agree with the spatial density of arm tracers, particularly OB-type stars. Other bright clouds in the Milky Way are primarily caused by rather low values of foreground extinction.

27.14.6 The Immediate Solar Neighborhood

The volume emission near the Sun as follows from the brightness distribution in the Milky Way amounts to 0.05 stars of absolute magnitude $M = +5$ per cubic parsec. The same value is obtained by integrating the luminosity function in the solar neighborhood. The luminosity function states how stars are distributed in absolute magnitudes. It should be noted that an accurate knowledge of the luminosity function at the faint end ($M > +10$) is unnecessary because the total luminosity is contributed to mostly by the intrinsically bright stars, despite their paucity.

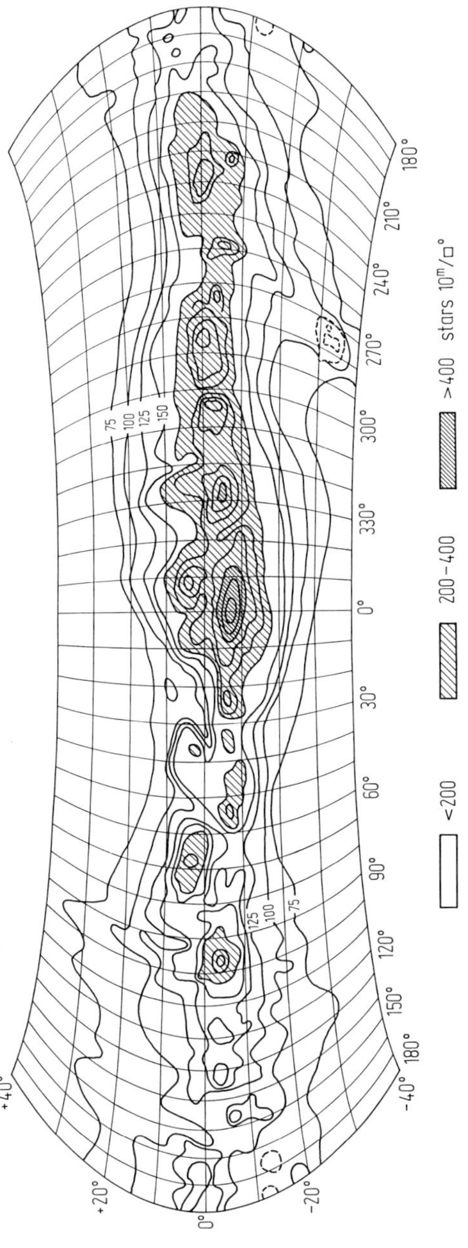

Fig. 27.43. Surface brightness of the Milky Way. From photoelectric measurements by H. Elsässer and U. Haug.

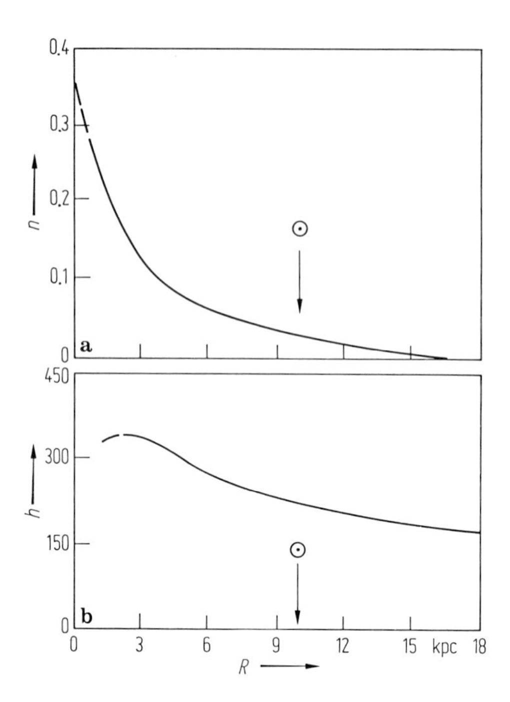

Fig. 27.44. The volume emission $N(R)$ of the galactic disk as a function of the distance from the center, and the thickness of the galactic disk expressed by the height above the galactic plane where the value of n has dropped to one-half of that at $z = 0$. After T. Neckel.

Astronomers have recorded all stars at or brighter than absolute magnitude $M = +10$ out to a radius of only about $r = 4$ pc from the Sun. Thereafter, the incompleteness at the faint end of the luminosity scale increases. Even high-luminosity OB-type stars with absolute magnitudes between -2.5 and -7 are known completely at best to $r = 300$ pc. Even within this range, their spatial distribution (see Fig. 27.45) shows significant details. First, there is a region of approximate dimensions 300 pc $\times$ 200 pc around the Sun which is conspicuously lacking in OB-type stars. A similar gap is shown by the dust distribution around the Sun. Among other things, this has the consequence that the extinction toward the galactic poles is nearly zero, and that the precise values of polar extinction, as derived assuming a homogeneous and plane-parallel dust distribution from counts of extragalactic nebulae, are wholly unreliable. This "hole" in the OB-star distribution presents a drawback: were such stars more smoothly distributed, the sky would be adorned with many more first-magnitude stars than is the case. Besides this "hole," Fig. 27.45 shows in the Scorpius–Centaurus direction, an aggregration which is typical of OB-type stars; it is the *Scorpio-Centaurus association.* At distances from the solar system ranging up to about 500 pc, the OB-type stars exhibit a strongly asymmetrical distribution relative to the galactic plane. In Orion and its immediate vicinity, almost all young stars lie on the average 20° south of the galactic equator, the Orion Trapezium group being no exception. On the opposite side as seen from the Sun, most young objects are found just as far north. Thus, young stars at distances up to $r = 500$ pc form a system tilted about 20° against the galactic plane called *Gould's Belt.*

The survey of OB-type stars is much less complete at larger distances, and moreover the absolute errors in their distances increase. Their spatial distribution up to $R = 3$ kpc was discussed in Sect. 27.14.4.

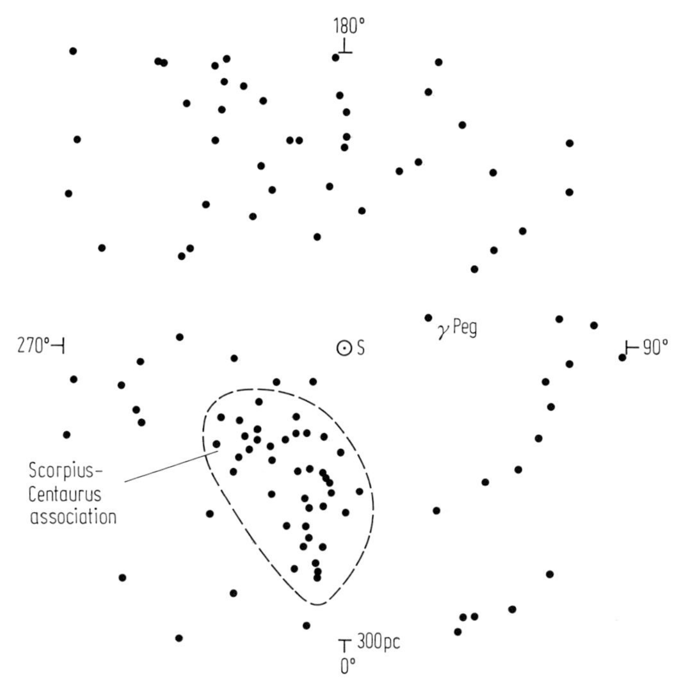

Fig. 27.45. In the immediate neighborhood of the Sun (denoted by an "S" at the center) OB-type stars are unusually rare, but they accumulate at a distance of about 200 pc in the Scorpio-Centaurus Association. After T. Neckel.

27.15 Hints for the Observer

Three technical developments in the past 20 years have opened up new and unforeseen opportunities for amateur astronomers. First, powerful telescopes with short tube lengths, such as those manufactured by Questar, Celestron, and Meade, are now available at affordable prices. The compact design allows even the amateur who lives in the downtown part of a light-polluted city to pack the telescope into the car at night and drive to a suitable site for observations. Very thorough and enthusiastic amateurs are known to travel to tall mountain peaks, say, to the Sierra Nevada range in Spain, for their observations.

Many amateur observers have been fascinated by the thought of taking photographs of nebulae, star fields, or galaxies. While photography of faint celestial objects was in former years quite laborious, the situation today has changed with the introduction of high-speed films, which have now been on the market for several years. For example, using ASA 400 film with a focal ratio $f/2.8$ at a site with dark night skies requires exposure times not exceeding one hour in order to obtain photographs limited by the sky background (see Chap. 6, Vol. 1).

State-of-the-art electronics equipment is also increasingly becoming a powerful tool of amateur observers. Systems which employ encoders at the hour and declination axes in conjunction with a computer have recently become available; these facilitate very rapid settings on large numbers of celestial objects. In this context it should also be noted that many amateur observers have equipped their telescopes with CCD cameras which, again, contributes to an enormous increase in the reach of their observational tools (see Chaps. 4 and 8 in Vol. 1).

27.15.1 Requirements of Site and Sky

The requirements of an observing site depend decisively upon the kinds of observations planned and the weather conditions. If one is searching for bright clusters, perhaps with binoculars, then the requisites are very few, since moonlight, scattered clouds, and moderate light pollution will not interfere much. The fainter the objects to be observed, the more a dark, haze-free sky will be desired. It is also possible to lessen the sky background by using suitable filters. Naturally, the best observing conditions exist at altitudes over 2000 m and located far from the urban "light sea." In fact, such an observing site *must* be selected if the goal is to photograph the faintest objects such as weak reflection nebulae or the gegenschein (see Chap. 22 in Vol. 2).

27.15.2 Visual Observations

Direct visual observations, i.e., those made with the eye directly at the eyepiece and not watching a television monitor, are now the exclusive domain of amateur observers. A professional astronomer only rarely has the opportunity to look through a telescope. An observer working on the coldest winter nights may certainly find it convenient to control his observations from a video terminal in a heated room, but the romance formerly associated with the work of the astronomer is thereby removed, and so is the pleasure of seeing Saturn or the Orion Nebula directly through a powerful telescope.

Visual observations of faint nebulae in particular require working in a very dark environment. There should be neither street lamps nor moonlight, as it is necessary to fully exploit the dark adaptation of the eye. This means that the eyes must be given enough time to adapt. The adaptation time varies from one individual to the next, and may amount to as much as one hour (see Sect. 4.8.1 in Vol. 1).

It is often stated that only bright nebulae are observable visually, while the fainter ones are accessible only photographically. Contrary to this statement, it is a fact that many faint nebulae have been detected visually; they include even the faintest galactic nebulae, including the Hagen clouds. Apparently the expectation of visual impressions through the eyepiece has been elevated so much by the many high-contrast photographs found in any standard astronomy textbook that many amateur astronomers (and *a fortiori* the professionals) do not have the patience to attempt to detect fainter nebulae visually.

The observation of emission nebulae, such as HII regions or planetary nebulae, is substantially expedited through the use of interference filters, which are transparent to the light of the primary emission lines but not to the major part of the continuum in between nor even to emission lines of the mercury-vapor lamps commonly used to illuminate city streets and parking lots. Since the brightest lines in most HII regions are Hα and neighboring lines of nitrogen, a filter which isolates just these red lines should be chosen. In planetary nebulae the forbidden oxygen ("nebulium") lines at 500.7 nm and 495.9 nm are strongest, and thus here a filter which is transparent to these green lines should be used.

The choice of a suitable magnification is important for observing faint nebulae. Extended objects appear brightest at the "normal magnification," which equals $f/8$ mm. At this value, the diameter of the exit pupil equals the diameter of the eye pupil, namely about 8 mm. A telescope with a focal length of 50 cm should thus be used with about 60× magnification to give fairly bright images of nebulae. On the other hand, for very large nebulae, such as the North American Nebula, this magnification is too big since then the field of view would include only a small fraction of the nebulosity. Extended objects are thus better observed with a small telescope or with binoculars.

All observations should be made as close to the zenith as possible. Unless observing at a high altitude, an extinction in the light of at least 20% to 25% in the direction of the zenith is incurred, and it increases with the secant of the zenith distance, reaching twice the zenith value at just 30° altitude.

27.15.3 Photographic Observations

Several tasks can be distinguished in the realm of photographic observation. Direct color photographs on high-speed 35-mm slide film will usually suffice to satisfy one's desire for impressive pictures. Here a standard single-lens reflex camera can achieve surprisingly good results. Using a wide-angle objective, large parts of the Milky Way can be combined into a single 35-mm slide. Large gaseous nebulae, such as the Orion Nebula, make excellent photographs with a good telephoto lens. The quality of modern objectives is so superior that they produce perfect, point-like stellar images even out to the corners of the photograph. This statement is based on the author's experiences with two separate $f/2.8$ Minolta objectives with focal lengths of 200 mm and 50 mm.

Smaller nebulae (e.g., most planetary nebulae) require large-aperture objectives with long focal lengths. The Celestron 14, for instance, provides excellent results, as numerous amateur photographs in *Sky & Telescope*, *Astronomy Now*, and other such magazines regularly show.

Extended exposures at longer-focus telescopes, of course, require a very careful adjustment of the polar axis of the telescope mount. Inadequate adjustment causes a rotation of the image field and elongated stellar images, particularly at the edges (see Sect. 5.13.1 in Vol. 1).

Black-and-white photographs are usually taken with special color filters in order to enhance specific features. For emission nebulae, a red filter, such as the Schott RG 630, which is transmissive from about 630 nm and above, is employed. A red-sensitive film or plate is also required. Reflection nebulae, however, are usually distinctly bluish, and therefore should be observed with a blue filter (see Chap. 6 in Vol. 1).

Quantitative work with photographs should use emulsion/filter combinations with which the Johnson *UBV* system can be realized. To the advanced amateur, this opens up the possibility of three-color photometry. Construction of color–magnitude diagrams of star fields or star clusters would be one interesting application.

Finally, the use of electronic CCD cameras has opened up unlimited possibilities. Photographing even the faintest galactic nebulae (say, the 900 nebulae north of declination $-33°$ in the *Atlas Galaktischer Nebel*) is certainly within reach of a Celestron 14, as is the photometry of stars as faint as the 17th magnitude.

References

27.1 Scheffler, H., Elsässer, H.: *Physics of the Galaxy and Interstellar Matter*, Springer-Verlag, Berlin Heidelberg 1987.
27.2 Scheffler, H.: *Interstellar Materie*, Vieweg, Braunschweig-Wiesbaden 1988.

Other relevant literature can be found in the Supplemental Reading List in this volume.

28 Extragalactic Objects

J.V. Feitzinger

28.1 Introduction

The building blocks of the observable universe on the scale of several megaparsecs (Mpc) are *galaxies*, stellar aggregrations similar to the Milky Way. The hierachical structure of the universe continues on an even grander scale as galaxies are ordered together into *clusters of galaxies*, the latter often connected together in the shape of chains or garlands. Between these chains of clusters of galaxies, there exist enormous voids.

Statistically, galaxies show a very large scatter with respect to their external morphologies, and this is also true of their physical parameters such as masses, diameters, rotational velocities, and luminosities. The energy output sometimes varies enormously from one galaxy to the next, and this fact allows astronomers to distinguish normal galaxies from the energetic active galaxies and quasars.

Galaxies are entities which inhabit the space exterior to the Milky Way system. The distances to them are therefore much larger than to objects within the Milky Way Galaxy. The remoteness of these objects generally means that they are necessarily faint and of small angular extension, a fact that the neophyte observer should keep in mind in order to prevent major disappointments. In the northern hemisphere, the Andromeda Nebula is the most suitable galaxy for the beginning observer, while in the southern hemisphere the Large Magellanic Cloud (LMC) is the best choice.

28.2 Catalogues and Photomaterials

Burnham's Celestial Handbook by R. Burnham [28.1] is, for good reason, subtitled *An Observational Guide for the Universe Beyond the Solar System.* This work provides a list of galaxies which are accessible to the amateur astronomer. Using data on apparent brightnesses and apparent diameters, one can decide which galaxies can easily be reached with a given telescope. As a good first exercise, one should try to photograph some of the 39 bright extragalactic objects in the Messier list (see Appendix Table B.33). In any event, *Burnham's Celestial Handbook* is the most detailed list of northern and southern hemisphere objects.

Individual names have been given to only a handful of galaxies, such as the Andromeda Galaxy and the Large Magellanic Cloud. Otherwise, galaxies are known only by a catalogue number and name. For example: Andromeda Galaxy = M31

Fig. 28.1. M51 and a companion galaxy. Photograph by K.-P Schröter and U. Kreller with the 36-cm Newtonian reflector at $f/6$, exposed 20^{m} on hypersensitized TP2415. The positive print was magnified $10\times$.

(Object No. 31 in the Messier list) = NGC 224 (Object No. 224 in the *New General Catalogue*; Triangulum Galaxy = M33 = NGC 598.

The most complete catalogue of galaxies for the northern skies is the *Uppsala General Catalogue of Galaxies* by Nilson [28.2], which employs information obtained using large Schmidt telescopes with apertures of 1 m and larger. For the southern skies, there is the *ESO/Uppsala Survery of the ESO (B) Atlas* by Lauberts [28.3] (ESO = European Southern Observatory). The *Second Reference Catalogue of Bright Galaxies* by de Vaucouleurs et al. compiles the data, including types, integrated magnitudes, and diameters, of the 6000 brightest galaxies.

Picture galleries of galaxies are offered in the following references: *The Hubble Atlas of Galaxies* by Sandage [28.5] and the *The Revised Shapley–Ames Catalogue* by Shapley and Tammann [28.6]. The latter includes, along with the pictorial material, an extensive catalogue section, while the *Hubble Atlas* describes the individual galaxies (189 pictures) in some detail. Galaxies with peculiarities, such as distortions or tails caused by tidal forces, are pictured in H. Arp's *Catalogue of Peculiar Galaxies* [28.7]; Arp and Madore [28.8] have also compiled a corresponding catalogue for the southern sky.

Fig. 28.2. NGC 4564. Photograph by K.-P. Schröter and U. Kreller. Technical data as in Fig. 28.1.

Fig. 28.3. M64. Photograph by K.-P. Schröter and U. Kreller. Technical data as in Fig. 28.1; exposure time 30^m.

Fig. 28.4. Double galaxy NGC 4485/4490; Photograph by A. Lindenmann with a Newtonian reflector (200/1500 mm) exposed 30^{m} on Kodak 103a0.

Fig. 28.5. M109, photographed by W. Münker with a 6-inch Newtonian reflector ($f/5$); exposed 20^{m} on hypersensitized TP2415.

Fig. 28.6. M81 and M82. Photograph by W. Baumann with a 10-inch Newtonian, exposed 52^{m} on hypersensitized TP2415. Positive magnified 6×.

Fig. 28.7. M33. Photograph by W. Baumann. Technical data as in Fig. 28.6; exposure time 40^m, magnified 13×.

The present chapter could have easily been embellished with spectacular photographs of galaxies gained with giant observatory telescopes. However, it was felt that it would be much more appropriate to provide pictorial material which had been obtained using equipment and techniques which are generally accessible to amateurs. The illustrative portion of the present chapter therefore uses photographs of galaxies obtained only by amateur astronomers using amateur instruments. The reader can easily find many and diverse examples of fine galaxy photographs by leafing through a few issues of the monthly magazines *Astronomy*, *Astronomy Now*, *Sky & Telescope*, and others. The beginner should first study some of these volumes and then try to duplicate the practical procedures presented therein. This will certainly spare him/her much disappointment and unnecessary duplication. The illustrations are here compiled

Fig. 28.8. NGC 2903. Photograph by W. Baumann. Technical data as in Fig. 28.6; exposure time 40^m, magnified 28×.

Fig. 28.9. M63. Photograph by W. Baumann. Technical data as in Figs. 28.6 and 28.8.

Fig. 28.10. M31 with companion galaxies. Photograph by B. Flach-Wilken using a flat-field camera by Lichtenknecker with $f = 500$ mm; effective focal ratio 4 (3.5/500), exposed 30^{m} on hypersensitized TP2415.

in Figs. 28.1 through 28.15. Completeness with respect to galaxy types, instruments, or photographic materials used has not been attempted. These pictures will hopefully provide incentive for the amateur's own work and to illustrate what is currently possible.

28.3 Classification of Galaxies

The richness of shapes of galaxies has always presented a challenge to observers. How can the various types be compiled into a scheme? The purpose of any classification system is to identify typical features in order to then progress to a deeper understanding of the systemic properties. One scheme of classification which is crude, but best suited for introductory purposes, was suggested by E. Hubble, and redefined by A. Sandage

Fig. 28.11. M101. Photograph by B. Flach-Wilken using a 300-mm focal length Schiefspiegler (Lichtenknecker optics) with effective f-ratio of 20, reduced via Shapley lens to 12 with effective focal length 3600 mm; exposed 180^m on hypersensitized TP2415. Positive magnified 9.5×.

Fig. 28.12. M66. Photograph by B. Flach-Wilken. Technical data as in Fig. 28.11, except with $f/12$, exposure 120^m. Positive magnified 15×.

Fig. 28.13. M99. Photograph by B. Flach-Wilken. Technical data as in Fig. 28.12.

Fig. 28.14. M82. Photograph by B. Flach-Wilken. Technical data as in Fig. 28.12; exposed 60^{m}. Compare with Fig. 28.6.

Fig. 28.15. Central region of the Virgo cluster of galaxies, containing the giant ellipticals M84, M86, M87, and M89. Photograph by B. Flach-Wilken. Technical data as in Fig. 28.10; exposure time 20^{m}, positive magnified 6.8×.

[28.9]. It distinguishes three classes of galaxies: *ellipticals*, *normal spirals*, and *barred spirals*. The scheme is illustrated in Fig. 28.16.

For elliptical systems, the number n attached to the symbol "E" expresses the ellipticity:

$$n = 10\,\frac{a-b}{a}, \tag{28.1}$$

where a and b are the semi-axes of the ellipse.

For spiral, or "S," galaxies, the letters a, b, c, d, m express a sequence of increasing openness of the spiral arms and/or of decreasing brightness of the central body relative to the disk. There is a gradual decrease in the continuity of the spiral pattern from a to m. When spiral arms are attached to a bar-like structure, the galaxy is classed as an "SB." Completing the classification are the irregular systems, which show no rotational symmetry. In between elliptical and spiral galaxies are the S0 systems; galaxies so classed contain a disk component but lack spiral structure and significant amounts of interstellar matter ($< 2\%$).

The luminosity class of a spiral galaxy can be estimated from the morphology of its spiral structure. Systems with well-developed, large-scale spiral arms are assigned luminosity class I, and systems with disrupted and ill-defined arms are classed V. The latter are also called *dwarf galaxies*. These are, on the average, smaller in brightness, diameter, and mass than normal galaxies by one or two orders of magnitude. Dwarf galaxies also exist among elliptical systems. The frequency percentages of the various galaxy types are shown in Table 28.1.

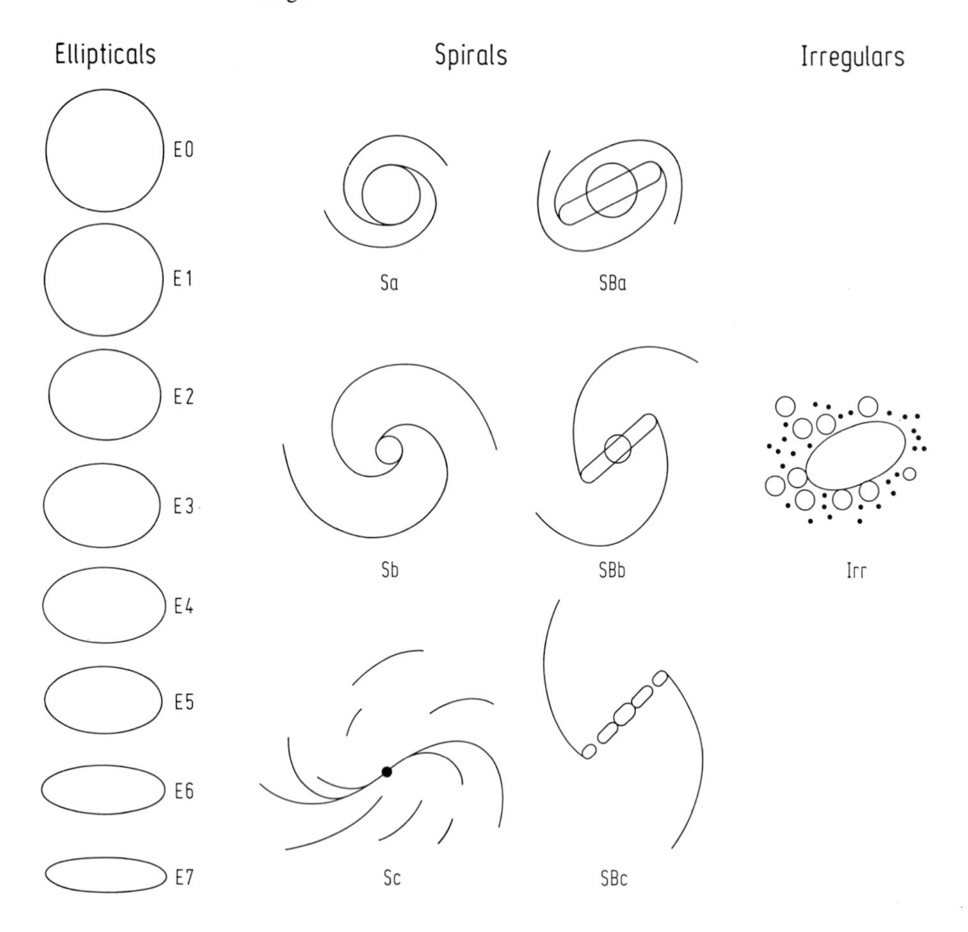

Fig. 28.16. The Hubble classification scheme for galaxies. Between E and S systems are the transition-type S0 galaxies. The S sequence a, b, c (d, m) proceeds from tightly wound to more open, loose, and disrupted structures.

Table 28.1. Apparent relative frequencies of galaxies according to Hubble type. After van den Bergh [28.36].

Type	%
E + S0	22.9
Sa + SBa	7.7
Sb + SBb	27.5
Sc + SBc	27.3
Ir	2.1
Other	12.5

28.3.1 Interacting and Peculiar Galaxies

The class of irregular galaxies is not to be confused with *peculiar galaxies*. Peculiar features could include warping of the systems, ring structures, matter bridges, or matter ejection in the form of gas clouds. Such features may also be identifiable in certain wavelength ranges, for instance, the radio luminosity is unusually high at the centers of some galaxies. These peculiar galaxies, also called active galaxies, include objects such as quasars, Seyfert galaxies, BL Lac objects, and radio galaxies (see Sect. 28.6). However, the limits of the classification presented here are somewhat flexible. The physical causes for the departures from the average properties of the corresponding galactic systems involve internal instabilities.

Tidal actions can also cause such peculiarities in galaxies. Distortions, warps, bridges, tails, or the triggering of bursts of vigorous star formation are mutually generated in pairs or multiples of galaxies. Galaxies undergoing tidal interaction show an enormous variety of shapes, and the reader should consult atlases of galaxies [28.7,8] to get an idea of the many diverse features of disturbances. Feitzinger [28.10] has described how the group of ring galaxies originates through tidal interaction.

Close groups of galaxies provide a natural laboratory for experiments on tidal interactions. Observations of these phenomena and attempts to explain them provide insight into the workings of these multi-component star systems. Figure 28.1 shows the interacting system M51 (NGC 5194/5), while Fig. 28.4 shows NGC 4485/90. Enormous amounts of interstellar or intergalactic matter have disturbed the galaxy M82 (Figs. 28.6 and 28.14); it is currently being debated whether or not matter has recently been ejected.

28.4 The Structure of Galaxies

A galaxy is composed of stars, interstellar matter (gas and dust) and high-energy cosmic particles (electrons and protons) tied to interstellar magnetic fields. These components organize themselves via equilibria between attractive and centrifugal forces, and according to the energy densities of the cosmic magnetic fields, galactic rotation, the turbulent motion of matter, and the entire radiation field. The three galactic constituents, interstellar matter, cosmic particles, and magnetic fields are distributed primarily in the plane of symmetry, the so-called *disk*, of spiral galaxies. A disk component is distinctly absent in elliptical systems, although they may well contain substantial amounts of interstellar matter.

Most galaxies contain a distinct nucleus. It can be the locus of active phenomena such as mass ejection, strong radio radiation, or X-ray emission. The nucleus is embedded in a central bulge, onto which the disk can be superimposed (Fig. 28.17; see also Fig. 28.2). Elliptical galaxies, lacking a disk component, resemble in shape the central spheroidal structures of spirals. Spiral galaxies may be illustrated by imagining

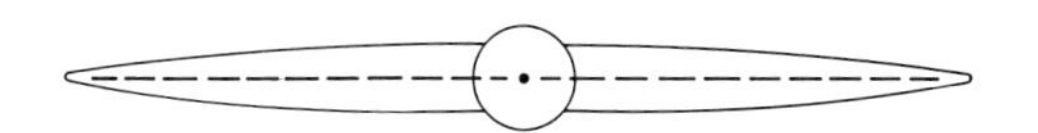

Fig. 28.17. Schematic graph of a galactic disk and central body; compare with Fig. 28.2.

a discus with a small spheroid the size of an egg clamped onto its center; a sand grain representing the nucleus lies at the center of the spheroid. The discus-shaped part can itself be structured by spiral arms. In barred spiral galaxies, the rotation spheroid of the central body is partially replaced by a cigar-shaped spheroid.

Elliptical galaxies are larger than the central spheroids of spiral galaxies; on the scale of the discus-egg-sand-grain model, an elliptical galaxy would correspond approximately in size and shape to an (American) football. Spiral galaxies are evidently also surrounded by a spherical halo, whose chemical composition is currently unknown. Its existence is inferred from the orbital velocities of stars and gas around the centers of galaxies. In this model, its size would be represented by a basketball. The basic components of a galactic system are not composed of stars of equal ages. Stellar populations of various ages, chemical compositions, and velocity distributions are the sub-units of multi-component galaxies. The stellar populations can also be found intermingled with each other. Basically two populations can be distinguished: Population I consists of young stars with ages substantially less than 10^{10} years, a mean velocity dispersion of about 20 km s^{-1}, and a heavy-element content of less than 4%; also belonging to Population I is the interstellar medium of gas and dust. Population II consists of stars with ages on the order of 10^{10} years, a very low abundance of heavy elements, and a mean velocity dispersion of over 40 km s^{-1}. Figure 28.18 illustrates the population components of a galaxy in schematic form. Population II dominates in the central bulge and, with decreasing strength, in the disk. It is steadily distributed over the entire galaxy, and also contains globular clusters. Population I has a discontinuous distribution, but is essentially concentrated in the disk. Stars, dust, and gas may appear preferentially along spiral arms, and the disk may appear to have a dearth of Population I objects around its center. This population has a steeper outward decline than does the old Population II.

The fact that different star groups have various ages means that they also possess different average colors. Population I objects are on the average hotter, and hence bluer, than Population II objects, which are redder. A qualitative separation of the two primary populations can be achieved by photographic means, such as Zwicky's "sandwich technique." Radio astronomical methods can register the gaseous interstellar matter.

28.4.1 General Properties of Galaxies

In order to deduce the properties of galaxies, their distances must first be known. True diameters, masses, and luminosities always depend on distance, as do the corresponding stellar parameters. The apparent angular size of a galaxy on the celestial sphere may be appreciable when the system is located near the Milky Way and very small when it resides far off in the depths of the Universe, irrespective of the true size. Methods of distance determination are discussed in Sect. 28.5.

After the distance has been ascertained, the measured quantities for a galaxy must be corrected for absorption and inclination effects. These rather complicated methods are described by de Vaucouleurs et al. [28.4]. Also, it is necessary to decide up to which isophote of surface brightness the extension of a galactic system should be

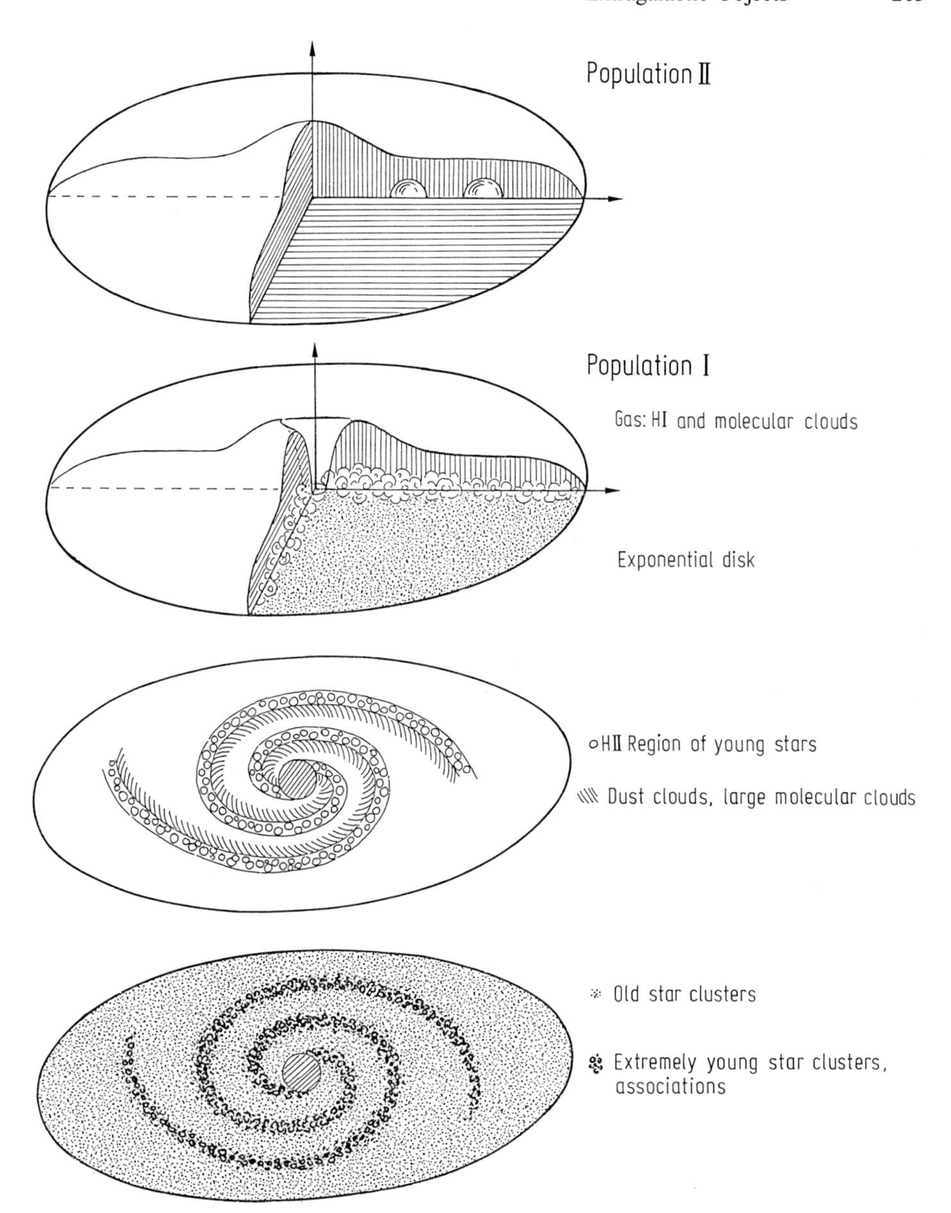

Fig. 28.18. Cutaway view of a galactic system showing its components. The graphs are to be visualized as symmetric to the central plane. *Top graph*: two "hoses" are hollowed out as spiral arms. *Bottom graph*: the uniform distribution of older clusters illustrates the steady transition to Population II objects. Globular clusters, as the oldest Pop. II constituents, are distributed with spherical symmetry with respect to the entire system.

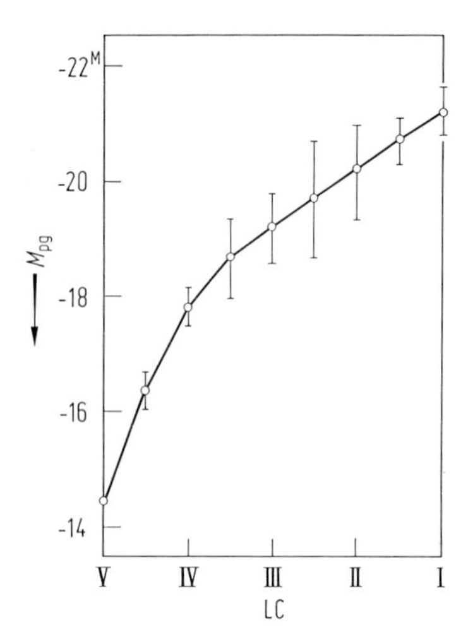

Fig. 28.19. Calibration of luminosity classes of galaxies. Absolute photographic magnitudes are corrected for absorption and inclination. Error bars show scatter around mean values. After Freeman [28.38].

counted. As a boundary, the *Holmberg radius*, which is defined as the major axis of the isophote of apparent photographic magnitude +26.5 per square arcsecond, is often used. With the angular diameter defined and the distance to the system known, the linear diameter is then determined.

The casual observer will certainly find it instructive to make diameter comparisons on a galaxy. Photographs taken on various days under different atmospheric conditions and also using different photographic materials and varying exposure times, will readily show that apparent differences in diameter can be found in a particular galaxy.

The range of linear diameters of galaxies extends from 0.1 kpc to 50 kpc, with spiral systems falling mostly between 10 and 30 kpc, and irregulars between 5 and 20 kpc. Dwarf elliptical galaxies typically have the smallest diameters. The Milky Way Galaxy is a large spiral system with a diameter of about 36 kpc.

Galaxies do not have a sharp boundaries. As with the definition of the diameter, the overall apparent brightness of a galaxy is defined in terms of the Holmberg radius. The sum of the individual brightness contributions within this radius forms the total apparent magnitude. Once the apparent magnitude has been determined, the absolute magnitude is obtained with the aid of the distance and two corrections. Absorption within the Milky Way itself must be considered, and the so-called *K-correction* applied. The latter is meant to guarantee that the same spectral range is always used in determining magnitudes, irrespective of the cosmological redshift a system may have (see Sects. 28.5 and 28.7). A galaxy's redshift may bring short-wavelength (UV) portions of the spectrum into the range defined for photographic observations, thus altering the isophotes.

The absolute magnitude M and the system radius R in kpc are closely related via the formula after Rubin [28.11],

$$\log R = -3.94 - 0.246M. \tag{28.2}$$

The absolute calibration of van den Bergh's [28.12] luminosity classes was performed by Sandage and Tamman [28.13], and is graphed in Fig. 28.19. Typical galaxies range

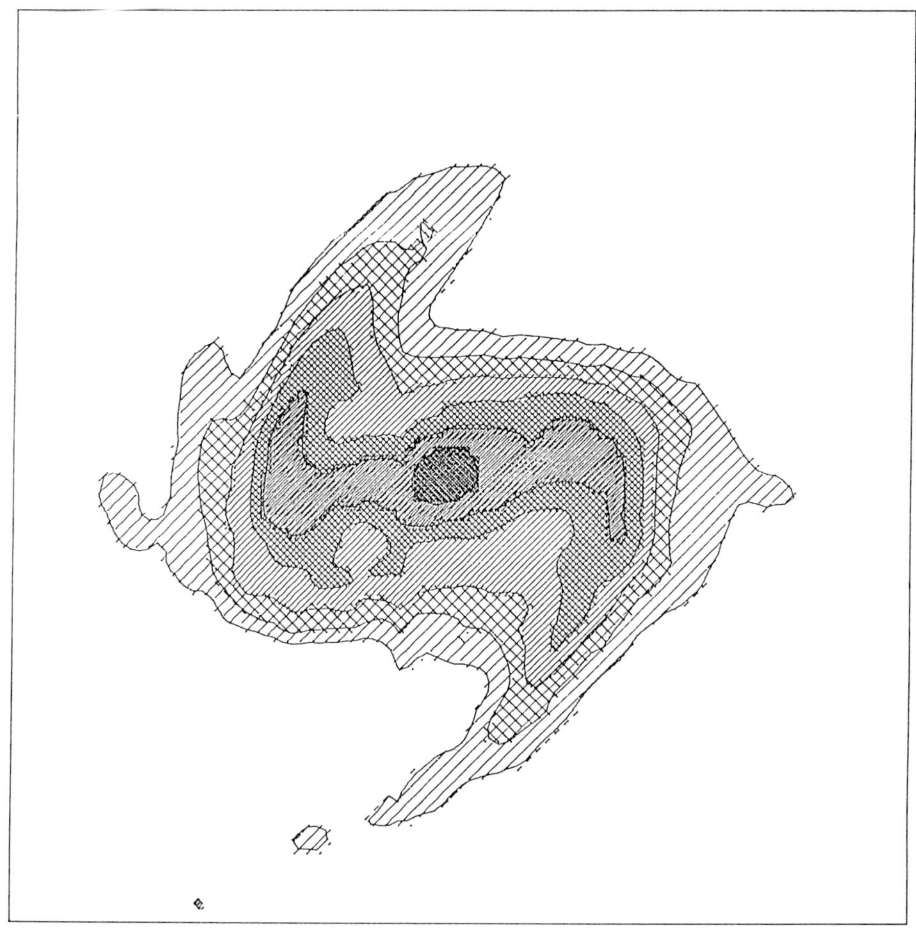

Fig. 28.20. Area photometry of the SBbI galaxy NGC 1365. Area graphed is about $3^\circ \times 3^\circ$. Isophotes are in relative units 50, 80, 110, 140, 170, 200. Image processing by Astronomical Institute of Ruhr University, Bochum.

between $M = -15$ to -20 in absolute magnitude. The range is larger for elliptical than for spiral systems. Again, the dwarf systems occupy the lower end of the brightness distribution, whereas some large deviations in the higher classes can be explained by peculiar cases.

The overluminous giant elliptical galaxies ($M = -23.3$) observed at the centers of clusters of galaxies are explained as mergers of several galaxies.

28.4.1.1 Brightness and Color. Surface brightness distributions (as graphed in Fig. 28.20 for barred SB types and in Fig. 28.21 for an Sc system) can be used to deduce the radial brightness distribution of a galaxy. The patterns of surface brightness follow a more or less universal law of luminosity which is characteristic for both ellipticals and spirals. The occurrence of a uniform law suggests that the star systems are structured according to similar "blueprints." The physical interpretation

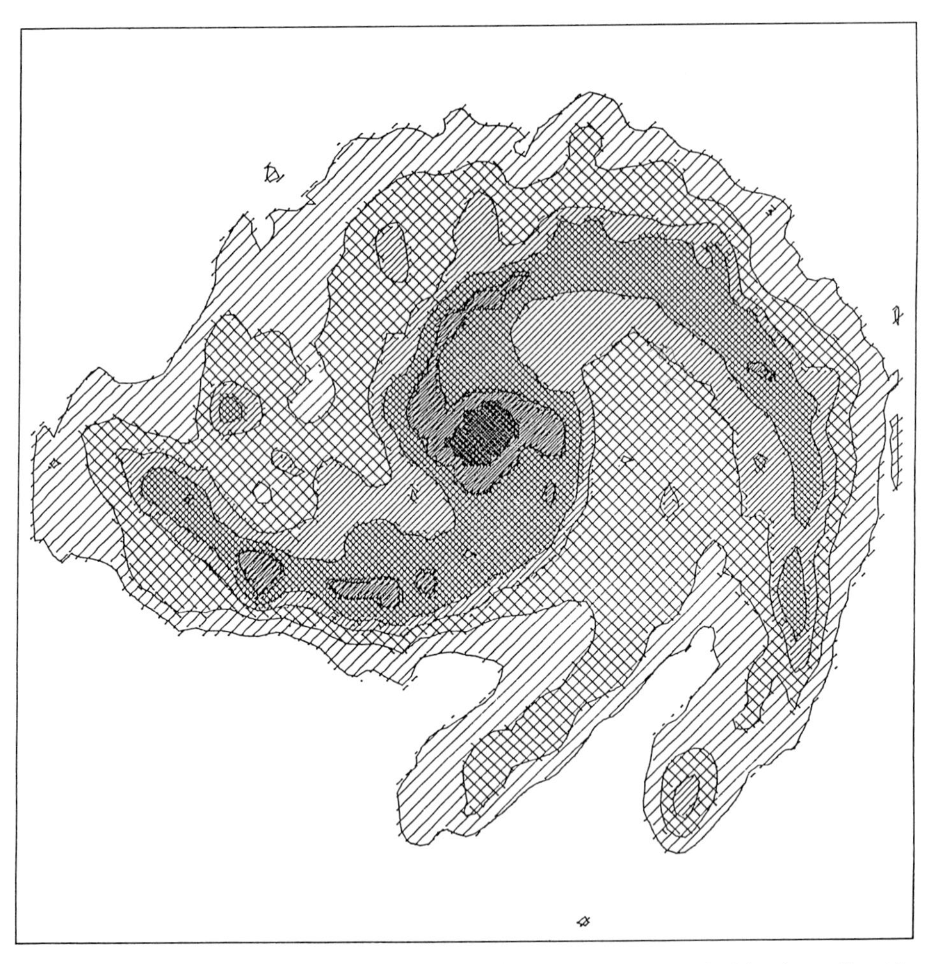

Fig. 28.21. Area photometry of the ScI-II galaxy NGC 5247. Area graphed is about 1° × 1°. Isophotes are in relative units 80, 110, 140, 160, 180, 200. Image processing by Astronomical Institute of Ruhr University, Bochum.

is that they should be in the same or at least in similar dynamical states. For elliptical systems, the empirical rule is

$$\log I(\rho) = -3.33 \left(\rho^{1/4} - 1\right), \qquad (28.3)$$

where R is the radial coordinate, R_e the effective radius of the isophote enclosing 50% of the total light, and $\rho = R/R_e$. Then $I(\rho)$ gives the surface brightness measured along the major axis as a function of distance from the center.

In spiral systems, Eq. (28.3) is augmented by a second term

$$I_2(R) = I_0 \, e^{-\alpha R}, \qquad (28.4)$$

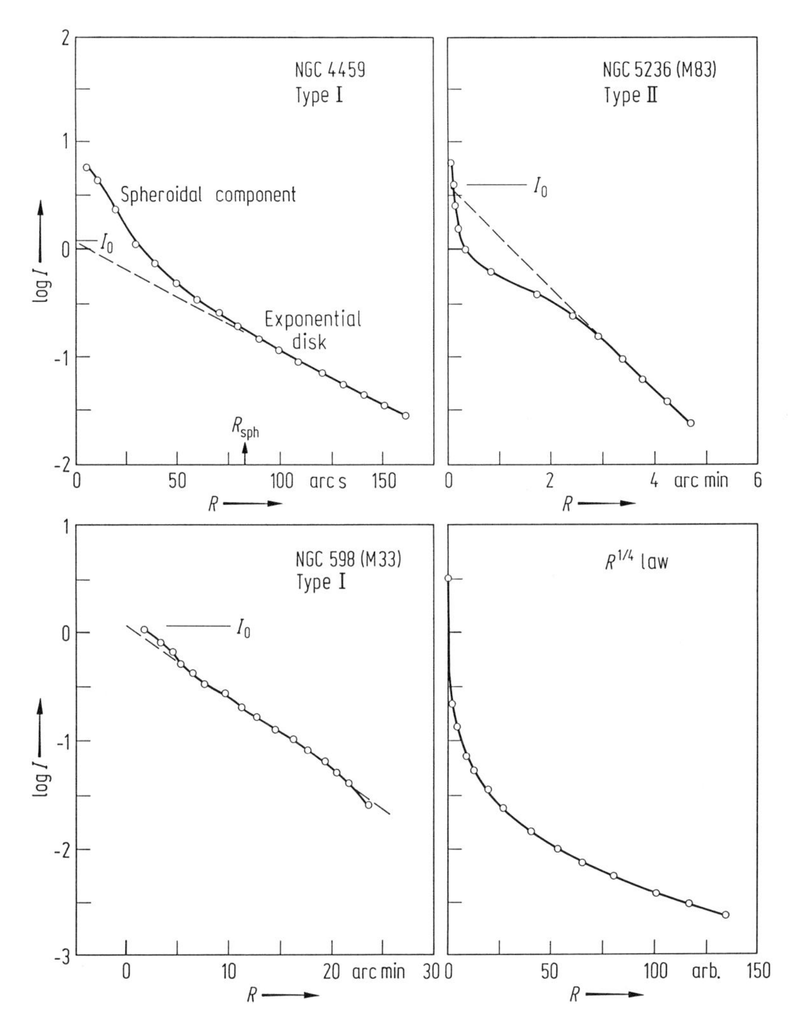

Fig. 28.22. Radial distribution of luminosity in three galaxies: NGC 4459 (S0), M83 (Sc), and M33 (Scd; cf. Fig. 28.7). Luminosity distributions of types I and II differ owing to different contributions from the spheroidal components. After Freeman [28.38].

which describes the exponentially declining brightness distribution in the disk. I_0 measures the brightness at the center, and α is the inverse scale of the brightness gradient. The quantity $1/\alpha$ can lie between 1.5 and 5 kpc, depending on galaxy type.

The central bulge of a spiral galaxy is characterized by Eq. (28.3), which applies to all elliptical systems. All disk systems show an exponential brightness decline in the radial direction, which means that the star population in the disk is always in a similar dynamical state. Within the Hubble classification of galaxies, the ratio of disk to central brightness is used as a classifying criterion. The distance at which the transition from the elliptical to the exponential brightness occurs is a quantitative measure of

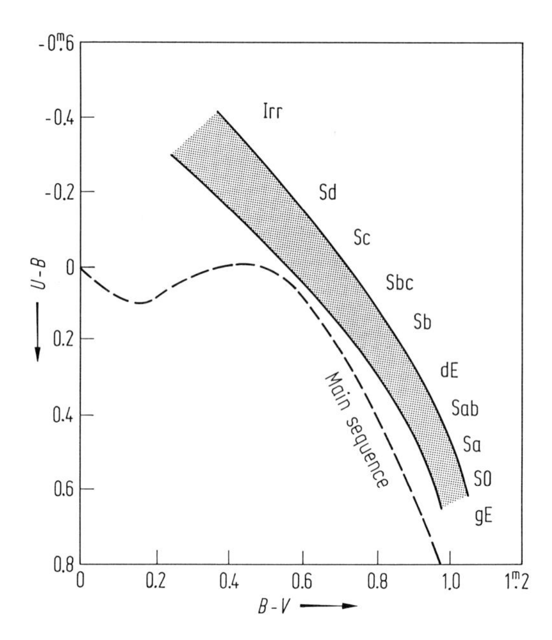

Fig. 28.23. Relation between *UBV* colors and galaxy types. After Mitton [28.39].

the light contributions in radial brightness distributions; the more advanced the type of the galaxy (c, d, m), the more inward are the transitions. Figure 28.22 illustrates radial brightness distibutions of varying contributions of spheroidal versus exponential components; a pure $R^{-1/4}$ curve is given for comparison. Brightness distributions in the bars of barred (SB) galaxies follow the $R^{-1/4}$ law.

Elliptical galaxies also may occasionally show disk structures. These, however, are either small, compact components deeply embedded in the spheroidal body, or else extended disk-shaped shells caused by tidal effects from neighboring galaxies (Kodaira et al. [28.14]).

From the apparent ellipticities of spheroidal components, the true frequency distribution of flattenings can be computed. If a and c are the true major and minor axes, respectively, then the ellipticity is

$$e = 1 - c/a = 1 - q. \tag{28.5}$$

A representative average is $e = 0.36$ for elliptical and $e = 0.75$ for spiral galaxies. The value $q = c/a$ decreases along the Hubble sequence from the E- to the S-type systems, and so does the ratio of brightnesses of central body to disk. Oblateness is not entirely conditioned by rotation. Since oblateness as a basic dynamical property cannot change on time scales of $\leqq 10^{12}$ years, such differences between E and S galaxies must have been imposed at the time when galaxies were formed. A galaxy cannot evolve from one type into another.

Surface photometry of a galaxy can be done practically in any color range (e.g., *U*, *B*, *V*). Depending on the wavelength chosen, different brightnesses are obtained which characterize the dominance of the different populations. The nuclear regions of spiral systems are reddish in color, as are elliptical systems throughout. From Sa to Sm, the outer parts exhibit a progressively bluer color as a result of stronger star-forming activity in the more irregular-type galaxies. A two-color diagram (Fig. 28.23) shows the galaxies arranged according to mean colors in a broad band above the typical stellar main sequence.

28.4.1.2 Stars and Gas. The fact that diverse stellar populations can coexist in space indicated early on that the distribution of matter in galaxies is not like a well-mixed dough. The various constituents of different ages, chemical compositions, and motion patterns are found to preferentially inhabit distinctly different regions of galaxies. Of course, galaxies consist of the same parts as the Milky Way. The entire spectrum of objects—normal stars, variables, novae, supernovae, planetary nebulae, and clusters—are found in galaxies, although the extreme values of the parameters of these constituents may have systematic differences. One striking example is the set of *young* (blue) globular clusters in the Large Magellanic Cloud.

The integral properties of galaxies are summed from the features of the constituents. The correlation of color indices with the morphological classification (Fig. 28.23) reflects the different mixture ratios of the populations. Galaxies become increasingly blue from type Sa to Sm, which expresses a steady increase in young stars and their maximum radiation output in the blue spectral range. Also increasing from Sa to Sm is the absolute luminosity of the brightest stars in the spiral arms and the size and number of HII regions. In addition, the percentage of gas and dust of the total mass increases, reaching about 15% to 20% in Sd and Sm systems, compared to 5% to 10% in Sa galaxies. Elliptical galaxies are not totally devoid of interstellar matter, but the mass contribution is less than about 3%.

Interstellar matter is strongly concentrated toward the central plane when the galaxy has a dominating central body. In cases of widely extended arms and the absence of a distinct central part, the plane of symmetry of the galaxy is enveloped in interstellar matter. Neutral atomic hydrogen in S-type systems has a typical extension of more than 2 Holmberg radii. Molecular gas components (CO, H_2) are concentrated predominantly in the optically visible galactic disk. Ring-shaped distributions can also be seen in molecular hydrogen as well as in atomic hydrogen.

More than 90% of the continuous radio radiation is generated within the disk of the spiral galaxy. It is radiated from free electrons in the interstellar medium, with the electrons traveling at high velocities along interstellar magnetic field lines. The magnetic fields are tied to interstellar matter, and it has been established from studies of a few nearby galaxies that they follow the spiral arms. The radio luminosity of the disk and nucleus or central bulge is proportional to optical luminosity.

The 21-cm radio line of neutral atomic hydrogen provides information on the distribution, mass fraction, and motion pattern (rotation curve) of gas, and hence on the internal-motion features of the entire galaxy. Every part of the electromagnetic spectrum yields information on certain parts of the galaxy. For example, infrared radiation may stem from the dust component of interstellar gas, or the X-ray emission from gas at a temperature of several 10^6 K in the central region of the galaxy. Composing all of this information, an overall picture of the galaxy may be stepwise constructed.

An optical spectrum recorded at moderate dispersion from an entire galaxy yields a composite (average) of the contributions of various populations, and, depending on

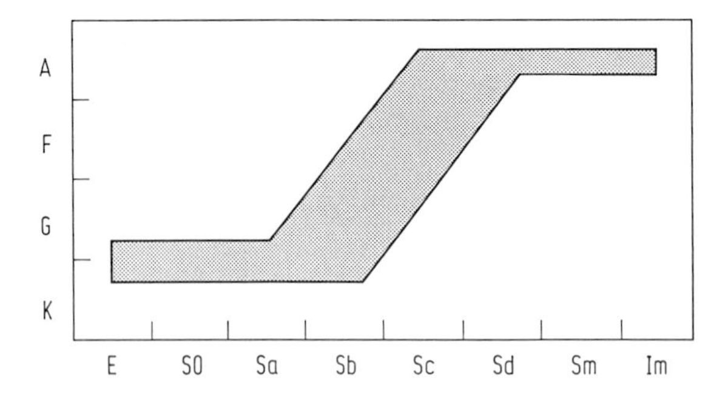

Fig. 28.24. Relation between average spectral type and galaxy type. After Voigt [28.40].

which population dominates, is characteristic for the type of galaxy. Galaxies may be assigned integrated spectral classes resulting from the blending of various spectral types. Figure 28.24 shows this relation. The sequence of spectral types thus runs parallel with an increase in younger population components from E to Sm.

28.4.1.3 Masses and the Mass–Luminosity Ratio. As is the case for stars, the mass of a galaxy is a fundamental parameter. The mass distribution then describes the structure and determines the kinematics and dynamics of the system. Mass determinations always require a theoretical model of the system according to which the calculations are laid out. The simplest model is the assumption of a point mass undergoing purely circular motion in the outer part of a galaxy around the total mass concentrated at the center. The actions of gravitational and centrifugal forces then balance and lead to a mass calculation.

There are six methods which can be used to determine the mass of a galaxy: (1) from the rotation curve, (2) from the width of the 21-cm emission of neutral hydrogen, (3) from the velocity dispersion of stars, or (4) of galaxies in a cluster of galaxies, (5) from the motion of a globular cluster or a dwarf elliptical system accompanying a large primary galaxy, and (6) from the motion of a pair of binary galaxies. The last two methods in principle are the same as used for mass determination in binary stars. The methods of velocity dispersions employ the *virial theorem*, which states that twice the amount of the average kinetic energy equals the gravitational energy when the system is considered invariant in time. The average kinetic energy is then determined from the average velocity dispersion.

Most recently, the mass determination of galaxies has used the correlation between absolute magnitude and total width of the 21-cm emission of neutral interstellar hydrogen. The line profile carries the information on the inner status of motion of the system, which, however, is governed by the mass distribution and, hence, by the total mass of the system. This correlation, named after Tully and Fischer [28.15] and which is graphed in Fig. 28.25, will be returned to in Sect. 28.5.

The rotation curve of a galaxy relates the rotational velocity of the system, as determined by the mass distribution, with the radius of the galaxy. The total mass of the galaxy can be calculated from the rotation curve, assuming circular orbits about the center, and by superposing various mass distributions. Figure 28.26 illustrates the schematic run of a rotation curve.

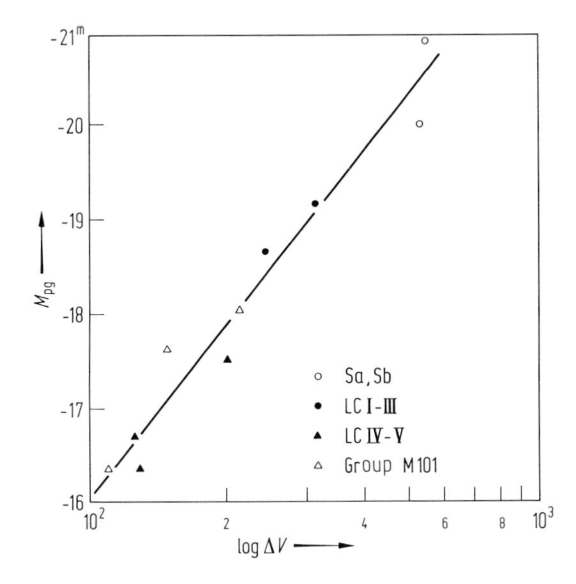

Fig. 28.25. Relation between the total width of the 21-cm hydrogen line (expressed as the log of the velocity dispersion) and absolute photographic magnitude of spiral galaxies, discovered by Brosche and named after Tully–Fischer. After Voigt and Huchtmeier [28.41].

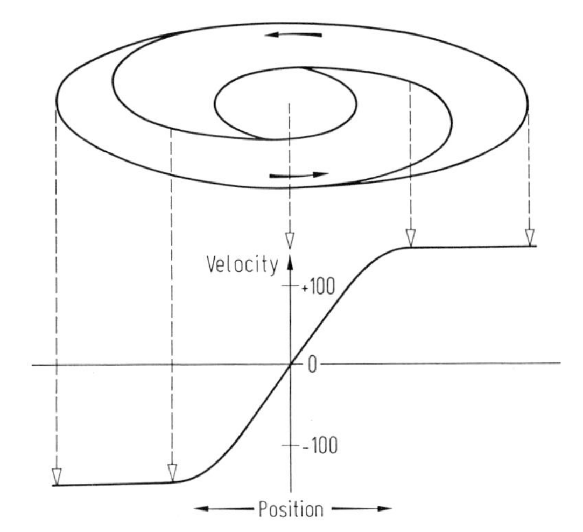

Fig. 28.26. The schematic rotation curve of a spiral galaxy shows a steep ascent in the inner part and a flat line of constant velocity in the outer part.

The *mass–luminosity ratio* combines two physical parameters. The luminosity is photometrically observed and the mass derived, as was stated previously, from kinematic data via a model. The mass–luminosity ratio describes how much luminosity per unit mass is developed in the system. Table 28.2 lists some typical masses and luminosities according to galaxy type, and Fig. 28.27 a,b illustrate the ratio of hydrogen mass to total mass and to blue-light luminosity as a function of spectral type. Underlying this connection is evidently a link between both the proportion of interstellar mass and the total luminosity with the type. The more interstellar matter present, the greater the number of stars which can form. Large numbers of young, hot stars

Table 28.2. Masses and mass–luminosity ratios for different types of galaxies. N = number of galaxies in sample. After Burbidge [28.37].

Type	N	$\mathfrak{M}(10^{10}\mathfrak{M}_{\odot})$	$\mathfrak{M}/L$
E/S0	9	0.36 ⋯ 350	10 ⋯ 80
Sa	2	1.9 ⋯ 20	3.6 ⋯ 7
Sb	15	1.2 ⋯ 34	1.2 ⋯ 8.4
Sc	24	0.13 ⋯ 27	0.4 ⋯ 20
Ir	8	0.07 ⋯ 13	2 ⋯ 11

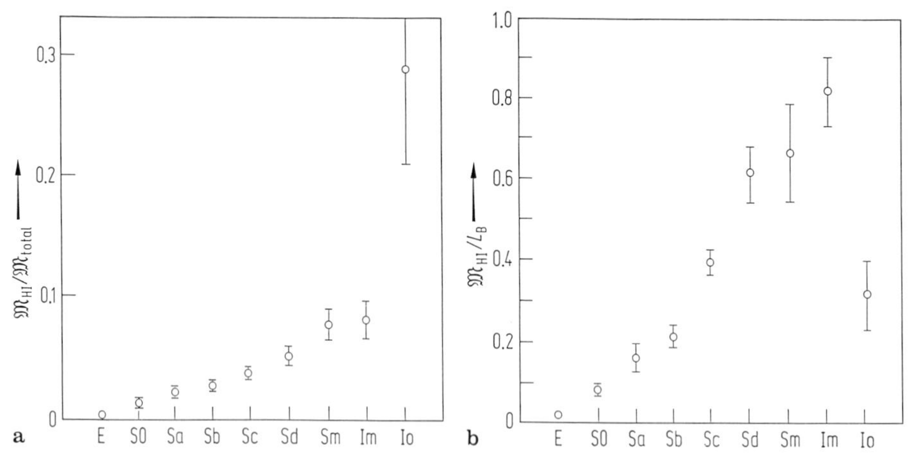

Fig. 28.27a, b. Mean hydrogen mass over total mass and over blue luminosity graphed against galaxy type. After Voigt and Huchtmeier [28.42].

in turn increase the blue luminosity. This explanation makes plausible some of the differences between types of galaxies.

28.4.2 Formation of Structures in Galaxies

The significance of the rotation curve for a galaxy has already been pointed out in Sect. 28.4.1.3. It will be readily understood that the large-scale motion in a galaxy, namely the rotation, has significant influence on the formation of structures, be they bar- or spiral-shaped. Rotation curves of galaxies are largely flat; after a steep increase in the inner part of a galaxy, the rotation speed does not drop but instead levels off to some constant value. This simple observed result has far-reaching consequences. If a galaxy does not exhibit a decline in rotational speed in its outer parts, as would be expected for a Keplerian orbit, an additional mass component must exist. Such a component has thus far not been detected in the optical, infrared, or radio-astronomical

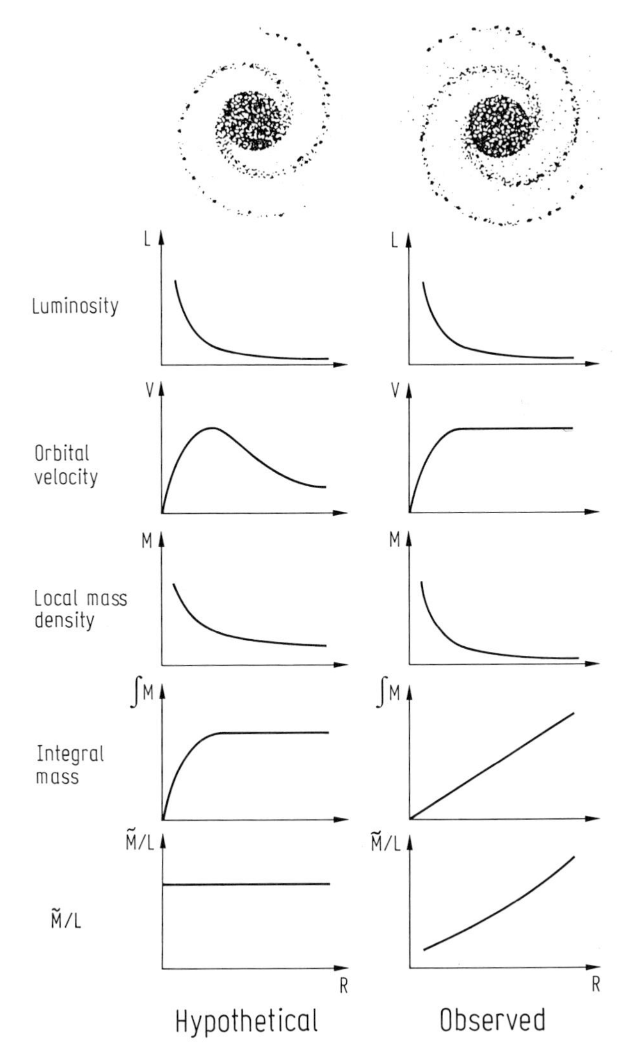

Fig. 28.28. A hypothetical and an observed galaxy. A halo is absent in the first, but the latter contains a massive nonluminous halo which changes all properties except optical appearance. $\tilde{M}$: total mass. After Rubin [28.16].

ranges. This additional mass component is called the *dark halo*, and it affects nearly all properties of a galaxy except its visual appearance. Figure 28.28 shows a comparison between the classical (but false) model and a real galaxy. Only the two luminosity distributions are identical. The classical model has no halo, the luminosity declines steeply, and so does the rotational velocity outside the central region. The local mass density runs parallel to luminosity, the integrated mass reaches a constant limiting value, and so does the mass–luminosity ratio. In a real galaxy, the rotational velocity remains high, the local mass density decreases but slowly, the integrated mass con-

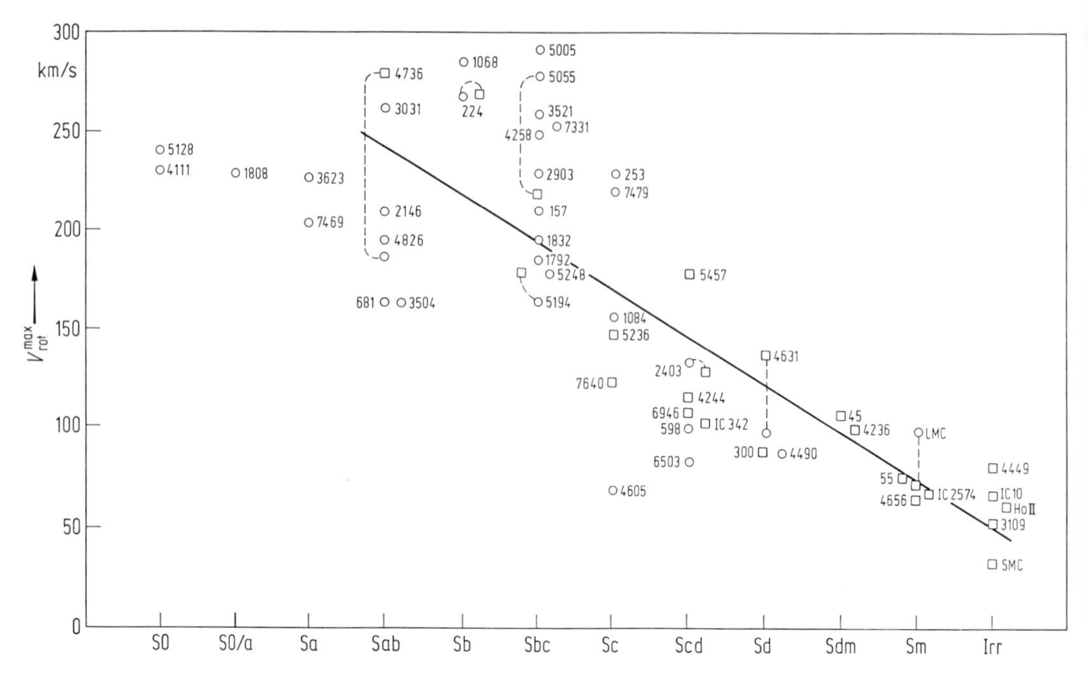

Fig. 28.29. Maximum rotational speeds of galaxies as a function of morphological type. The more tightly wound the spiral, the faster the rotation. *Circles* = optical data; *squares* = 21 cm radio data. Galaxies are identified by NGC numbers. After Brosche [28.42].

tinuously increases, and so does the mass–luminosity ratio. This result is caused by the influence of the halo component, which increases radially outward, combined with the brightness decline in the disk. What kind of matter composes the dark halo? Only one thing is currently certain: this matter must have an $\mathfrak{M}/L$ ratio of 100 or higher (Rubin [28.16]).

The maximum rotational velocity of a galaxy, represented by the mean of the flat part of the rotation curve, is correlated with the morphological type of the galaxy. The slower the rotational velocity, the more open is the spiral structure. It dissipates in the slowly rotating Sm spirals ($V_{max} = 70$ km s^{-1}) in some spiral arm filaments. The relation between type and rotational velocity is shown in Fig. 28.29.

Figure 28.30 illustrates that luminosities for a given spiral type (here Sc) range over a factor of 100, and that the luminosities increase with the rotational speed of the galaxy. A comparison between Sa and Sc galaxies shows that, at equal luminosity, the rotational speed of an Sa system is always higher, so they must contain more mass per unit luminosity and volume than Sc galaxies (Fig. 28.31).

In contrast to spiral galaxies, the structure of elliptical systems (including globular clusters and the central bulges of galaxies) is determined by the velocity dispersion of the individual stars. Elliptical systems rotate much more slowly, if at all, than the degree of ellipticity might suggest. V_{max} is less than 100 km s^{-1}. Velocity dispersions of the stars composing the systems differ in the directions of the three major axes.

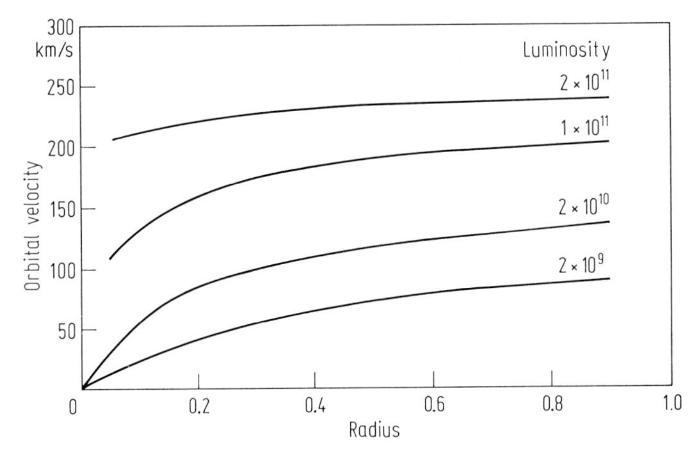

Fig. 28.30. Schematic rotation curves of Sc galaxies of different luminosities; radii normalized so that the edge of optical disk = 1. After Rubin [28.16].

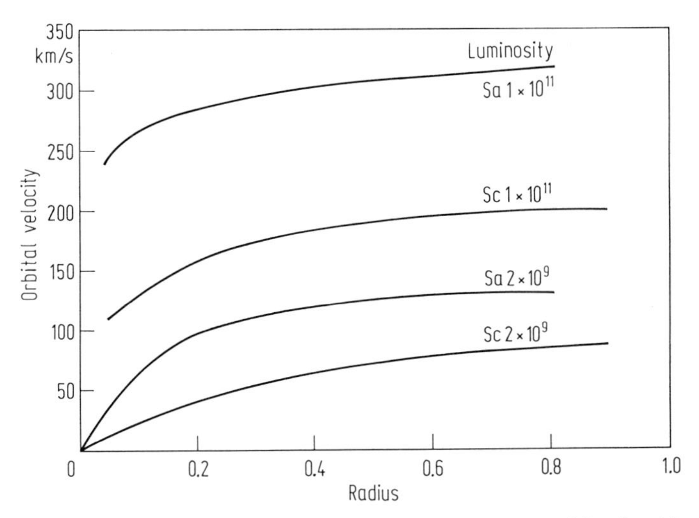

Fig. 28.31. Comparison of Sa and Sc galaxies. At equal luminosities, the Sa systems show higher rotational velocities. After Rubin [28.16].

Indeed, theory suggests that elliptical galaxies may be tri-axial galaxies with three axes of different lengths. The relation between shape (oblateness), central velocity dispersion σ, and luminosity L is a basic connection of dynamical and photometric parameters in elliptical systems:

$$L \sim \sigma^n \qquad 4 < n < 6. \tag{28.6}$$

The mean values of velocity dispersions are between 200 and 300 km s^{-1}.

Lacking pronounced rotation, elliptical systems yield only traces of rotation curves, and these are certainly not suitable for mass determinations. In such cases, the velocity dispersion in the center or its change among major axes may be used to calculate the mass. Velocity profiles measured along a major axis can thus serve as a surrogate for the rotation curve and are the most important observed data for describing the dynamics of such systems, and hence for interpreting the structure of elliptical galaxies. The typical decrease in the velocity dispersion from the inner to the outer regions of an elliptical galaxy is by a factor of two.

28.4.2.1 Spiral Structure. Because of differential rotation, any accumulation of matter within the galactic disk experiences a shear. The more distant a star is from the center of a galaxy, the longer is the period of revolution about the center; the inner parts overtake the outlying regions. Shear causes spiral structures to be trailed as a natural consequence of galactic rotation. This fact, however, actually presents a major dilemma: such rotation would quickly wind up any spiral structure so that it would effectively disappear. What mechanism prevents this winding up? Two theories purport to provide an explanation: the density-wave theory, and the theory of self-propagating, stochastic star formation.

The density-wave theory begins with the assumption that in the disk a spiral-shaped, symmetric perturbation of density forms and then rotates rigidly with fixed velocity. This can be illustrated by analogy with a moving construction area on a highway. The flow of cars (stars) condenses in a particular region, and this region slowly moves along with the progress of the construction work. The cause of these jam-ups in the traffic flow (or star flow) is the narrowed road. In nature, an external force (perhaps tidal action of a neighboring galaxy) or an instability within the disk itself may create such a mass concentration, which then acts as a region of enhanced gravitational attraction on the surrounding gaseous interstellar medium. The interstellar gas which is pulled into the region will be decelerated and compressed, thus initiating intense star formation in the cool molecular clouds. The outcome, in the form of young stars and HII regions, then manifests itself as a visible spiral arm. Thus it is understood why spiral arms consist so predominantly of young, recently formed stars and cool interstellar clouds (optically visible as dust lanes). The books by Scheffler and Elsässer [28.17] and by Bowers and Deeming [28.18] present the relevant mathematical formalisms.

The difficulty of this theory lies in providing an explanation for the origin of the stellar density wave and for its lifetime. As is true of any wave, the den-

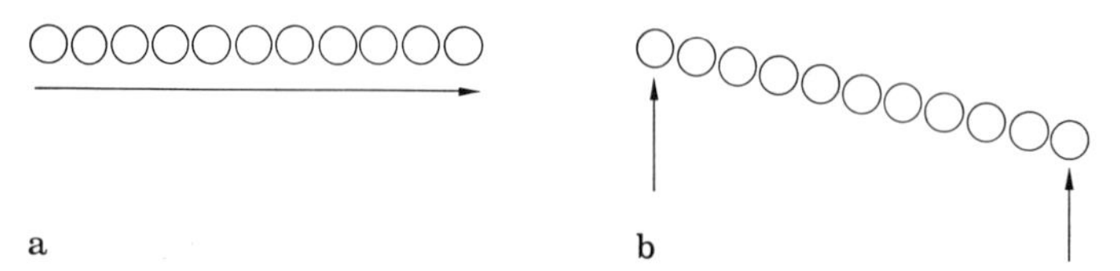

Fig. 28.32. a Sequential star formation with probability P and **b** shear by differential rotation.

sity wave propagates at some typical velocity, moving outward until it exits the galactic disk. The spiral structure dissappears on a time scale of a few 10^8 years, which corresponds to just one galactic rotation period. The spiral structure created by density waves is therefore very transient (Feitzinger and Schmidt-Kaler [28.19]).

The density-wave theory requires a perturbation in the disk density as well as an interaction of the interstellar medium with this disturbance in order to explain the spiral structure. By contrast, the theory of self-propagating star formation is much simpler; here, it is the interaction of interstellar medium with the rotation of the galaxy which shapes the structure.

This theory begins with the notion that the formation of structure (the origin of a spiral arm) is linked to the random event of star formation. This idea is observationally founded upon spatially propagating starbirth.

The formation of stars is a process of extremely low efficiency. Only a few percent of the mass of a cool interstellar cloud will condense into stars, the rest becoming heated and dissipated by the newly formed hot stars. The leftover gas, which constitutes at least 85% of a cloud, is then shifted into adjacent regions, causing the gas already existing there to condense, triggering more star formation, or it can cool, merge with other clouds, and thus form the birthplaces for later generations of stars. It has previously been mentioned that the angular speed of a galaxy decreases with increasing distance from the center, and that this causes shears. The rotating galaxy necessarily generates, via shears in stellar nurseries, a transient spiral structure. The spiral pattern changes with time as it is hallmarked by continuously forming stars. The pattern itself, however, is self-propagating, as the new stars continue to be born.

Assuming that each star-forming cell is a region with an extent of 200 pc, and imagining that a galaxy is subdivided into such cells, then each star-forming cell will, over the course of time, come into contact with various other neighboring cells by way of the differential rotation of the galaxy. The spark of starbirth then can spill over into the neighboring cells. Intense star formation in one cell increases the probability of such activity in neighboring cells. The probability for generating stars depends, of course, on the contents of the cell, i.e., on its previous history. If the cell has not experienced star formation for a long time, it will contain many cool, dense molecular clouds, and will be quite ripe for starbirth activity. If, on the other hand, a cell has recently produced fresh stars, it will require some time to "recover" before the birthing process can begin anew.

The processes described above outline a self-propagating chain-reaction of star formation which changes the contents of a star-forming cell with time. Also, the encroachment into neighboring cells triggers a pattern of structure. This encroachment, physically called a *percolation process*, enhances the probability that additional star formation will occur, and can be illustrated by the model of a forest fire.

Assume that a forest contains a single tree which is on fire. It is now desired to calculate the probability that the fire will spread to the next tree. This probability is dependent on such factors as the distance between trees, the undergrowth, wind direction, humidity, and type of tree. In the astrophysical analogue, the burning tree is replaced by a star-forming cell, and the probability for the fire (star formation) spreading to other trees (cells) is influenced by, for instance, the density of the molec-

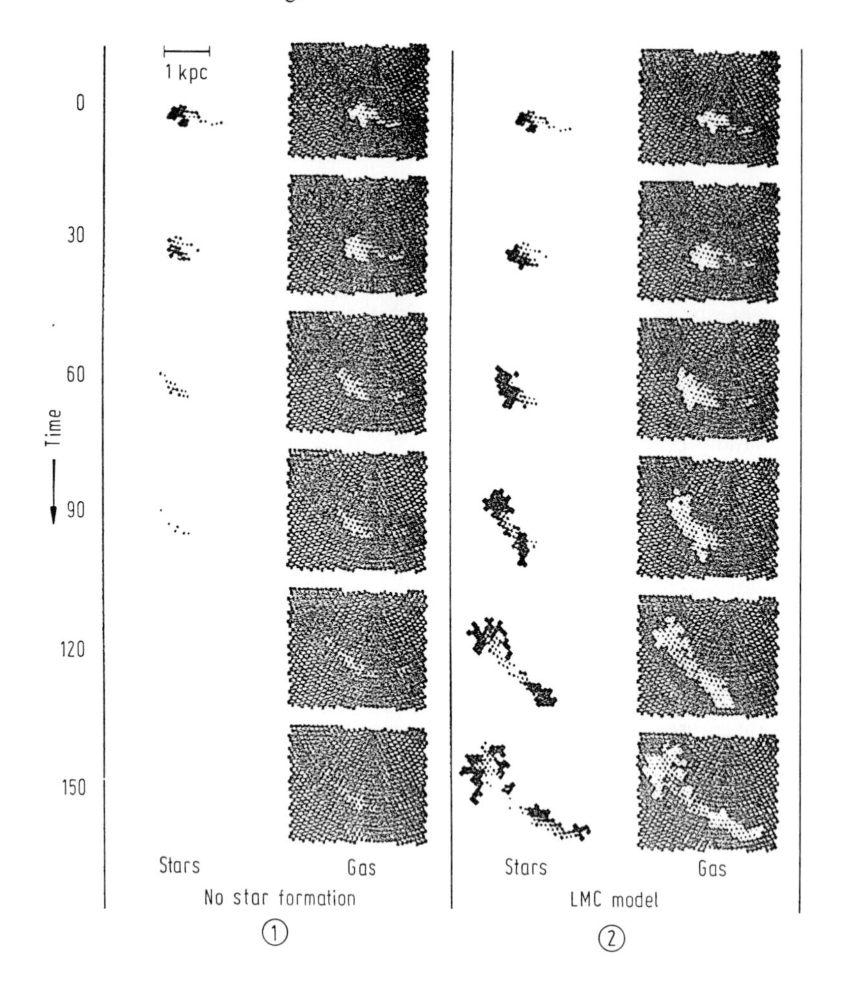

ular cloud, rotational speed, matter flows, gas temperatures, and dust content. At a probability $P = 1$, the fire will spread to all other trees (cells), while for $P = 0$, the fire will not spread at all. The crucial point in percolation theory is that the transition from $P = 1$ to $P = 0$ is nonlinear. This nonlinearity (cf. Sect. 28.4.2.2) encompasses the feedback between star formation, development of structure, and the state of the interstellar medium as the carrier of these processes.

The pattern of cells into which the star formation spreads is now sheared by differential rotation. The final result is a spiral arm. See Fig. 28.32.

The spiral arm may appear to possess a filamentary and disjointed structure, and may have a short coherence length. Or, it can form continuously with large coherence lengths and present a large-scale pattern. Coherence lengths are the regions of contiguous connections, and are determined by the dynamics and thermodynamics of the interstellar medium.

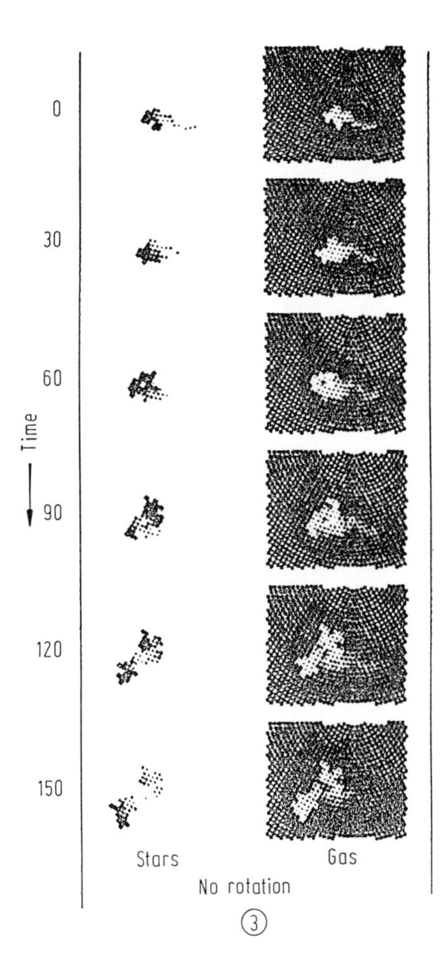

Fig. 28.33. Interaction of differential rotation and star formation, graphing the evolution of the gaseous disk (*dark*) and of a star-forming region (*dotted*) with time in units of 10^6 years. Where stars have formed the gas in the disk has been removed. When star formation is turned off, the region dies out and the disk cells refill (*left-hand sequence*). Without differential rotation, star formation virtually stagnates in place as no fresh gas cells are moved past the active cells (*right-hand sequence*). Only the presence of both effects (*center sequence*) can generate a spiral-arm filament.

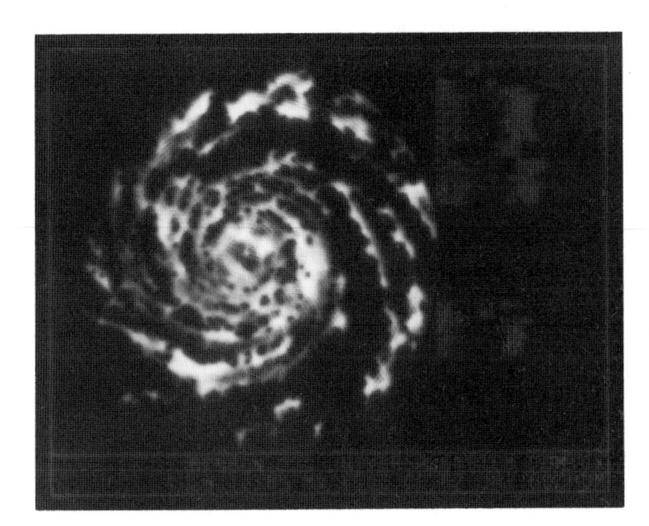

Fig. 28.34. A computer-generated, artificial galaxy with diameter 20 kpc, 7400 cells, and rotation speed 250 km s^{-1}. The slightly unsharp reproduction suppresses cell "graininess."

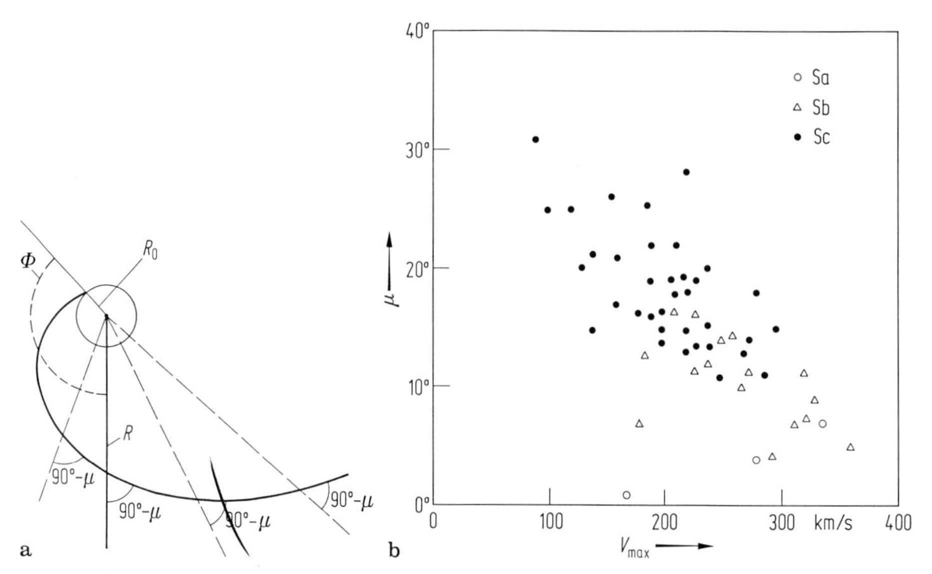

Fig. 28.35. a A logarithmic spiral with winding angle Φ and slope angle μ. **b** Relation between slope angle and maximum rotational speed. After Kennicutt [28.22].

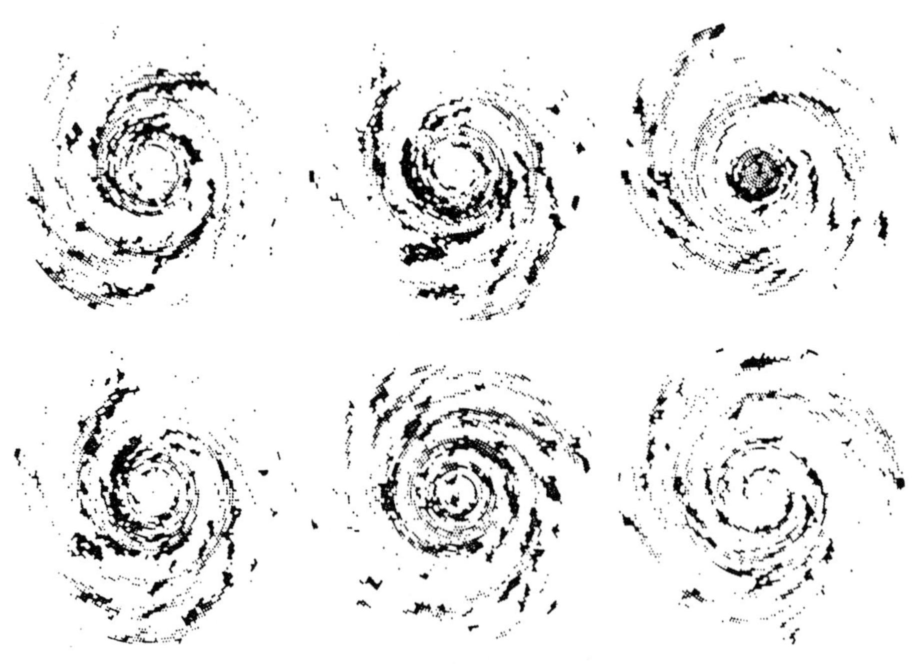

Fig. 28.36. Evolutionary sequence of a galaxy model at time steps 152, 154, 155, 161, 175, and 231 ($\times 10^7$ yr). Star-forming regions are graphed by dots whose sizes decrease with age. The smallest dots represent a maximum age of 10^8 yr.

The interaction of differential rotation and propagating star formation is illustrated in Fig. 28.33. The formation of the spiral-arm filament is graphed for three cases: star formation with and without differential rotation, and differential rotation without star formation. Star-forming regions create a hole in the gaseous disk (each right-side column, black) as the gas becomes transformed into stars. When star formation dies out, the gas disk is gradually replenished, and the cells thus regenerate. The ages of stars are represented by the sizes of dots graphed: the smaller the point, the older the stars represented. The time steps are marked in units of 10^6 years. In a nonrotating system (right column), no fresh star-forming cells move past active regions; the star formation hardly moves on. Only if rotation is added (central column) does a spiral filament begin to form.

In order to compare such computer-generated, artificial galaxies in appearance with real galaxies, Fig. 28.34 shows an unsharp image of such an artificially generated galaxy (type Sb). It is not difficult to find matching photographs in atlases of galaxies; for instance, if the image is tilted by about 25°, it looks like M81 = NGC 3031. Not only stellar morphology but also the distribution of the various gaseous components (neutral hydrogen, molecular clouds, and the hot coronal phase of the interstellar medium) can be generated in models and compared with observations.

This theory of spiral structure encounters difficulties in explaining the symmetry of spiral arms. The predominance of two-armed spirals can be achieved only awkwardly with a very random theory [28.20,21]. On the other hand, the differential rotation leads naturally to a spiral pattern, which can be anywhere between a logarithmic spiral (of constant slope) and a hyperbolic one (linear relation between slope angle and radius). Measurements by Kennicutt [28.22] of 113 spiral galaxies confirmed this morphological ambiguity. Figure 28.35 graphs the relation between slope angle μ, winding angle Φ, and the radius R in a logarithmic spiral.

Finally, Fig. 28.36 displays the time evolution of a spiral pattern. Though the substructures change, the large-scale shape is preserved over a long time.

28.4.2.2 Cosmic Cycles and Energy Equipartitions. The basic question raised by a classification of galaxies is what it signifies physically. The effects of two different processes, namely the formation as well as the evolution of galaxies, will here be linked. The basic correlations associated with the morphological classification will help clarify this connection.

The continuously decreasing ratio of central bulge diameter to disk diameter from elliptical to spiral systems is tied to the origin of galaxies. The elliptical central bulges of various extents had very early on undergone star formation, together with the final shaping of the galaxy; this process was connected with the turbulent state of matter in the early Universe as it cooled following the Big Bang. The basic structure of galaxies was already imprinted on them at the time of their birth.

The second primary correlation is the increasing influence of interstellar matter and of young stars upon the sequence of classification. For spirals, the decrease of maximum rotational speed along the sequence provides a further correlation. Here, also, certain initial conditions seem to have been impressed upon the galaxies since the time of their formation.

The difference between E and S systems is the result of certain dynamically different initial states which shaped their morphologies. Within the framework of these premisses, the birth of a galaxy can ensue, and this always means first a *chemical* evolution. The interstellar medium is continually enriched with heavy elements (atomic numbers > 2) as new generations of stars are born and die. The basic structure of the galaxy itself seems to remain unchanged on cosmic time scales where, however, as is the case for spiral structure, the shaping constituents continuously change and rejuvenate. The prerequisite for this cosmic cycle—interstellar matter, star birth, evolution, and death, return of matter into the interstellar medium—is the equilibrium of energy sources and energy reservoirs which govern the evolution of a galaxy.

Energy sources include the energies of differential rotation (E_{rot}), of cosmic radiation (E_{cos}), and of stellar radiation (E_{str}). The interstellar medium harbors three kinds of energy: (1) the thermal energy of gas particles (E_{thr}), described by the kinetic temperature T, (2) the turbulence energy (E_{tur}) of regions moving relative to one another, and ultimately (3) the magnetic energy (E_{mag}) coupled with the hot gas. The sources for the radiation energy are gravitation, which, to begin with, initiates the collapse of a gas cloud to form stars, and nuclear fusion in stellar interiors. The radiation emitted into intergalactic space constitutes the energy loss of a galaxy.

Typical energy densities E of the various mechanisms are

$$\begin{aligned} E_{str} &= 7 \times 10^{-13} \text{ erg cm}^{-3}, \\ E_{cos} &= 10 \times 10^{-13} \text{ erg cm}^{-3}, \\ E_{rot} &= 7 \times 10^{-13} \text{ erg cm}^{-3}. \end{aligned}$$

The energy densities in the interstellar medium amount to

$$\begin{aligned} E_{thr} &= 4.5 \times 10^{-13} \text{ erg cm}^{-3}, \\ E_{tur} &= 3 \times 10^{-13} \text{ erg cm}^{-3}, \\ E_{mag} &= 3.5 \times 10^{-13} \text{ erg cm}^{-3}. \end{aligned}$$

The striking equipartition between these energies had already been pointed out by S. von Hoerner in 1974:

$$E_{thr} \approx E_{tur} \approx E_{mag} \approx E_1, \tag{28.7}$$

$$E_{str} \approx E_{cos} \approx E_{rot} \approx E_2, \tag{28.8}$$

$$E_1 \approx E_2. \tag{28.9}$$

These equipartitions seem to contain the secret of structure formation and of cyclic processes in galaxies. The interstellar medium is described as a system of energy reservoirs, couplings, and feeding and releasing mechanisms (Fig. 28.37). Stellar and cosmic radiations contribute, via absorption, to the increase in the thermal radiation. The differential galactic rotation feeds energy into the turbulence via the friction between different gas clouds. Turbulence and thermal energy are coupled by way of spatial differences of densities, velocities, and temperatures. Equipartition is reached through the conversion of turbulence energy, which appears as an increase in the mean thermal energy. On the other hand, local heating (e.g., through radiation or the formation of new stars) causes inhomogeneities in the temperature distribution and hence gradients in pressure and velocity, which in turn increase the turbulent energy.

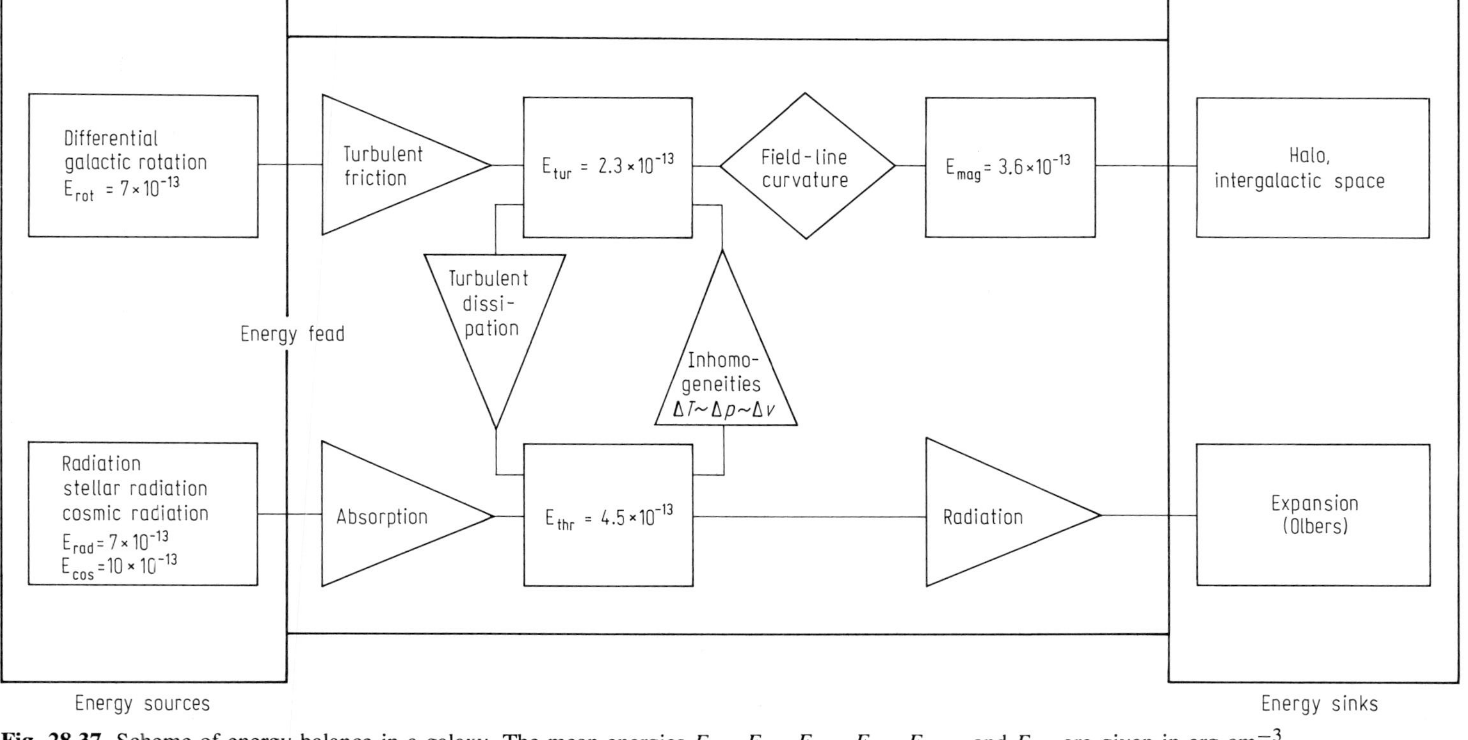

Fig. 28.37. Scheme of energy balance in a galaxy. The mean energies E_{rot}, E_{str}, E_{cos}, E_{tur}, E_{mag}, and E_{thr} are given in erg cm^{-3}.

Magnetic energy can be converted, through turbulence, via an extension of field lines. Conversely, turbulence is generated when field lines are curved so much at the local level that the magnetic pressure exceeds the gas pressure. Of course, energy also escapes from the system via radiation, and it is assumed that, owing to the expansion of the Universe, the photon density in intergalactic space does not increase substantially, because the photons are simultaneously reduced in energy through that expansion.

The equipartition of energy in galaxies (distorted and active galaxies excepted, of course) and thus the operation of long-term evolutionary and cyclic processes can be achieved in the following way: the time constants of the coupling mechanisms between energy reservoirs must be very much smaller than those in the energy feeding and releasing processes. Turbulent friction and imbalances in temperatures, pressures, and velocities will be smoothed out almost instantaneously, so that the energy reservoirs almost always maintain equal energy contents. If the energy contents of the reservoirs are all equal, there must also be equality in the contents of the energy *sources*. Per unit time, no more energy can flow into a system than flows out. Equipartition between energy sources is coupled via the origin and evolution of stars. It has been seen how stochastic star formation is coupled with differential rotation. However, rotational energy and the rate of star formation seem to be proportional to one another, and star formation provides the energy source of stellar radiation and, ultimately, via the deaths of massive stars, for cosmic radiation. Stars exploding as supernovae feed, through interstellar magnetic fields, the energy source of cosmic radiation. The two named equipartitions are the most difficult to understand and also the least investigated.

Star formation is the cause of galactic structure. The randomly progressing process of starbirth is a trigger for the formation of large-scale organization in the interstellar medium as manifested in the spiral arms (see also Chap. 24).

28.5 Determination of Distances

A basic prerequisite for an understanding of the structure of the Universe is the determination of distances to galaxies. A typical description of the state of the Universe could be derived solely from the knowledge of cosmic distance scales. In the determination of distances, primary and secondary methods are distinguished. The primary procedures come from studies of the Milky Way Galaxy, and have been calibrated on objects contained within it. Secondary methods employ the extragalactic systems themselves, and have been calibrated on nearby galaxies. The principle of determining distances is to observe a class of objects, either from the Milky Way or from other galaxies, whose brightness or geometrical size is known. This further assumes that the physics of these objects is the same everywhere in the Universe. A system of scale must always be constructed first, for instance by measuring the distances and diameters of globular clusters in the Milky Way. Such a scale can then be linked to the values measured for globular clusters in other galaxies.

The connection between angular extension α and the distance D of an object follows the simple relation

$$D = \frac{\overline{R}}{\tan\alpha}, \qquad (28.10)$$

where $\overline{R}$ is the calibrated diameter of the object. The connection with magnitudes is affected by the relation between the apparent magnitude m and the absolute magnitude M of the object,

$$\log D = 0.2\,(m - M) + 1, \qquad (28.11)$$

where absorption effects must be allowed for. At very large distances, a correction depending on the particular model of the Universe employed also enters.

Primary distance indicators can be any of the following:

- the brightest stars (supergiants);
- Cepheids, using their period–luminosity relation, or other kinds of variable stars;
- the brightest stars in globular clusters;
- integrated brightness or diameters of globular clusters;
- novae and supernovae;
- planetary nebulae;
- diameters of HII regions.

Counted as secondary distance indicators are:

- overall brightness of galaxy types (large range);
- brightest member of a cluster of galaxies;
- diameters of galaxies;
- luminosity classes of galaxies.

Recently, a correlation between absolute magnitude and maximum rotational velocity of a galaxy, first described by Brosche [28.23] and named after Tully and Fischer [28.15,24], has proven to be a valuable method of distance determination. The width Δv_0 of the 21-cm line of the total neutral hydrogen in a galaxy, as measured by radio observations, is proportional to the galaxy's absolute magnitude in blue light according to the relation

$$M_B = -8.2 \log \Delta v_0. \qquad (28.12)$$

This line width again is proportional to $2V_{\max}$, and thus depends on the galaxy type. Large radio telescopes permit the measurement of Δv_0 fairly easily and quickly. When combined with the knowledge of the apparent magnitude of the galaxy, a distance determination then becomes possible.

The various methods of distance determination have different reliabilities and limiting ranges. A combination of various methods will result in determining dependable distances, which then can be used to fix the value for the Hubble constant. In ranges over 20 Mpc, the distances can be bounded only to within a factor of two. Figure 28.38 illustrates the distance ladder.

All of the procedures for finding distances contribute to the determination of the constant H, named after E. Hubble, which connects the general recessional motion of

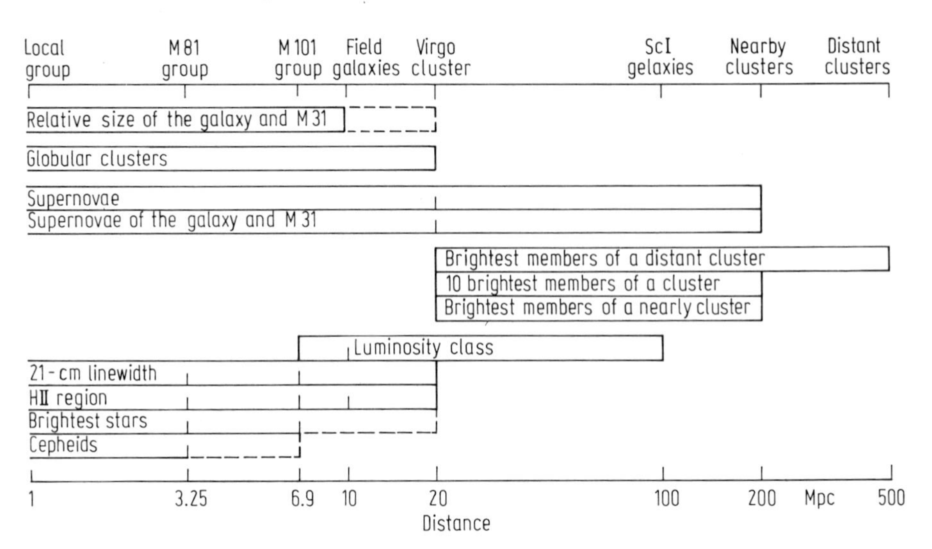

Fig. 28.38. The distance ladder and the determination of the Hubble constant H. After Tammann [28.43].

the galaxies (expansion of the Universe) with their distances. The larger the distance of a galaxy, the higher the velocity of recession. The measured redshift of spectral lines is interpreted as a Doppler effect due to a receding motion. This velocity v is proportional to the distance D via

$$v = HD. \tag{28.13}$$

The value of H lies in the range 50 to 100 km s^{-1} Mpc^{-1}. This substantial uncertainty is caused by a combination of various inaccuracies entering into the method of distance determination. A weighted mean H gives 65 km s^{-1} Mpc^{-1} (Rowan-Robinson [28.25]).

As the largest cosmical distances are measured via redshifts, that is, from the relative line shift $z = \Delta\lambda/\lambda$ from one or several spectral lines in the spectrum of an extragalactic object, it is this latter quantity which is often quoted in the literature instead of the distance; z is well defined, while the deduced distance depends on the chosen value of H. The relation between the recessional velocity and the distance follows from the Doppler equation in its simplest (nonrelativistic) approximation:

$$\begin{aligned} v &= cz, \\ v &= c\Delta\lambda/\lambda = HD, \\ z &= HD/c, \end{aligned} \tag{28.14}$$

where c is the speed of light in a vacuum and $z = \Delta\lambda/\lambda$ is the redshift.

28.6 Active Galaxies and Quasars

The class of *active galaxies* comprises objects of various types, all of them showing an energy output which cannot be explained by normal evolution and radiation of stars. This energy output is limited to a very small region at the centers of these objects. Among active galaxies are counted Seyfert galaxies, radio galaxies with active nuclei, BL Lac objects (named after the first-known object of its kind, which was originally classed as a variable star), LINER galaxies (LINER = low ionization nuclear emission-line region), and quasars (QSOs = quasi-stellar objects).

A galaxy falls into one of these various group if at least one, but preferably two, of the following criteria apply:

1. A very compact nuclear region, much brighter than the cores of galaxies with the same morphological type;
2. Emission lines in the nuclear region, indicating an origin different from the radiation mechanisms operating in normal stellar atmospheres;
3. Continuum radiation and/or emission-line radiation which is variable with time;
4. Continuum radiation from the nuclear region is nonthermal.

LINER systems show properties (2) and often (1) or (4), while BL Lac objects exhibit features (1), (3), and (4). Many Seyfert galaxies, radio galaxies, and quasars comply with all four criteria.

Two groups of objects, namely Seyfert galaxies and quasars, will be considered here in more detail. Seyfert systems account for 1% of all bright spirals. The very bright nuclei of these galaxies exhibit all four of the characteristics mentioned. Luminosities of their nuclear regions range from 10^{42} to 10^{45} erg s^{-1}. Two different classes are distinguished: Seyfert-1 systems show very broad, forbidden emission lines indicating speeds of around 5000 km s^{-1} in their nuclei. Seyfert-2 systems exhibit emission lines (both forbidden and permitted) with line widths ranging from 300 to 1000 km s^{-1}. The gas generating these emission lines lies at various distances from a central, unknown energy source and consequently displays various states of excitation and velocity. Seyfert galaxies are relatively nearby and therefore quite accessible to most observers. It is hoped that in Seyferts a first glance can be cast on what is the central powerhouse of galaxy activity.

By contrast with Seyferts, quasars are among the most distant objects in the observable Universe. With their energy output in the range 10^{45} to 10^{48} erg s^{-1}, quasars represent the most luminous active nuclei, corresponding to absolute magnitudes in the range $-31 < M_V < -24$. Quasars appear as starlike points on photographic plates, but under very good seeing and on very deep photographic exposures, a galaxy disk can be photometrically detected around some quasars. A quasar is the nucleus of an elliptical galaxy or of a spiral galaxy of extremely low apparent magnitude. The smallest redshifts in quasars are about $z = 0.1$, while the largest measured values of z are currently 4 and up. The brightest and nearest quasar is 3C 273 at the 1950 coordinates $(\alpha, \delta) = (12^h26^m33^s, +02°20')$. It has an apparent magnitude $m_V = 12.8$ and a redshift of $z = 0.158$.

The spectral features, such as emission lines, and the continuous energy distribution at optical, infrared, radio, X-ray, γ-ray wavelengths need to be explained by a uniform

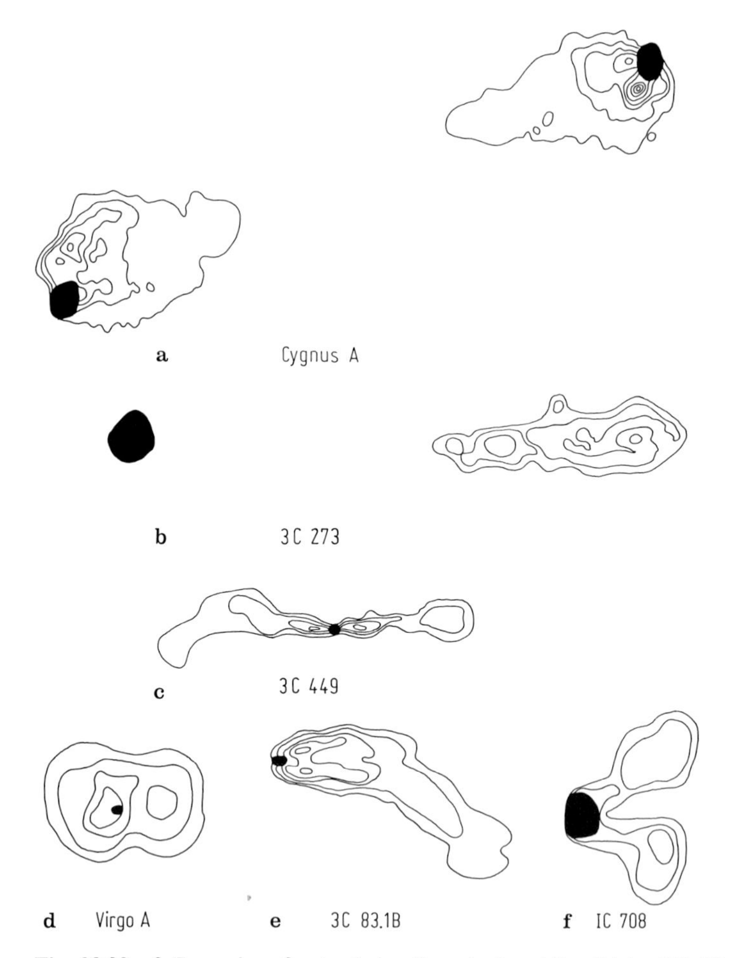

Fig. 28.39a–f. Examples of extended radio galaxies. After Fricke [28.44].

model for active galaxies. This model could then also provide information on the masses and the evolution of massive galactic nuclei. The nucleus might contain an extremely dense star cluster, or a rotating magnetic superstar with from 10^8 to 10^{10} $\mathfrak{M}_\odot$, or a black hole. The third possibility is the one most discussed in the literature. Matter flowing into the black hole liberates the energy required to power an active galaxy. In quasars, about 10 $\mathfrak{M}_\odot$ per year must be consumed, while for Seyfert galaxies, only about 0.1 $\mathfrak{M}_\odot$ per year is required to account for the radiated energy. The different observed types of active galactic nuclei can presumably be distinguished in terms of various distributions of matter around a black hole (accretion disk, hot and cool gas clouds, steep or flat density gradients) and the rate of mass influx.

Such active nuclei within their mother galaxies are also sources of radio-astronomically and partially optically identified *jets* and of the symmetrically structured, far-extended radio emission regions. Most of the extended radio sources are symmetric with an optically visible galaxy or a quasar (Fig. 28.39). The jets reach into the extended radio emission. The extension of these radio structures ranges from 100 kpc into the Mpc range. Optical jets are 3 kpc to 400 kpc long. The symmetry and the straight-line arrangement of the jet, measured from the innermost core to the outermost radio lobes, holds to within a few degrees. The jets are often found to be continuously curved, and the extended radio tails are displaced relative to one another. This suggests that the central source of activity rotates, and this motion is reflected in the ejecta. Taken together with estimates of the lifetime (about 10^5 years) of the radio-emitting regions, this fact suggests that active galactic nuclei eject matter and constantly resupply energy. Only in this fashion can these enormous cosmic structures be created and maintained. The radio-astronomically identified very compact structures in the range of sizes of active nuclei (a few pc) sometimes show expansion velocities with superluminal speeds of $5 < v/c < 15$. This, however, is a relativistic illusion, for which Feitzinger and Rohlfs [28.36] provide an illustrative description.

28.7 The Universe

The entire cosmos is hierarchically structured. Stars are ordered into stellar associations, clusters, and galaxies. Galaxies are ordered into groups and clusters, and even clusters of galaxies are apparently linked up with a still higher order. Field galaxies, i.e., single or isolated stellar systems, exist only in low numbers.

An agglomerate of galaxies must comply with three conditions in order to be ranked as a cluster:

1. There should be no connection between apparent brightness and redshift (distance). If such a relation does exist, then the group in reality consists of randomly aligned objects in space.
2. The number density per unit area in the group should, from a statistical standpoint, be clearly above random fluctuations of surrounding galaxy numbers.
3. The crossing time of the cluster diameter for the individual galaxies of the possible group must be less than 1/3 of the age of the Universe, so that a stable equipartition of kinetic energy within the group has had time to form. This energy distribution guarantees a good mixture in the group. The central region is often occupied by a large dominant galaxy.

The Milky Way is a member of the *Local Group*, which consists of about 24 members, including three large galaxies (the Milky Way, the Andromeda Nebula, and M33) and 21 dwarf galaxies. A detailed description of the Local Group is given by van den Bergh [28.27]. The Local Group is apparently a small appendage of a local *supercluster*, whose center lies in the Virgo cluster of galaxies. If the Local Group is taken as a representative sample, then the overwhelming majority of galaxies in the Universe must be dwarf objects, which at larger distances will be beyond the reach of observation. Catalogues of dwarf galaxies have been compiled by Feitzinger and Galinski [28.28] for the southern hemisphere, and by van den Bergh [28.29] for the northern hemisphere.

In clusters of galaxies, as in the case of groups, the compliance with the criteria given must also be carefully adhered to in order to identify true clusters. The best-known and nearest clusters of galaxies are the Coma and Virgo clusters. The Virgo cluster has about 3000 certain members, and lies at a distance of 24 Mpc; the Coma cluster has 11 000 members (estimated) and a distance of 140 Mpc. Dwarf galaxies are not included in the membership numbers.

Clusters of galaxies show diverse structures, with the central concentration being the primary distinguishing criterion. Those clusters with many members show a strong concentration of the brighter members toward the center. The richer and more compact a cluster of galaxies, the higher the fraction of elliptical systems it contains. On the other hand, the more loosely packed a cluster is assembled, the more spiral galaxies it contains. These distributions are presumably connected with internal interactions of galaxies within the cluster. During collisions between galaxies, interstellar matter is swept out of the systems, a frequent occurrence in compact clusters. But without interstellar matter, no structure can form in galaxies.

The intergalactic gas in these clusters was first identified by its X-ray emission. Since it has been found to have nearly the same chemical composition as stellar and interstellar matter, it must come from the galaxies themselves. Its temperature of 10^7 K to 10^8 K corresponds to the kinetic energy of cluster galaxies from whose energy supply in motion and radiation the energy is fed into the intergalactic medium. Average relative velocities of galaxies in clusters are in the range of 1000 to 2000 km s^{-1}.

The problem of the "hidden mass" appears in individual galaxies as well as in the clusters to which they belong. The mass as estimated from the state of motion of a cluster is considerably higher than that derived from the optical luminosity. The fact that the observed brightness underestimates the mass may be explained by noting that a substantial part of the mass is contained in the invisible dwarf galaxies, in the dark haloes of galaxies, and in intergalactic matter.

Apparently, there are no spherical superclusters composed of well-defined clusters of galaxies. The latter appear rather to be arranged in *strings*. These strings form a kind of lattice with larger voids between its nodes. Superclusters thus are fluctuations on the vastest size scale in the distribution of cosmic matter.

Knowledge about the origin of the Universe is derived from the distribution of matter as accessible from the recessional motion of galaxies, from elemental abundances, and from the cosmic background radiation. The latter, the microwave background in the wavelength range 0.1 to 6 cm, arrives uniformly from all directions in the sky. It is interpreted as the remnant of blackbody radiation emitted during the initial phase of the Universe. This remnant of the radiation from the cosmic Big Bang flooded into the expanding Universe, and it now has the properties which would be expected after about 1.5×10^{10} years. Next to the recessional motion of galaxies, it is the most convincing evidence for an evolving Universe which started with a Big Bang.

The microwave background arrives isotropically from all directions, and galaxies recede similarly isotropically into all directions with their velocities increasing with the distance. This symmetry with respect to the terrestrial point of observation does not mean that our Milky Way is at the center of the Universe. The cosmos offers the same view from every point. Writing the Hubble relation as a vector equation, it takes on the form

$$\mathbf{v} = H\mathbf{D}. \tag{28.15}$$

Here the origin of the coordinates is the Milky Way Galaxy. Another galaxy may have, relative to the Milky Way, a position $\mathbf{D}_1$ and a velocity $\mathbf{v}_1$:

$$\mathbf{v}_1 = H\mathbf{D}_1. \tag{28.16}$$

With the aid of a coordinate transformation into the system of the other galaxy, a Hubble relation again results:

$$\mathbf{v} - \mathbf{v}_1 = H(\mathbf{D} - \mathbf{D}_1), \tag{28.17}$$

$$\Delta\mathbf{v} = H\Delta\mathbf{D}. \tag{28.18}$$

The fact expressed by this transformation, that is, by the spatial uniformity of the Universe, is called the *cosmological principle.*

Cosmologies are scientific theories regarding the large-scale structure of the Universe, and purport to explain the structure of space, the origin, expansion, and age of the Universe, and the origin and evolution of the objects within it. The books by R. Kippenhahn [28.30], H. Fritsch [28.31], and S. Weinberg [28.32] treat this question at a generally accessible level.

The age of the Universe and its current expansion (as measured from the velocities of galaxies) are interrelated. Since the expansion must have begun at the singularity of the "Big Bang," the expansion formula also contains implicitly the age of the Universe, that is, of its present state. The simplest assumption is a constant recessional velocity of galaxies. The age T of the Universe is then immediately found from the reciprocal Hubble constant:

$$T = 1/H = 1/50 \text{ Mpc s km}^{-1} = 6 \times 10^{17} \text{ s} = 2 \times 10^{10} \text{ yr}. \tag{28.19}$$

The theoretical boundary of the observable Universe is reached when the recessional velocity becomes equal to the speed of light c. For $v = c$, the radius of the Universe comes out to be $R = 6000$ Mpc.

The enormous fascination of present theories of cosmologies lies in the fact that an understanding of the origin of the Universe is also an aid in determining the physics of atoms, protons, neutrons, and electrons. The physical descriptions of elementary particles, which form the just-named classical constituents of atoms, trace their roots back to the first seconds of the birth of the Universe.

The birth and evolution of objects in the Universe is tied to its origin. Galaxies were created during a specific phase of expansion by the contraction of denser regions of pre-galactic gas clouds. The degree of turbulence in these clouds seems to have determined the type of galaxy which formed. The distribution of angular momentum resulting from the turbulence is the basic parameter for the later dynamical evolution of the galaxies, in particular with respect to interstellar matter.

28.8 Amateur Techniques and Projects

It was stated in the first edition of this *Handbook* that where extragalactic astronomy begins the possibilities for amateurs end. Certainly this is still true now when dealing with visual observations of galaxies at the telescope. On the other hand, modern high-speed photographic films, as well as techniques for processing before and after exposure, have opened entirely new avenues for photography of galaxies. The spectral ranges from the UV to gamma rays and those from the near-infrared to radio wavelengths (see, however, Chap. 9 in Vol. 1) are largely inaccessible to the amateur, but active working groups whose personal computers have sufficient storage capacity and printing equipment should not be discouraged, as they can obtain from professional scientists data on the inaccessible wavelength ranges. Thus, radio charts of nearby galaxies scaled to the optical photographs will allow the observer to work on many interesting astrophysical problems. Astronomical databases may also be of help in data acquisition, but here it will be necessary to provide evidence of serious intent.

Making photographic surveys of supernovae, novae, and variables in galaxies in the Local Group is a virtually unending project. Here, the first virtue of the observer is perseverance and continuity. The same galaxies, photographic materials, and exposure and developing times should always be used in order to achieve success, which, however, may be realized only after years of patient work. Working groups of amateur associations can provide much and helpful instruction in this regard. This area of work will certainly expand in the years to come as optoelectronics becomes less expensive and more widely accessible.

A photographic study of galaxies may approach various astrophysical concepts; such an investigation may address the following possibilities:

1. Testing various photographic materials for speed and color, and, in combination with the telescope, for resolution and color;
2. Taking photographs centered on different wavelengths through suitable instrument/filter/film combinations (e.g., red and blue), which will qualitatively show the different star populations in galaxies;
3. Employing Zwicky's Sandwich Technique [28.33]: superposing a positive onto a negative of different wavelength range enhances color contrast and also allows one to qualitatively separate populations, thus elucidating the structure of a galaxy;
4. Using invariable photographic procedures, the apparent magnitude of a galaxy can be recorded as depending on seeing and on sky brightness;
5. Construct one's own catalogue of galaxies, including, if possible, all morphological types.

Two reports by members of the Working Group on Astrophotography of the German *Vereinigung der Sternfreunde* (Riepe et al. [28.34]) present the techniques for photography of galaxies in detail; other advice is given by B. Gordon in his book *Astrophotography* [28.35]. Neither very large instruments nor expensive equipment is required to achieve success in photographing galaxies; telescopes with focal lengths in the range 750 mm to 1500 mm can produce images of high quality. All photographs in this chapter were taken by amateurs using instruments which are small in comparison with professional telescopes. It is hoped that they will provide incentive for the reader to make his/her own attempts in this realm.

The primary requirements for extragalactic astrophotography are:

1. The night sky background should be as dark as possible, the atmospheric transparency quite good, and the air turbulence negligibly small. High contrast and rich detail are reached in adequate quality only at excellent sites and under favorable conditions.
2. For long-focus astrophotography, stable and precise tracking is required. Long exposures (one hour or more) demand accurate guiding. Very fine galaxy photographs will be obtained only with a mechanically stable instrument.
3. It is always necessary to use high-contrast, high-speed photographic material. Deep-cooled, hypersensitized, or spectroscopic emulsions satisfy these requirements. Film, telescope resolving power, and required exposure times may be considered as forming a unit. With the addition of certain filter combinations, optimal exposure values will have to be found through carefully arranged test sequences.
4. Precise focusing is always a problem in long-focus focal photography; the knife-edge method (see Sect. 4.4.4 in Vol. 1) is recommended for satisfying results. During the night the focus should be reexamined after each exposure, as temperature and mechanical changes of the optics or of telescope flexure may affect the sharpness of the image.

What instrument is most suited for extragalactic astrophotography? Actually, any observer, irrespective of what size or kind of telescope he/she has, can attempt it. The best results, however, are obtained by using a stably mounted, long-focus instrument, and by taking long exposures with careful guiding. Respectable pictures can be secured with focal lengths of only $f > 1$ m. As the light-gathering power increases with the square of the aperture A, the brightness of extended objects is thus given by squaring the aperture ratio (A/f). Galaxy photography is necessarily something for the right instrument and for the right observing crew—in order to finance the equipment jointly, as well as to share the various labors, problems, and joys. (See also Chap. 6, Vol. 1.)

References

28.1 Burnham, R.: *Celestial Handbook I–III*, Celestial Handbook Publ., Flagstaff 1977.

28.2 Nilson, P.: *Uppsala General Catalogue of Galaxies*, Uppsala Astron. Obs. Ann. **6** (1973).

28.3 Lauberts, A.: *The ESO/Uppsala Survey of the ESO (B) Atlas*, European Southern Observatory, Munich 1982.

28.4 de Vaucouleurs, G., de Vaucouleurs, A., Corwin, H.: *Second Reference Catalogue of Bright Galaxies*, University of Texas Press, Austin 1976.

28.5 Sandage, A.: *The Hubble Atlas of Galaxies*, Carnegie Institution of Washington, Washington 1961.

28.6 Sandage, A., Tammann, G.A.: *A Revised Shapley–Ames Catalogue of Bright Galaxies*, Carnegie Institution of Washington, Washington 1981.

28.7 Arp, H.: *Atlas of Peculiar Galaxies*, California Institute of Technology, Pasadena 1966.

28.8 Arp, H., Madore, B.F.: *A Catalogue of Southern Peculiar Galaxies and Associations, Vols. I, II*, Cambridge University Press, Cambridge 1987.

28.9 Sandage, A.: In *Stars and Stellar Systems, Vol. IX*, G.P. Kuiper (ed.), University of Chicago Press, Chicago 1975, p. 1.

28.10 Feitzinger, J.V.: Ring Galaxies. *Sterne und Weltraum* **17**, 287 (1978).
28.11 Rubin, V.C.: In *Internal Kinematics and Dynamics of Galaxies, IAU Symp. 100*, E. Athanassoula (ed.), Reidel Publ. Co., Dordrecht 1983, p. 3.
28.12 van den Bergh, S.: *Astrophys. J.* **131**, 215, 558 (1960).
28.13 Sandage, A., Tammann, G.A.: *Astrophys. J.* **194**, 569 (1974).
28.14 Kodaira, K., Watanabe, M., Okamura, S.: *Astrophys. J. Suppl.* **62**, 703 (1986).
28.15 Tully, R.B., Fischer, J.R.: *Astron. Astrophys.* **54**, 661 (1977).
28.16 Rubin, V.C.: *Scientific American* **6**, June 1983.
28.17 Scheffler, H., Elsässer, H.: *Bau und Physik der Galaxis*, B.F. Wissenschaftsverlag, Mannheim 1982.
28.18 Bowers, R., Deeming, T.: *Astrophysics I, II*, Iones and Bartlett Publ., Boston 1984.
28.19 Feitzinger, J.V., Schmidt-Kaler, T.: *Astron. Astrophys.* **88**, 41 (1980).
28.20 Seiden, P.E., Gerola, H.: *Fund. Cosm. Phys.* **7**, 241 (1982).
28.21 Seiden, P.E., Schulman, L.S., Feitzinger, J.V.: *Astrophys. J.* **253**, 91 (1982).
28.22 Kennicut, R.C.: *Astron. J.* **86**, 1847 (1981).
28.23 Brosche, P.: *Astron. Astrophys.* **23**, 259 (1973).
28.24 Roberts, M.S.: *Astron. J.* **83**, 1026 (1978).
28.25 Rowan-Robinson, M.: *The Cosmological Distance Ladder*, Freeman and Co., New York 1985.
28.26 Feitzinger, J.V., Rohlfs, K.: *Sterne und Weltraum* **80**, 133 (1980).
28.27 van den Bergh, S.: *J. Roy. Astron. Soc. Canada* **62**, No. 4, 1 (1968).
28.28 Feitzinger, J.V., Galinski, T.: *Astron. Astrophys. Suppl.* **61**, 503 (1985).
28.29 van den Bergh, S.: *Publ. David Dunlap Obs.*, Vol. II, No. 5 (1959).
28.30 Kippenhahn, R.: *One Hundred Billion Suns*, Basic Books Inc., New York 1983.
28.31 Fritsch, H.: *Vom Urknall zum Zerfall*, Piper Verlag, Munich 1983.
28.32 Weinberg, S.: *The First Three Minutes*, Macmillan, New York 1977.
28.33 Zwicky, F.: *Morphological Astronomy*, Springer, Berlin 1957.
28.34 Riepe, P., Flach-Wilken, B., Bosse, F., Baumann, W., Celnik, W.E.: *Sterne und Weltraum* **26**, pp. 34 and 155 (1987).
28.35 Gordon, B.: *Astrophotography*, Willman-Bell Inc., Richmond 1985.
28.36 van den Bergh, S.: In Landolt-Börnstein, *Astronomie und Astrophysik*, Vol. 1, Springer, Berlin 1965, p. 664.
28.37 Burbidge, E.M., Burbidge, G.: In *Stars and Stellar Systems Vol. IX*, G.P. Kuiper (ed.), University of Chicago Press, Chicago 1975.
28.38 Freeman, K.C.: *Galaxies*, Swiss Society of Astronomy and Astrophysics, Saas Fee 1976, p. 1.
28.39 Mitton, S.: *Exploring the Galaxies*, C. Scribner Sons, New York 1976.
28.40 Voigt, H.H.: *Abriß der Astronomie*, B.I. Wissenschaftsverlag, Mannheim 1980.
28.41 Voigt, H.H., Huchtmeier, W.K.: In Landolt-Börnstein, *Astronomy and Astrophysik*, Vol. 2c, Springer, Berlin 1982, p. 254.
28.42 Brosche, P.: *Astron. Astrophys.* **13**, 293 (1971).
28.43 von Hoerner, S.: *Vorlesungsmitschrift*, Bonn 1975.
28.44 Tammann, G.A.: Décalages vers le rouge et expansion de l'universe, Edition du CNRS, No. 263, Paris 1977, p. 43.
28.45 Fricke, H.J., Witzel, A.: In Landolt-Börnstein, *Astronomy and Astrophysik*, Vol. 2c, Springer, Berlin 1982, p. 315.

Appendix B: Astronomical Data

Greek Alphabet

A, α	Alpha	H, η	Eta	N, ν	Nu	T, τ	Tau
B, β	Beta	Θ, θ	Theta	Ξ, ξ	Xi	Y, υ	Upsilon
Γ, γ	Gamma	I, ι	Iota	O, o	Omicron	Φ, ϕ	Phi
Δ, δ	Delta	K, κ	Kappa	Π, π	Pi	X, χ	Chi
E, ε	Epsilon	Λ, λ	Lambda	$P, \rho(\varrho)$	Rho	Ψ, ψ	Psi
Z, ζ	Zeta	M, μ	Mu	Σ, σ	Sigma	Ω, ω	Omega

Table B.0. Abbreviations and symbols often used in astronomy, mathematics, and physics.

Å	ångstrom unit $= 10^{-10}$ m
α, RA	right ascension
AU (or AUD)	astronomical unit (distance) $= 1.495\,979 \times 10^{11}$ m
A	azimuth
a	semimajor axis of (= mean distance in) orbit
a	altitude (angular: above horizon)
b, B, β	latitude (in various spherical coordinate systems)
c	speed of light $= 2.997\,924\,58 \times 10^{8}$ m
CM	central meridian
Δ	distance from Earth (of a planet, comet)
δ	declination
ε	obliquity of ecliptic
e	eccentricity
e	base of natural logarithm $= 2.718\,2818$
E	eccentric anomaly
f, ν	frequency
f, F	focal length
G	(Newtonian) constant of gravitation
g	gravitational acceleration at Earth's surface
H, α	Hubble factor (km s^{-1} Mpc^{-1})
H	altitude (linear: above sea level)
h	hour angle
h	Planck's constant
i	inclination (of an orbital plane)
J	joule, SI unit of energy (replacing the erg)
JD	Julian date
Jy	jansky $= 10^{-26}$ J m^{-2} Hz^{-1} s^{-1}
K	kelvin (temperature scale)
k	Gaussian gravitational constant

Table B.0. (continued)

kpc	kiloparsec, 10^3 pc
L	luminosity (J s^{-1}, but usually given in solar units)
LY	light year = 9.46×10^{12} km
l, L, λ	longitude (in various spherical coordinate systems)
log	base 10 logarithm
ln	natural or base e logarithm
λ	wavelength
M	absolute magnitude
M	mean anomaly
Mpc	megaparsec, 10^6 pc
$\mathfrak{M}, m$	mass
m	apparent magnitude
m_{pv}	photovisual magnitude
m_{pg}	photographic magnitude
μ, pm	proper motion (arcsec/yr)
μ	mean motion (usually in degrees per day)
μm	micrometer, $= 10^{-6}$ m
nm	nanometer, $= 10^{-9}$ m
N	newton, SI unit of force
ω	angular distance of pericenter (perihelion) from node
ω	angular rotation rate; also angular frequency ($= 2\pi f$)
P	period (of revolution)
p, θ	position angle
Pa	pascal ($=$ N m^{-2}), SI unit of pressure
pc	parsec, $= 3.0857 \times 10^{13}$ km
π, p	parallax (arcsec)
π	$= 3.141\ 5927$, ratio circumference/diameter of circle
φ	geographic latitude
q	perihelion distance (in parabolic and hyperbolic orbits)
R, R_0	refraction constant
R, r	radius (in orbits: distance from the Sun)
ρ	(equatorial) radius of Earth
ρ, s	angular separation (of binary stars)
ρ	density
s,m,h,d,a	(superscripts) second, minute, hour, day, year
T	time of pericenter (perihelion) passage
T	absolute temperature
t	time (sometimes also hour angle)
τ	sidereal time
TDT	terrestrial dynamical time
UT	universal time
v	true anomaly
X, Y, Z	rectangular solar coordinates in equatorial coordinate system
x, y, z	heliocentric rectangular coordinates (of a planet, etc.); also galactic rectangular coordinates
ξ, η, ζ	geocentric rectangular coordinates
z	zenith distance

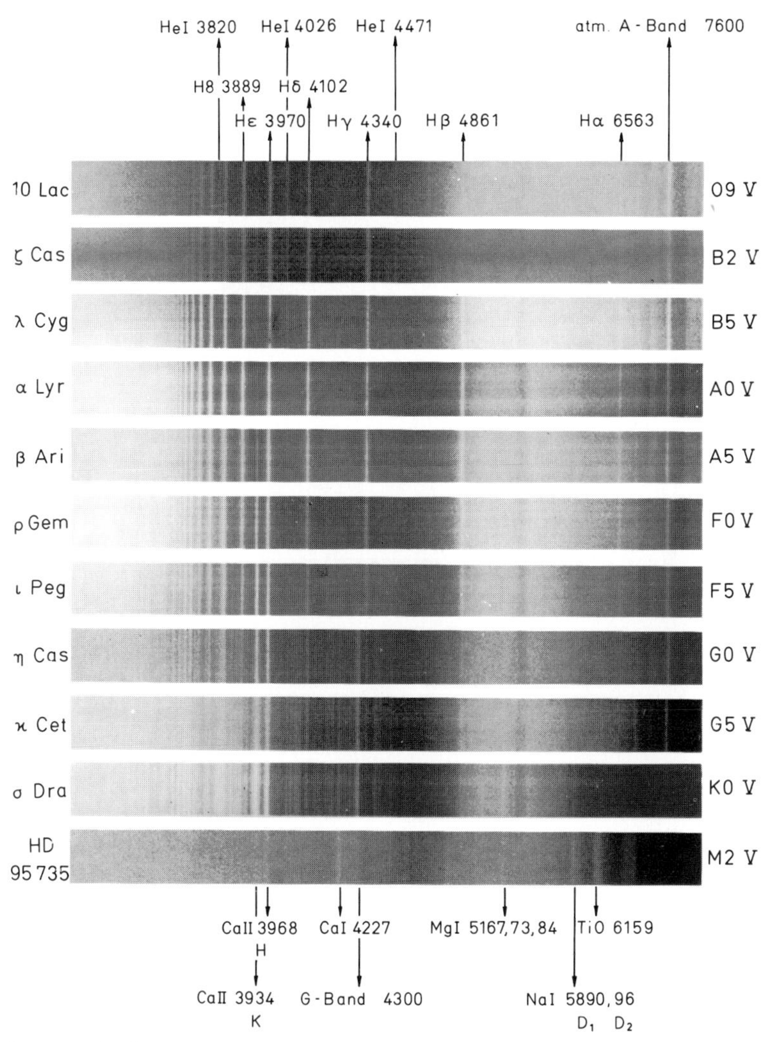

Fig. B.1. Main-sequence (V) spectral classes, represented by objective-prism spectra of typical stars (W. Seitter, Observatory Hoher List). Lower sensitivity of plates ("green gap") causes the apparently weakened continuum between 500 and 650 nm.

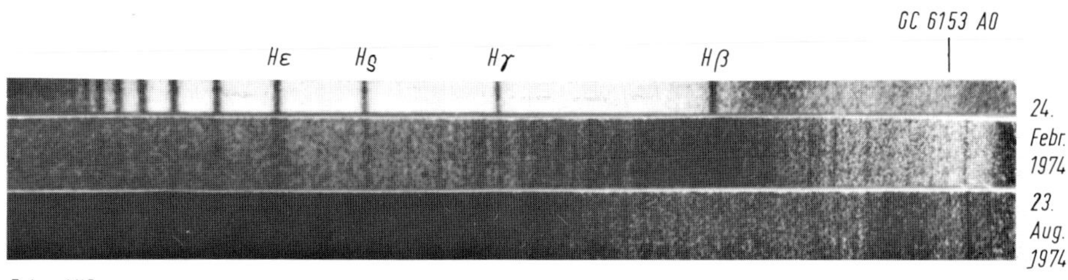

Fig. B.3. Spectra of ζ Aur (K4 Ib + B6 V), with comparison spectrum of GC6153 (A0) on top. Photo: C. Albrecht, 300 mm Newtonian with 60° crown prism spectrograph.

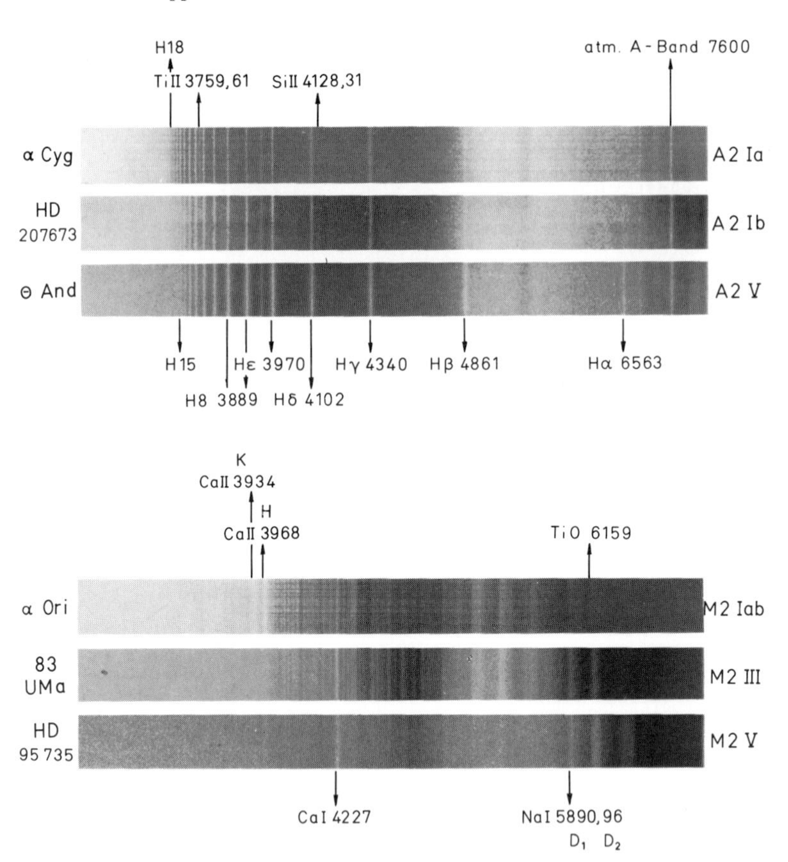

Fig. B.2. Objective-prism spectra of different luminosity classes for stars of identical Harvard types, A2 and M2 respectively (W. Seitter).

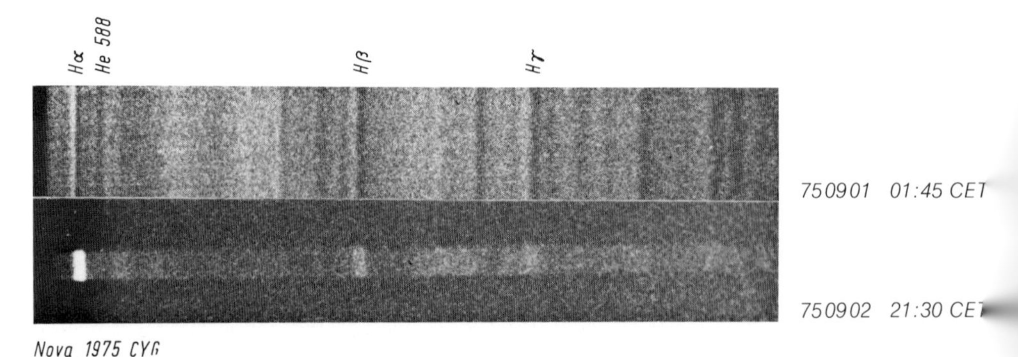

Fig. B.4. Spectra of Nova Cygni 1975. Photo: C. Albrecht, 110 mm Newtonian 1:4 and 30° crown prism, Kodak Recording Film.

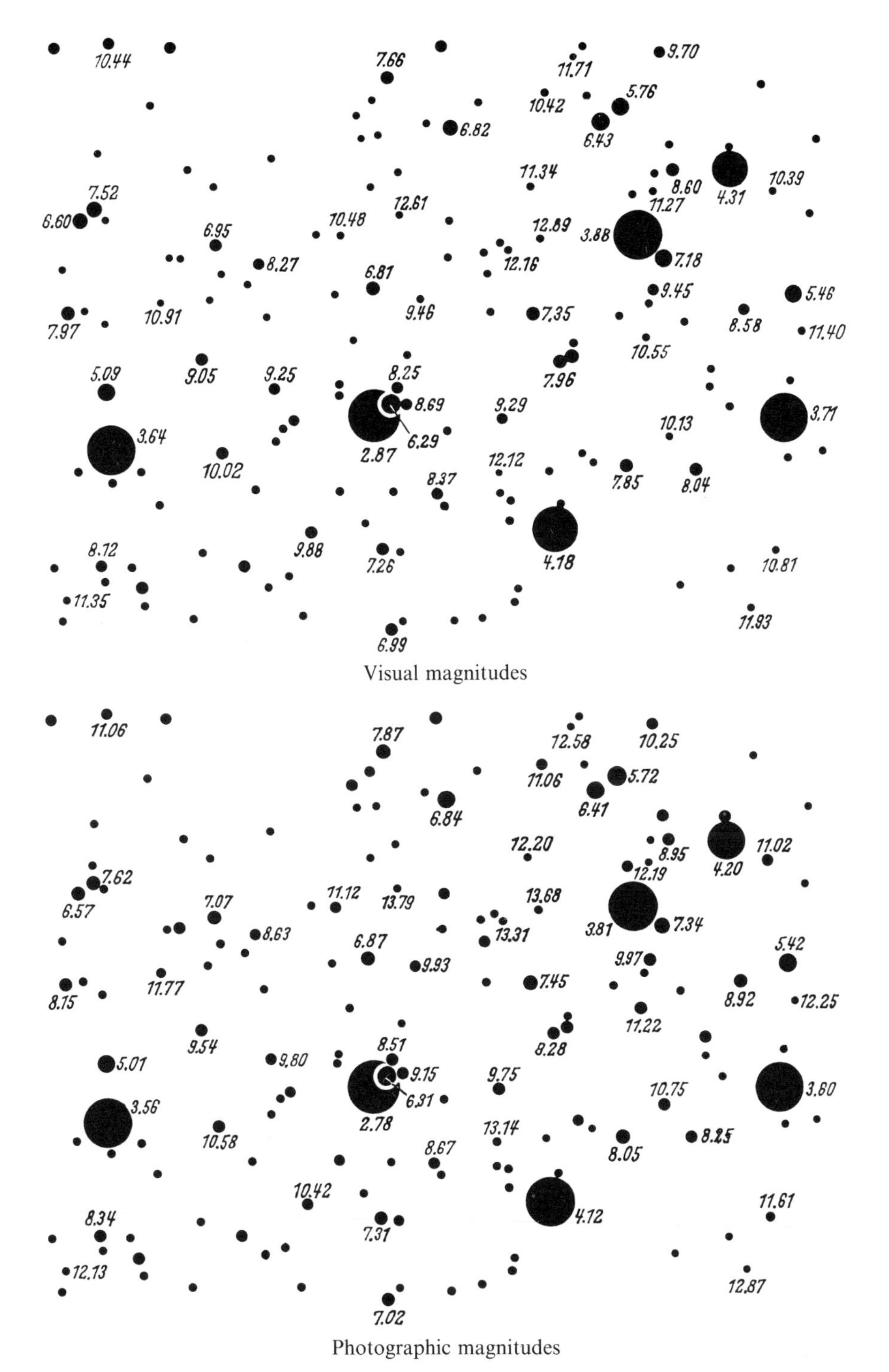

Fig. B.5. Magnitudes of selected objects at the center of the Pleiades. At the top are the visual magnitudes, at the bottom photographic magnitudes. The data, which are taken from H.L. Johnson and R.J. Mitchell: *Ap. J.* **128**, 31 (1958), are in the *UBV* system. This map will suffice for most amateur astronomers. For greater detail, reference should be made to the *Atlas der Kapteynschen Areas* (Selected Areas) by A. Brun and H. Vehrenberg (Treugesell-Verlag, Düsseldorf 1971).

Tables B.1–3. Atmospheric transmission.

z = zenith distance; λ = wavelength (here in μm).

Empirical zenith extinction k_r by Rayleigh scattering:

$$k_r = 0.00906\ \lambda^{-4}.$$

λ	0.4	0.55	0.7
k_r	0.367	0.099	0.037

Zenith extinction by haze:

$$k_h = 1.086\beta\lambda^{-\alpha}.$$

β is proportional to the density of scattering particles ($\beta = 0.10$ at normally good transparency) while the exponent α depends on the particle size d. Normally $\alpha = 1.3$ when $d \approx 1\ \mu$m for larger particles (dust) $\alpha = 0.5$, and for clouds $\alpha = 0$ with large β.

$$\text{Air mass}\quad M(z) = \sec(z - \Delta z)$$

with the approximation (holding for $z < 87°$)

$$\Delta z \approx 0.''05 z/(93° - z)$$

or, in arc,

$$\Delta z \approx 0.00087 z/(1.62 - z).$$

The transmission $T(z)$ then follows from

$$\log T(z) = -0.4(k_r + k_h)M(z).$$

Table B.1. Transmission $T(0)$ at the zenith. Photometry often uses only the "reduction to zenith" (Table B.3). The total transmission, if needed, is obtained by multiplication of the values from Tables B.1 and B.3.

Zenithal transmission β	Wavelength		
	650 nm (red)	550 nm (yellow-green)	400 nm (violet)
0.05 (exceptionally clear)	0.875	0.819	0.605
0.10 (normal, good)	0.802	0.735	0.513
0.20 (marginal transparency)	0.673	0.592	0.369

Table B.2. Approximate air masses $M(z)$ and reductions Δm_V to zenith in visual magnitudes.

z	$M(z)$	Δm_V
0°	1.000	0.00
10	1.015	0.00
20	1.064	0.01
30	1.154	0.03
40	1.304	0.06
50	1.553	0.12
60	1.995	0.23
70	2.904	0.45
75	3.816	0.65
80	5.60	0.99
85	10.40	1.77
87	15.36	2.61

For **Table B.3** see. p. 252

Table B.4. Normal refraction (for standard atmosphere at 1013 mb and 0°C).

z_{obs}	R	z_{obs}	R
0°	0′	70°	2′45″
10°	0′11″	75	3′42″
20°	0′22″	80	5′31″
30°	0′35″	85	10′15″
40°	0′51″	88	19′07″
50°	1′11″	89	25′36″
60°	1′45″	90	36′38″

Table B.3. Extinction tables (relative to zenith), computed with $\alpha = 1.3$.

Zenith Distance	Wavelength								
	0.065 μm (red)			0.55 μm (green)			0.40 μm (blue)		
	$\beta = 0.05$	0.10	0.20	$\beta = 0.05$	0.10	0.20	$\beta = 0.05$	0.10	0.20
0°	1.000	1.000	1.000	1.000	1.000	1.000	1.000	1.000	1.000
10°	0.998	0.997	0.994	0.997	0.995	0.992	0.992	0.990	0.985
20°	0.992	0.986	0.975	0.987	0.980	0.967	0.968	0.958	0.938
25°	0.986	0.977	0.960	0.980	0.969	0.947	0.950	0.934	0.903
30°	0.980	0.966	0.941	0.970	0.954	0.922	0.925	0.902	0.858
35°	0.971	0.953	0.916	0.957	0.934	0.891	0.895	0.864	0.803
40°	0.960	0.935	0.886	0.941	0.910	0.852	0.858	0.816	0.739
45°	0.946	0.913	0.849	0.921	0.880	0.805	0.812	0.759	0.663
50°	0.929	0.885	0.803	0.895	0.843	0.748	0.757	0.692	0.576
52°	0.920	0.872	0.782	0.883	0.826	0.721	0.732	0.661	0.539
54°	0.911	0.857	0.758	0.870	0.806	0.693	0.704	0.628	0.499
56°	0.901	0.841	0.733	0.855	0.785	0.662	0.674	0.593	0.458
58°	0.889	0.823	0.705	0.838	0.762	0.629	0.642	0.555	0.415
60°	0.875	0.803	0.674	0.820	0.736	0.592	0.607	0.515	0.371
62°	0.861	0.780	0.641	0.799	0.707	0.554	0.569	0.473	0.327
64°	0.844	0.755	0.604	0.775	0.675	0.511	0.527	0.427	0.281
66°	0.824	0.726	0.564	0.749	0.640	0.467	0.483	0.381	0.237
68°	0.802	0.694	0.519	0.718	0.600	0.419	0.436	0.332	0.192
70°	0.775	0.656	0.470	0.683	0.556	0.367	0.384	0.281	0.150
72°	0.744	0.614	0.417	0.643	0.506	0.313	0.330	0.229	0.111
74°	0.708	0.564	0.359	0.596	0.450	0.256	0.272	0.178	0.076
75°	0.686	0.536	0.328	0.569	0.419	0.227	0.243	0.153	0.060
76°	0.663	0.507	0.296	0.541	0.387	0.198	0.213	0.128	0.047
77°	0.637	0.474	0.263	0.510	0.353	0.170	0.184	0.105	0.035
78°	0.609	0.440	0.229	0.476	0.317	0.141	0.155	0.084	0.025
79°	0.577	0.402	0.195	0.439	0.280	0.114	0.126	0.064	0.016
80°	0.541	0.362	0.162	0.399	0.242	0.089	0.099	0.046	0.010
81°	0.500	0.318	0.128	0.355	0.202	0.065	0.074	0.032	0.006
82°	0.456	0.272	0.097	0.309	0.163	0.045	0.052	0.020	0.003
83°	0.405	0.224	0.068	0.258	0.127	0.028	0.033	0.011	0.001
84°	0.348	0.174	0.044	0.206	0.087	0.016	0.019	0.005	0.000
85°	0.285	0.125	0.024	0.153	0.055	0.007	0.009	0.002	0.000
86°	0.217	0.080	0.011	0.101	0.029	0.002	0.003	0.001	0.000
87°	0.147	0.042	0.003	0.057	0.012	0.001	0.001	0.000	0.000

Table B.5. Semi-diurnal arc (including normal refraction).

δ \ φ	+ 30°	+ 32°	+ 34°	+ 36°	+ 38°	+ 40°	+ 42°	+ 44°	+ 46°	+ 48°	+ 50°
°	h m	h m	h m	h m	h m	h m	h m	h m	h m	h m	h m
− 30	4 45.4	4 38.8	4 31.8	4 24.4	4 16.5	4 8.1	3 58.9	3 48.9	3 37.9	3 25.7	3 11.8
29	4 48.6	4 42.3	4 35.6	4 28.6	4 21.1	4 13.0	4 4.3	3 54.9	3 44.5	3 33.0	3 20.1
28	4 51.7	4 45.7	4 39.3	4 32.6	4 32.6	4 17.8	4 9.6	4 0.7	3 50.9	3 40.1	3 28.0
27	4 54.7	4 49.0	4 42.9	4 36.5	4 29.8	4 22.5	4 14.7	4 6.2	3 57.0	3 46.9	3 35.5
26	4 57.7	4 52.2	4 46.5	4 40.4	4 33.9	4 27.1	4 19.7	4 11.7	4 3.0	3 53.4	3 42.8
25	5 0.6	4 55.4	4 49.9	4 44.2	4 38.0	4 31.5	4 24.5	4 16.9	4 8.7	3 59.7	3 49.7
24	5 3.5	4 58.5	4 53.3	4 47.8	4 42.0	4 35.8	4 29.2	4 22.0	4 14.3	4 5.8	3 56.5
23	5 6.3	5 1.6	4 56.6	4 51.4	4 45.9	4 40.1	4 33.8	4 27.0	4 19.7	4 11.8	4 3.0
22	5 9.0	5 4.6	4 59.9	4 55.0	4 49.7	4 44.2	4 38.3	4 31.9	4 25.0	4 17.5	4 9.3
21	5 11.7	5 7.5	5 3.1	4 58.4	4 53.5	4 48.3	4 42.7	4 36.7	4 30.2	4 23.2	4 15.4
− 20	5 14.4	5 10.4	5 6.2	5 1.8	4 57.2	4 52.3	4 47.0	4 41.3	4 35.3	4 28.7	4 21.4
19	5 17.0	5 13.3	5 9.3	5 5.2	5 0.8	4 56.2	4 51.2	4 45.9	4 40.2	4 34.0	4 27.3
18	5 19.6	5 16.1	5 12.4	5 8.5	5 4.4	5 0.0	4 55.4	4.50.4	4 45.1	4 39.3	4 33.0
17	5 22.2	5 18.9	5 15.4	5 11.7	5 7.9	5 3.8	4 59.9	4 54.9	4 49.9	4.44.5	4.38.6
16	5 24.7	5 21.6	5 18.4	5 14.9	5 11.4	5 7.5	5 3.5	4 59.2	4 54.6	4 49.5	4 44.1
15	5 27.2	5 24.3	5 21.3	5 18.1	5 14.8	5 11.2	5 7.5	5 3.5	4 59.2	4 54.5	4 49.5
14	5 29.7	5 27.0	5 24.2	5 21.3	5 18.2	5 14.9	5 11.4	5 7.7	5 3.7	4 59.5	4 54.8
13	5 32.1	5 29.7	5 27.1	5 24.4	5 21.5	5 18.5	5 15.3	5 11.9	5 8.2	5 4.3	5 0.0
12	5 34.6	5 32.3	5 29.9	5 27.4	5 24.8	5 22.1	5 19.1	5 16.0	5 12.6	5 9.0	5 5.1
11	5 37.0	5 34.9	5 32.7	5 30.5	5 28.1	5 25.6	5 22.9	5 20.1	5 17.0	5 13.7	5 10.2
− 10	5 39.4	5 37.5	5 35.5	5 33.5	5 31.3	5 29.1	5 26.7	5 24.1	5 21.4	5 18.4	5 15.2
9	5 41.7	5 40.1	5 38.3	5 36.5	5 34.6	5 32.5	5 30.4	5 28.1	5 25.7	5 23.0	5 20.2
8	5 44.1	5 42.6	5 41.1	5 39.5	5 37.8	5 36.0	5 34.1	5 32.1	5 29.9	5 27.6	5 25.1
7	5 46.4	5.45.2	5 43.8	5 42.4	5 41.0	5 39.4	5 37.8	5 36.0	5 34.2	5 32.2	5 30.0
6	5 48.8	5 47.7	5 46.6	5 45.4	5 44.1	5 42.8	5 41.4	5 40.0	5 38.4	5 36.7	5 34.9
5	5 51.1	5 50.2	5 49.3	5 48.3	5 47.3	5 46.2	5 45.1	5 43.9	5 42.6	5 41.2	5 39.7
4	5 53.4	5 52.7	5 52.0	5 51.2	5 50.4	5 49.6	5 48.7	5 47.8	5 46.8	5 45.7	5 44.5
3	5 55.8	5 55.2	5 54.7	5 54.1	5 53.6	5 53.0	5 52.3	5 51.6	5 50.9	5 50.1	5 49.3
2	5 58.1	5 57.7	5 57.4	5 57.1	5 56.7	5 56.3	5 55.9	5 55.5	5 55.1	5 54.6	5 54.1
− 1	6 0.4	6 0.2	6 0.1	6 0.0	5 59.8	5 59.7	5 59.5	5 59.4	5 59.2	5 59.0	5 58.9
0	6 2.7	6 2.7	6 2.8	6 2.9	6 2.9	6 3.0	6 3.1	6 3.2	6 3.4	6 3.5	6 3.6
+ 1	6 5.0	6 5.2	6 5.5	6 5.8	6 6.1	6 6.4	6 6.7	6 7.1	6 7.5	6 7.9	6 8.4
2	6 7.3	6 7.7	6 8.2	6 8.7	6 9.2	6 9.8	6 10.3	6 11.0	6 11.6	6 12.4	6 13.2
3	6 9.6	6 10.3	6 10.9	6 11.6	6 12.3	6 13.1	6 14.0	6 14.8	6 16.8	6 16.8	6 18.0
4	6 11.9	6 12.8	6 13.6	6 14.5	6 15.5	6 16.5	6 17.6	6 18.7	6 20.0	6 21.3	6 22.8
5	6 14.3	6 15.3	6 14.6	6 17.5	6 18.6	6 19.9	6 21.2	6 22.6	6 24.2	6 25.8	6 27.6
6	6 16.6	6 17.8	6 19.1	6 20.4	6 21.8	6 23.3	6 24.9	6 26.6	6 28.4	6 30.4	6 32.5
7	6 19.0	6 20.4	6 21.8	6 23.4	6 25.0	6 26.7	6 28.6	6 30.5	6 32.6	6 34.9	6 37.4
8	6 21.3	6 22.9	6 24.4	6 26.4	6 28.2	6 30.2	6 32.3	6 34.5	6 36.9	6 39.5	6 42.3
9	6 23.7	6 25.5	6 27.4	6 29.4	6 31.4	6 33.7	6 36.0	6 38.5	6 41.2	6 44.1	6 47.3
10	6 26.1	6 28.1	6 30.2	6 32.4	6 34.7	6 37.2	6 39.8	6 42.5	6 45.6	6 48.8	6 52.3
+ 11	6 28.5	6 30.7	6 33.0	6 35.4	6 38.0	6 40.7	6 43.6	6 46.6	6 49.9	6 53.5	6 57.4
12	6 31.0	6 33.4	6 35.9	6 38.5	6 41.3	6 44.3	6 47.4	6 50.8	6 54.4	6 58.3	7 2.5
13	6 33.4	6 36.0	6 38.8	6 41.6	6 44.7	6 47.9	6 51.3	6 54.9	6 58.9	7 3.1	7 7.8
14	6 35.9	6 38.7	6 41.7	6 44.8	6 48.0	6 51.5	6 55.2	6 59.2	7 3.4	7 8.0	7 13.1
15	6 38.4	6 41.4	6 44.6	6 47.9	6 51.5	6 55.2	6 59.2	7 3.5	7 8.1	7 13.0	7 18.5
16	6 41.0	6 44.2	6 47.6	6 51.2	6 54.9	6 58.9	7 3.2	7 7.8	7 12.7	7 18.1	7 23.9
17	6 43.5	6 47.0	6 50.6	6 54.4	6 58.5	7 2.7	7 7.3	7 12.2	7 17.5	7 23.3	7 29.5
18	6 46.1	6 49.8	6 53.7	6 57.7	7 2.0	7 6.6	7 11.5	7 16.7	7 22.4	7 28.5	7 35.3
19	6 48.8	6 52.7	6 56.8	7 1.1	7 5.7	7 10.5	7 15.7	7 21.3	7 27.4	7 33.9	7 41.1
20	6 51.5	6 55.6	6 59.9	7 4.5	7 9.4	7 14.5	7 20.1	7 26.0	7 32.4	7 39.4	7 47.1
+ 21	6 54.2	6 58.6	7 3.1	7 8.0	7 13.1	7 18.6	7 24.5	7 30.8	7 37.6	7 45.1	7 53.3
22	6 56.9	7 1.6	7 6.4	7 11.5	7 17.0	7 22.8	7 29.0	7 35.7	7 42.9	7 50.9	7 59.6
23	6 59.8	7 4.6	7 9.7	7 15.1	7 20.9	7 27.0	7 33.6	7 40.7	7 48.4	7 56.8	8 6.1
24	7 2.6	7 7.7	7 13.1	7 18.8	7 24.9	7 31.3	7 38.3	7 45.8	7 54.0	8 2.9	8 12.9
25	7 5.6	7 10.9	7 16.6	7 22.6	7 29.0	7 35.8	7 43.1	7 51.1	7 59.8	8 9.3	8 19.9
26	7 8.5	7 14.2	7 20.1	7 26.4	7 33.2	7 40.4	7 48.1	7 56.5	8 5.7	8 15.8	8 27.1
27	7 11.6	7 17.5	7 23.8	7 30.4	7 37.5	7 45.0	7 53.2	8 2.1	8 11.8	8 22.6	8 34.7
28	7 14.7	7 20.9	7 27.5	7 34.4	7 41.9	7 49.9	7 58.5	8 7.9	8 18.2	8 29.7	8 42.6
29	7 17.9	7 24.4	7 31.3	7 38.6	7 46.4	7 54.8	8 3.9	8 13.9	8 24.8	8 37.1	8 51.0
+ 30	7 21.2	7 28.0	7 35.2	7 42.9	7 51.1	7 59.9	8 9.5	8 20.1	8 31.7	8 44.8	8 59.7

Table B.5. (Continued)

δ \ φ	+ 50°	+ 51°	+ 52°	+ 53°	+ 54°	+ 55°	+ 56°	+ 57°	+ 58°	+ 59°	+ 60°
°	h m	h m	h m	h m	h m	h m	h m	h m	h m	h m	h m
− 30	3 11.8	3 4.1	2 55.8	2 46.8	2 36.9	2 25.9	2 13.5	1 59.3	1 42.4	1 21.0	0 49.7
29	3 20.1	3 12.9	3 5.3	2 57.0	2 48.0	2 38.1	2 27.1	2 14.7	2 0.4	1 43.4	1 21.9
28	3 28.0	3 21.3	3 14.2	3 6.6	2 58.3	2 49.3	2 39.4	2 28.4	2 15.9	2 1.6	1 44.5
27	3 35.5	3 29.3	3 22.7	3 15.7	3 8.0	2 59.8	2 50.8	2 40.8	2 29.8	2 17.3	2 2.9
26	3 42.8	3 37.0	3 30.8	3 24.2	3 17.2	3 9.6	3 1.4	2 52.4	2 42.4	2 31.3	2 18.8
25	3 49.7	3 44.3	3 38.6	3 32.4	3 25.9	3 18.9	3 11.3	3 3.1	2 54.1	2 44.1	2 33.0
24	3 56.5	3 51.4	3 46.0	3 40.3	3 34.3	3 27.8	3 20.8	3 13.2	3 5.0	2 56.0	2 46.0
23	4 3.0	3 58.2	3 53.2	3 47.9	3 42.3	3 36.2	3 29.8	3 22.8	3 15.3	3 7.1	2 58.0
22	4 9.3	4 4.9	4 0.2	3 55.2	3 50.0	3 44.3	3 38.4	3 31.9	3 25.0	3 17.5	3 9.3
21	4 15.4	4 11.3	4 6.9	4 2.3	3 57.4	3 52.2	3 46.6	3 40.7	3 34.3	3 27.4	3 19.9
− 20	4 21.4	4 17.5	4 13.5	4 9.1	4 4.6	3 59.8	3 54.6	3 49.1	3 43.2	3 36.9	3 30.0
19	4 27.3	4 23.7	4 19.9	4 15.8	4 11.6	4 7.1	4 2.3	3 57.2	3 51.8	3 45.9	3 39.6
18	4 33.0	4 29.6	4 26.1	4 22.3	4 18.4	4 14.2	4 9.8	4 5.1	4 0.1	3 54.7	3 48.9
17	4 38.6	4 35.4	4 32.1	4 28.7	4 25.0	4 21.1	4 17.0	4 12.7	4 8.1	4 3.1	3 57.8
16	4 44.1	4 41.2	4 38.1	4 34.9	4 31.5	4 27.9	4 24.1	4 20.1	4 15.9	4 11.3	4 6.4
15	4 49.5	4 46.8	4 43.9	4 41.0	4 37.8	4 34.5	4 31.0	4 27.4	4 23.4	4 19.3	4 14.8
14	4 54.8	4 52.3	4 49.7	4 46.9	4 44.1	4 41.0	4 37.8	4 34.4	4 30.8	4 27.0	4 22.9
13	5 0.0	4 57.7	4 55.3	4 52.8	4 50.2	4 47.4	4 44.5	4 41.4	4 38.1	4 34.6	4 3.9
12	5 5.1	5 3.0	5 0.9	4 56.3	4 56.2	4 53.7	4 51.0	4 48.2	4 45.2	4 42.0	4.38.7
11	5 10.2	5 8.3	5 6.4	5 4.3	5 2.1	4 59.8	4 57.4	4 54.9	4 52.2	4 49.3	4 46.3
− 10	5 15.2	5 13.5	5 11.8	5 9.9	5 7.9	5 5.9	5 3.7	5 1.5	4 59.1	4 56.5	4 53.8
9	5 20.2	5 18.7	5 17.1	5 15.2	5 13.7	5 11.9	5 10.0	5 8.0	5 5.8	5 3.6	5 1.2
8	5 25.1	5 23.8	5 22.4	5 21.0	5 19.5	5 17.9	5 16.2	5 14.4	5 12.5	5 10.6	5 8.5
7	5 30.0	5 28.9	5 27.7	5 26.4	5 25.1	5 23.8	5 22.3	5 20.8	5 19.2	5 17.5	5 15.7
6	5 34.9	5 33.9	5 32.9	5 31.8	5 30.7	5 29.6	5 28.4	5 27.1	5 25.7	5 24.3	5 22.8
5	5 39.7	5 38.9	5 38.1	5 37.2	5 36.1	5 35.4	5 34.4	5 33.4	5 32.2	5 31.1	5 29.9
4	5 44.5	5 43.9	5 43.3	5 42.6	5 41.9	5 41.2	5 40.4	5 39,6	5 38.7	5 37.8	5 36.9
3	5 49.3	5 48.9	5 48.4	5 47.9	5 47.4	5 46.9	5 46.3	5 45.8	5 45.2	5 44.5	5 43.8
2	5 54.1	5 53.8	5 53.5	5 53.3	5 52.9	5 52.6	5 52.3	5 52.0	5 51.6	5 51.2	5 50.8
− 1	5 58.9	5 58.8	5.58.7	5.58.6	5 58.4	5 58.3	5.58.2	5.58.1	5 58.0	5.57.9	5 57.7
0	6 3.6	6 3.7	6 3.9	6 4.0	6 4.1	6 4.8	6 4.2	6 4.3	6 4.4	6 4.5	6 4.7
+ 1	6 8.4	6 8.6	6 8.9	6 9.2	6 9.5	6 9.8	6 10.1	6 10.4	6 10.8	6 11.2	6 11.6
2	6 13.2	6 13.6	6 14.0	6 14.5	6 15.0	6 15.5	6 16.0	6 16.6	6 17.2	6 17.8	6 18.5
3	6 18.0	6 18.6	6 19.2	6 19.8	6 20.5	6 21.2	6 22.0	6 22.8	6 23.6	6 24.6	6 25.5
4	6 22.8	6 23.5	6 24.4	6 25.2	6 26.1	6 27.0	6 28.0	6 29.0	6 30.1	6 31.3	6 32.5
5	6 27.6	6 28.6	6 29.6	6 30.6	6 31.7	6 32.8	6 34.0	6 35.3	6 36.6	6 38.1	6 39.6
6	6 32.5	6 33.6	6 34.8	6 36.0	6 37.3	6 38.7	6 40.1	6 41.6	6 43.2	6 44.9	6 46.7
7	6 37.4	6 38.7	6 40.0	6 41.5	6 43.0	6 44.6	6 46.2	6 48.0	6 49.8	6 51.8	6 53.9
8	6 42.3	6 43.8	6 45.3	6 47.0	6 48.7	6 50.5	6 52.4	6 54.4	6 56.5	6 58.8	7 1.2
9	6 47.3	6 48.9	6 50.7	6 54.5	6 54.5	6 56.5	6 58.7	7 0.9	7 3.3	7 5.9	7 8.6
10	6 52.3	6 54.1	6 56.1	6 58.2	7 0.3	7 2.6	7 7.5	7 7.5	7 10.2	7 13.1	7 16.2
+ 11	6 57.4	6 59.4	7 1.6	7 3.9	7 6.3	7 8.8	7 11.4	7 14.2	7 17.2	7 20.4	7 23.8
12	7 2.5	7 4.8	7 7.2	7 9.7	7 12.3	7 15.1	7 18.0	7 21.1	7 24.3	7 27.8	7 31.5
13	7 7.8	7 10.2	7 12.8	7 15.5	7 18.4	7 21.4	7 24.6	7 28.0	7 35.4	7 35.4	7 39.5
14	7 13.1	7 15.7	7 18.6	7 21.5	7 24.6	7 27.9	7 31.4	7 35.1	7 39.0	7 43.2	7 47.7
15	7 18.5	7 21.4	7 24.4	7 27.6	7 31.0	7 34.6	7 38.3	7 42.4	7 46.3	7 51.2	7 56.1
16	7 23.9	7 27.1	7 30.4	7 33.8	7 37.5	7 41.4	7 45.4	7 49.8	7 54.4	7 59.4	8 4.7
17	7 29.5	7 32.9	7 36.5	7 40.2	7 44.1	7 48.3	7 52.7	7 57.4	8 2.5	8 7.9	8 13.7
18	7 35.3	7 38.9	7 42.7	7 46.7	7 50.9	7 55.4	8 0.2	8 5.3	8 10.8	8 16.6	8 23.0
19	7 41.1	7 45.0	7 49.1	7 53.4	7 57.9	8 2.8	8 7.9	8 13.4	8 19.4	8 25.7	8 32.6
20	7 47.1	7 51.3	7 55.6	8 0.3	8 5.2	8 10.4	8 15.9	8 21.9	8 28.3	8 35.2	8 42.8
+ 21	7 53.3	7 57.7	8 2.4	8 7.3	8 12.6	8 18.2	8 24.2	8 30.7	8 37.6	8 45.2	8 53.5
22	7 59.6	8 4.3	8 9.4	8 14.7	8 20.3	8 26.4	8 32.8	8 39.8	8 47.4	8 55.7	9 4.8
23	8 6.1	8 11.2	8 16.6	8 22.3	8 28.3	8 34.9	8 41.9	8 49.5	8 57.7	9 6.8	9 16.9
24	8 12.9	8 18.3	8 24.0	8 30.2	8 36.7	8 43.8	8 51.4	8 59.6	9 8.7	9 18.8	9 30.0
25	8 19.9	8 25.7	8 31.8	8 38.4	8 45.5	8 53.1	9 1.4	9 10.5	9 20.5	9 31.7	9 44.4
26	8 27.1	8 33.4	8 40.0	8 47.0	8 54.7	9 3.0	9 12.1	9 22.1	9 33.2	9 45.9	10 0.6
27	8 34.7	8 41.4	8 48.5	8 56.1	9 4.4	9 13.5	9 23.5	9 34.6	9 47.3	10 1.9	10 19.5
28	8 42.6	8 49.8	8 57.5	9 5.8	9 14.8	9 24.8	9 35.9	9 48.5	10 3.1	10 20.5	10 42.9
29	8 51.0	8 58.7	9 7.0	9 16.1	9 26.0	9 37.1	9 49.6	10 4.1	10 21.5	10 43.7	11 18.1
30	8 59.7	9 8.1	9 17.2	9 27.1	9 38.2	9 50.7	10 5.1	10 22.3	10 44.4	11 18.5	– –

Table B.6. Astronomical constants.

Defining Constants		
Gaussian gravitational constant	k	$= 0.017\ 202\ 098\ 95$
Speed of light	c	$= 299\ 792\ 458$ m s^{-1}
Primary Constants		
Light travel time for unit distance	τ	$= 499.004\ 782$ s
Equatorial radius of Earth	$a_\oplus$	$= 6\ 378\ 140$ m
Dynamical form factor of Earth	J_2	$= 0.001\ 082\ 63$
Geocentric gravitational constant	$G_\oplus$	$= 3.986\ 005 \times 10^{14} m^3\ s^{-2}$
(Newtonian) gravitational constant	G	$= 6.672 \times 10^{-11} m^3\ s^{-2}\ kg^{-1}$
Mass ratio Moon/Earth	$\mathfrak{M}_{☾}/\mathfrak{M}_\oplus$	$= 0.012\ 300$
General precession in longitude per Julian century[a] (epoch zero)	μ	$= 5029''.0966$
Obliquity of ecliptic (2000.0)	ε	$= 23^\circ 26' 21''.4$
Selected Other Constants[b]		
Astronomical unit[c]	$A = c\tau$	$= 1.495\ 9787 \times 10^{11}$ m
Solar parallax	$= \arcsin(a_\oplus/A)$	$= 8''.794\ 148$
Mass ratio Sun/Earth	$\mathfrak{M}_\odot/\mathfrak{M}_\oplus$	$= 332\ 946.0$
Mass of Sun	$\mathfrak{M}_\odot$	$= 1.9891 \times 10^{30}$ kg
Parsec (pc)	$= 2.062\ 648 \times 10^5 A$	$= 3.086 \times 10^{13}$ km
Light year[d] (ly)	$= 6.3240 \times 10^4 A$	$= 0.3066$ pc
Tropical year	$= 365^{d}.242\ 190$	
Sidereal year	$= 365^{d}.256\ 363$	
Anomalistic year	$= 365^{d}.259\ 635$	
Tropical month	$= 27^{d}.321\ 582$	
Sidereal month	$= 27^{d}.321\ 662$	
Anomalistic month	$= 27^{d}.554\ 550$	
Synodic month	$= 29^{d}.530\ 589$	
Heliocentric gravitational constant ($G_\odot$)	$= A^3 k^2 d^{-2}$	$= 1.327\ 1244 \times 10^{20}\ m^3\ s^{-2}$

[a] The astronomical unit of time is the day (d) of 86 400 s, the mean unit is the mass of the Sun ($\mathfrak{M}_\odot$). (The Gaussian constant originally meant the mean motion of the Earth in radians per day, but the definition is now reversed in that the astronomical unit distance A expresses the distance at which a massless object would circle the Sun with mean motion given by k.)

[b] Following from, but not part of, the primary constants.

[c] The Julian century is an interval of 36 525 days.

[d] The light year is not used by professional astronomers.

Table B.7. Beginning and end times of twilight at 20-day intervals for different latitudes.

a Beginning and end of nautical twilight (center of Sun 12° below horizon).

Latitude		+20°		+30°		+40°		+45°		+50°		+55°	
Jan	0	5 43	18 23	5 59	18 07	6 17	17 49	6 27	17 39	6 39	17 28	6 53	17 14
	20	5 47	18 35	6 01	18 22	6 15	18 07	6 23	17 59	6 33	17 51	6 43	17 40
Feb	9	5 44	18 45	5 52	18 37	5 59	18 29	6 05	18 25	6 09	18 20	6 15	18 15
Mar	1	5 33	18 53	5 35	18 51	5 36	18 50	5 35	18 51	5 34	18 52	5 22	18 54
	21	5 16	18 59	5 12	19 04	5 04	19 12	4 58	19 17	4 52	19 24	4 42	19 34
Apr	10	4 58	19 05	4 47	19 17	4 30	19 34	4 18	19 46	4 04	20 00	3 45	19 25
	30	4 42	19 13	4 23	19 32	3 57	19 59	3 39	20 16	3 17	20 40	2 44	21 13
May	20	4 30	19 23	4 06	19 48	3 31	20 23	3 07	20 48	2 33	21 21	1 38	22 18
Jun	9	4 26	19 32	3 58	20 01	3 18	20 42	2 48	21 12	2 04	21 55	—	
	29	4 30	19 37	4 01	20 06	3 19	20 47	2 49	21 18	2 04	22 02	—	
Jul	19	4 38	19 35	4 12	20 00	3 36	20 36	3 10	21 04	2 33	21 38	1 30	22 43
Aug	8	4 47	19 24	4 26	19 44	3 58	20 12	3 39	20 31	3 13	20 57	2 37	21 34
	28	4 54	19 08	4 41	19 21	4 22	19 40	4 09	19 53	3 52	20 09	3 29	20 31
Sep	17	4 59	18 49	4 53	18 56	4 43	19 05	4 36	19 12	4 27	19 21	4 15	19 33
Oct	7	5 04	18 31	5 05	18 31	5 03	18 32	5 02	18 33	4 55	18 36	4 55	18 40
	27	5 11	18 17	5 17	18 11	5 23	18 04	5 27	18 01	5 29	17 58	5 32	17 55
Nov	16	5 19	18 10	5 32	17 58	5 44	17 45	5 51	17 33	5 59	17 24	6 08	17 12
Dec	6	5 30	18 11	5 46	17 56	6 03	17 39	6 13	17 29	6 24	17 18	6 37	17 04
	26	5 41	18 20	5 58	18 03	6 16	17 46	6 26	17 35	6 37	17 24	6 52	17 10

b Beginning and end of astronomical twilight (center of Sun 18° below horizon).

Latitude		+20°		+30°		+40°		+45°		+50°		+55°	
Jan	0	5 16	18 50	5 30	18 36	5 45	18 22	5 52	18 15	6 00	18 07	6 08	17 58
	20	5 21	19 01	5 31	18 51	5 41	18 39	5 48	18 34	5 54	18 28	6 00	18 23
Feb	9	5 17	19 11	5 24	19 04	5 29	19 00	5 31	18 58	5 32	18 57	5 33	18 57
Mar	1	5 07	19 18	5 07	19 19	5 04	19 22	5 01	19 25	4 57	19 29	4 51	19 33
	21	4 51	19 25	4 44	19 32	4 32	19 44	4 23	19 52	4 12	20 04	3 57	20 20
Apr	10	4 32	19 32	4 17	19 47	3 56	20 09	3 40	20 24	3 20	20 45	2 51	21 15
	30	4 14	19 41	3 52	20 03	3 20	20 37	2 55	21 01	2 22	21 36	1 33	22 45
May	20	4 01	19 52	3 33	20 22	2 48	21 06	2 14	21 01	1 15	22 42	—	
Jun	9	3 56	20 02	3 22	20 36	2 30	21 30	1 44	22 16	—		—	
	29	4 00	20 07	3 27	20 41	2 31	21 35	1 44	22 33	—		—	
Jul	19	4 09	20 04	3 38	20 34	2 51	21 21	2 14	21 58	1 02	23 06	—	
Aug	8	4 19	19 52	3 55	20 16	3 19	20 51	2 52	21 18	2 14	21 56	0 51	23 15
	28	4 28	19 34	4 11	19 51	3 47	20 15	3 29	21 18	3 06	20 56	2 32	21 48
Sep	17	4 34	19 15	4 25	19 24	4 10	19 38	4 00	19 48	3 46	20 01	3 27	20 20
Oct	7	4 39	19 00	4 37	19 03	4 33	19 10	4 27	19 15	4 21	19 23	4 20	19 33
	27	4 45	18 43	4 49	18 38	4 52	18 36	4 52	18 35	4 52	18 35	4 50	18 36
Nov	16	4 53	18 37	5 02	18 26	5 12	18 17	5 16	18 13	5 20	18 08	5 25	18 04
Dec	6	5 04	18 39	5 17	18 25	5 30	18 11	5 37	18 05	5 44	17 57	5 53	17 49
	26	5 14	18 47	5 28	18 33	5 43	18 18	5 50	18 11	5 58	18 02	6 07	17 54

Times given (in hours and minutes) apply for the longitude of the standard meridian of a time zone. For an observer located away from the standard meridian, add the longitude difference (+ west, − east) between the meridian of the observer and that to which the time zone refers. Southern hemisphere observers: Enter table with date ±183 days. Errors caused by the equation of time will be less than $\pm 5^{m}$.

Table B.8. Julian day numbers.

a Number of days elapsed since beginning of era on January 0, 12^h noon UT.

Year A.D.	1000	1100	1200	1300	1400	1500	1600	1700	1800	1900
	20	21	21	21	22	22	23	23	23	24
0	86 307	22 832	59 357	95 882	32 407	68 932	05 447	41 971[1]	78 495[1]	15 019[1]
4	87 768	24 293	60 818	97 343	33 868	70 393	06 908	43 432	79 956	16 480
8	89 229	25 754	62 279	98 804	35 369	71 854	08 369	44 893	81 417	17 941
12	90 690	27 215	63 740	00 265	36 790	73 315	09 830	46 354	82 878	19 402
16	92 151	28 676	65 201	01 726	38 251	74 776	11 291	47 815	84 339	20 863
20	93 612	30 137	66 662	03 187	39 712	76 237	12 752	49 276	85 800	22 324
24	95 073	31 598	68 123	04 648	41 173	77 698	14 213	50 737	87 261	23 785
28	96 534	33 059	69 584	06 109	42 634	79 159	15 764	52 198	88 722	25 246
32	97 995	34 520	71 045	07 570	44 095	80 620	17 135	53 659	90 183	26 707
36	99 456	35 981	72 506	09 031	45 556	82 081	18 596	55 120	91 644	28 168
40	00 917	37 442	73 967	10 492	47 017	83 542	20 057	56 105	93 105	29 629
44	02 378	38 903	75 428	11 953	48 578	85 103	21 518	58 042	94 566	31 090
48	03 839	40 364	76 889	13 414	49 939	86 464	22 979	59 503	96 027	32 551
52	05 300	41 825	78 350	14 875	51 400	87 925	24 440	60 964	97 488	34 012
56	06 761	43 286	79 811	16 336	52 861	89 386	25 901	62 425	98 949	35 473
60	08 222	44 747	81 272	17 797	54 322	90 847	27 362	63 886	00 410	36 934
64	09 683	46 208	82 733	19 258	55 783	92 308	28 823	65 347	01 871	38 395
68	11 144	47 669	84 194	20 719	57 244	93 769	30 284	66 808	03 332	39 856
72	12 605	49 130	85 655	22 180	58 705	95 230	31 745	68 269	04 793	41 317
76	14 066	50 591	87 116	23 641	60 166	96 691	33 206	69 730	06 254	42 778
80	15 527	52 052	88 577	25 102	61 627	98 152	34 667	71 191	07 715	44 239
84	16 988	53 513	90 038	26 563	63 088	99 603	36 128	72 652	09 176	45 700
88	18 449	54 974	91 499	28 024	64 549	01 064	37 589	74 113	10 637	47 161
92	19 910	56 435	92 960	29 485	66 010	02 525	39 050	75 574	12 098	48 622
96	21 371	57 896	94 421	30 946	67 471	03 986	40 511	77 035	13 559	50 053
100	22 832	59 357	95 882	32 407	68 932	05 447	41 971[1]	78 495[1]	15 019[1]	51 544[1]
	20	21	21	21	22	22	23	23	23	24

b Number of days elapsed on day 0, 12^h UT, of each month since beginning of the year in section **a**.

Year	Jan 0	Feb 0	Mar 0	Apr 0	May 0	Jun 0	Jul 0	Aug 0	Sep 0	Oct 0	Nov 0	Dec 0
0	0[1]	31[2]	60	91	121	152	182	213	244	274	305	335
1	366	397	425	456	486	517	547	578	609	693	670	700
2	731	762	790	821	851	882	912	943	974	1004	1035	1065
3	1096	1127	1155	1186	1216	1247	1277	1308	1339	1369	1400	1430

[1] Valid for January −1.

[2] Add 1 for the years 1700, 1800, and 1900.

The table follows the Julian Calendar until 1582 Oct 15, and the Gregorian Calendar thereafter. Decrease tabular numbers by 10 for Gregorian dates from 1582 Oct 15 to 1583 Dec 31.

Table B.9. Conversion of days into decimal fractions of the Julian Year.

Day	Jan	Feb	Mar	Apr	May	Jun	Jul	Aug	Sep	Oct	Nov	Dec
1	0.000	0.085	0.162	0.246	0.329	0.413	0.496	0.580	0.665	0.747	0.832	0.914
2	003	088	164	249	331	416	498	583	668	750	835	917
3	005	090	167	252	334	419	501	586	671	753	838	920
4	008	093	170	255	337	422	504	589	674	756	841	923
5	011	096	172	257	340	424	507	591	676	758	843	925
6	014	099	175	260	342	427	509	594	679	761	846	928
7	016	101	178	263	345	430	512	597	682	764	849	931
8	019	104	181	266	348	433	515	600	684	767	851	934
9	022	107	183	268	350	435	517	602	687	769	854	936
10	025	110	186	271	353	438	520	605	690	772	857	939
11	027	112	189	274	356	441	523	608	693	775	860	942
12	030	115	192	277	359	444	526	611	695	778	862	945
13	033	118	194	279	361	446	528	613	698	780	865	947
14	036	120	197	282	364	449	531	616	701	783	868	950
15	038	123	200	285	367	452	534	619	704	786	871	953
16	041	126	203	287	370	454	537	622	706	789	873	956
17	044	129	205	290	372	457	539	624	709	791	876	958
18	047	131	208	293	375	460	542	627	712	794	879	961
19	049	134	211	296	378	463	545	630	715	797	882	964
20	052	137	214	298	381	465	548	632	717	799	884	966
21	055	140	216	301	383	468	550	635	720	802	887	969
22	057	142	219	304	386	471	553	638	723	805	890	972
23	060	145	222	307	389	474	556	641	726	808	893	975
24	063	148	225	309	392	476	559	643	728	810	895	977
25	066	151	227	312	394	479	561	646	731	813	898	980
26	068	153	230	315	397	482	564	649	734	816	901	983
27	071	156	233	318	400	485	567	652	736	819	904	986
28	074	159	235	320	402	487	569	654	739	821	906	988
29	077		238	323	405	490	572	657	742	824	909	991
30	079		241	326	408	493	575	660	745	827	912	994
31	082		244		411		578	663		830		997

The tabulated fraction is for the beginning of the date in universal time (0^h UT). The error caused by the leap cycle can be minimized by applying the following correction:

- Subtract 0.001 for a leap year before leap day and for the preceding year (e.g., 1991 Jan to 1992 Feb).
- Add 0.001 for a leap year after leap day and for the following year (e.g., 1992 Mar to 1993 Dec).

Table B.10. Annual precession P_α in right ascension and P_δ in declination.

α \ δ	P_α +60°	+50°	+40°	+30°	+20°	+10°	0°	−10°	−20°	−30°	−40°	−50°	−60°	P_δ
h	s	s	s	s	s	s	s	s	s	s	s	s	s	″
0	3.07	3.07	3.07	3.07	3.07	3.07	3.07	3.07	3.07	3.07	3.07	3.07	3.07	+ 20.0
1	3.67	3.48	3.36	3.27	3.20	3.13	3.07	3.01	2.95	2.87	2.78	2.66	2.47	+ 19.4
2	4.23	3.87	3.63	3.46	3.32	3.19	3.07	2.95	2.83	2.69	2.51	2.28	1.92	+ 17.4
3	4.71	4.20	3.97	3.62	3.42	3.24	3.07	2.91	2.73	2.53	2.28	1.95	1.44	+ 14.2
4	5.08	4.45	4.04	3.74	3.49	3.28	3.07	2.87	2.65	2.41	2.10	1.69	1.07	+ 10.0
5	5.31	4.61	4.16	3.82	3.54	3.30	3.07	2.84	2.60	2.33	1.99	1.53	0.84	+ 5.2
6	5.39	4.67	4.19	3.84	3.56	3.31	3.07	2.84	2.59	2.30	1.95	1.48	0.76	0.0
7	5.31	4.61	4.16	3.82	3.54	3.30	3.07	2.84	2.60	2.33	1.99	1.53	0.84	− 5.2
8	5.08	4.45	4.04	3.74	3.49	3.28	3.07	2.87	2.65	2.41	2.10	1.69	1.07	− 10.0
9	4.71	4.20	3.87	3.62	3.42	3.24	3.07	2.91	2.73	2.53	2.28	1.95	1.44	− 14.2
10	4.23	3.87	3.63	3.46	3.32	3.19	3.07	2.95	2.83	2.69	2.51	2.28	1.92	− 17.4
11	3.67	3.48	3.36	3.27	3.20	3.13	3.07	3.01	2.95	2.87	2.78	2.66	2.47	− 19.4
12	3.07	3.07	3.07	3.07	3.07	3.07	3.07	3.07	3.07	3.07	3.07	3.07	3.07	− 20.0
13	2.47	2.66	2.78	3.87	2.95	3.01	3.07	3.13	3.20	3.37	3.36	3.48	3.67	− 19.4
14	1.92	2.28	2.51	2.69	2.83	2.95	3.07	3.19	3.32	3.46	3.63	3.87	4.23	− 17.4
15	1.44	1.95	2.28	2.53	2.73	2.91	3.07	3.24	3.42	3.62	3.87	4.20	4.71	− 14.2
16	1.07	1.69	2.10	2.41	2.65	2.87	3.07	3.28	3.49	3.74	4.04	4.45	5.08	− 10.0
17	0.84	1.53	1.99	2.33	2.60	2.84	3.07	3.30	3.54	3.82	4.16	4.61	5.31	− 5.2
18	0.76	1.48	1.95	2.30	2.59	2.84	3.07	3.31	3.56	3.84	4.19	4.67	5.39	0.0
19	0.84	1.53	1.99	2.33	2.60	2.84	3.07	3.30	3.54	3.82	4.16	4.61	5.31	+ 5.2
20	1.07	1.69	2.10	2.41	2.65	2.87	3.07	3.28	3.49	3.74	4.04	4.45	5.08	+ 10.0
21	1.44	1.95	2.28	2.53	2.73	2.91	3.07	3.24	3.42	3.62	3.87	4.20	4.71	+ 14.2
22	1.92	2.28	2.51	2.69	2.83	2.95	3.07	3.19	3.32	3.46	3.63	3.97	4.23	+ 17.4
23	2.47	2.66	2.78	2.87	2.95	3.01	3.07	3.13	3.20	3.27	3.36	3.48	3.67	+ 19.4
24	3.07	3.07	3.07	3.07	3.07	3.07	3.07	3.07	3.07	3.07	3.07	3.07	3.07	+ 20.0

Table B.11. Solar data.

Diameter in km	$1.392\ 530 \times 10^6$
Diameter in Earth diameter (equator)	109.164
Oblateness	0
Surface Area in Earth areas	1.1957×10^4
Volume in Earth volumes	$1.307\ 465 \times 10^6$
Mass in Earth masses	$3.329\ 50 \times 10^5$
Mass in kg	1.9891×10^{30}
Density in kg m^{-3}:	
Mean (over entire Sun)	1.410×10^3
Interior (center of Sun)	1.6×10^5
Surface (photosphere)	1.0×10^1
Central pressure in Pa	6×10^{14}
Rotation period in days:	
Sidereal (at solar equator)	25.03
As seen from Earth:	
Solar equator	26.8
At 75° heliographic latitude	31.8
Inclination of solar equator against the ecliptic	7.25°
Magnetic flux density in tesla:	
Polar field	10^{-4}
Sunspot	0.3
Photospheric abundances of the five most common chemical elements, by percentage mass:	
Hydrogen	78.4
Helium	19.8
Oxygen	0.863
Carbon	0.395
Iron	0.140
Temperature in K:	
Center	1.5×10^7
Surface (photosphere)	6050
Umbra of sunspot	4240
Solar radiation output:	
Total luminosity in W	3.83×10^{26}
Flux in W m^{-2}:	
At the solar surface	6.29×10^7
At the distance of Earth	1.368×10^3
Photometric quantities:	
Radiated power in cd	3.07×10^{27}
Luminance in cd m^{-2}:	
Average over solar disk	2.01×10^9
Middle of solar disk	2.53×10^9
Illuminance of light (in lx) in the perpendicular direction at the average distance between Earth and Sun	1.37×10^5

Table B.12. Position angle P of the solar axis and heliographic latitude B_0 of the apparent center of disk at 5-day intervals. (Variations due to the leap cycle do not exceed 0°.3 in P and 0°.1 in B_0.)

Day		P	B_0	Day		P	B_0	Day		P	B_0
Jan	1	+ 2.4°	− 3.0°	May	1	− 24.3°	− 4.2°	Aug	29	+ 20.1°	+ 7.1°
	6	− 0.1	3.6		6	23.4	3.7	Sep	3	21.5	7.2
	11	2.5	4.1		11	22.3	3.1		8	22.6	7.3
	16	4.9	4.7		16	21.0	2.6		13	23.6	7.2
	21	7.2	5.1		21	19.5	2.0		18	24.5	7.2
	26	9.4	5.6		26	17.9	1.4		23	25.2	7.0
	31	11.5	6.0		31	16.1	0.8		28	25.8	6.9
Feb	5	13.6	6.3	Jun	5	14.2	− 0.2	Oct	3	26.1	6.6
	10	15.5	6.6		10	12.2	+ 0.4		8	26.3	6.4
	15	17.2	6.8		15	10.3	1.0		13	26.3	6.0
	20	18.9	7.0		20	8.0	1.6		18	26.2	5.7
	25	20.3	7.2		25	5.7	2.2		23	25.8	5.2
Mar	2	21.7	7.2		30	3.5	2.7		28	25.3	4.8
	7	22.9	7.3	Jul	5	− 1.2	3.3	Nov	2	24.5	4.3
	12	23.9	7.2		10	+ 1.1	3.8		7	23.6	3.8
	17	24.6	7.1		15	3.3	4.3		12	22.4	3.2
	22	25.4	7.0		20	5.5	4.8		17	21.1	2.6
	27	25.9	6.8		25	7.7	5.2		22	19.6	2.0
Apr	1	26.2	6.6		30	9.7	5.6		27	17.9	1.4
	6	26.4	6.3	Aug	4	11.7	6.0	Dec	2	16.0	0.8
	11	26.3	5.9		9	13.7	6.3		7	14.0	+ 0.1
	16	26.1	5.5		14	15.5	6.6		12	11.9	− 0.5
	21	25.7	5.1		19	17.1	6.8		17	9.6	1.1
	26	− 25.1	− 4.7		24	+ 18.7	+ 7.0		22	7.3	1.8
									27	+ 4.9	− 2.4

Table B.13. Coordinates of lunar formations in the Mucke–Rükl system. ξ is positive toward east (astronautical, toward Mare Crisium), η positive toward north, ζ positive toward Earth. λ, β = selenographic longitudes and latitudes.

No.	Name	ξ	η	ζ	λ	β
1	Lohrmann A	−0.888	−0.013	+0.460	−62.5	−0.7
2	Damoiseau E	−0.847	−0.091	+0.524	−58.2	−5.2
3	Byrgius A	−0.816	−0.416	+0.401	−63.8	−24.6
4	Billy	−0.744	−0.239	+0.624	−50.0	−13.8
5	Aristarchus	−0.676	+0.402	+0.618	−47.6	+23.7
6	Mersenius C	−0.676	−0.338	+0.655	−45.9	−19.8
7	Gassendi α	−0.654	−0.316	+0.687	−43.6	−18.4
8	Kepler	−0.609	+0.141	+0.781	−38.0	+8.1
9	Encke B	−0.598	+0.041	+0.800	−36.8	+2.3
10	Bessarion	−0.585	+0.256	+0.770	−37.2	+14.8
11	Brayley	−0.561	+0.356	+0.747	−36.9	+20.9
12	Lansberg D	−0.508	−0.052	+0.860	−30.6	−3.0
13	Milichius	−0.495	+0.174	+0.851	−30.2	+10.0
14	Euclides	−0.488	−0.128	+0.863	−29.5	−7.4
15	Lansberg B	−0.470	−0.043	+0.882	−28.1	−2.5
16	Dunthorne	−0.454	−0.501	+0.737	−31.6	−30.1
17	Sharp A	−0.456	+0.738	+0.497	−42.5	+47.6
18	Agatharchides A	−0.437	−0.395	+0.808	−28.4	−23.3
19	Foucault	−0.409	+0.770	+0.491	−39.8	+50.4
20	Darney	−0.386	−0.252	+0.887	−23.5	−14.6
21	Kies A	−0.340	−0.474	+0.812	−22.7	−28.3
22	Pytheas	−0.329	+0.351	+0.877	−20.6	+20.5
23	Gambart A	−0.321	+0.017	+0.947	−18.7	+1.0
24	La Condamina A	−0.292	+0.813	+0.504	−30.1	+54.4
25	Maupertuis A	−0.265	+0.772	+0.578	−24.6	+50.5
26	Guericke C	−0.196	−0.200	+0.960	−11.5	−11.5
27	Birt	−0.137	−0.380	+0.915	−8.5	+22.3
28	Tycho C.P.[a]	−0.141	−0.685	+0.715	−11.2	−43.2
29	Alpetragius B	−0.115	−0.261	+0.958	−6.8	−15.1
30	Pico	−0.106	+0.717	+0.689	−8.7	+45.8
31	Archimedes A	−0.098	+0.470	+0.877	−6.4	+28.0
32	Mösting A	−0.090	−0.056	+0.994	−5.2	−3.2
33	Maginus H	−0.106	−0.793	+0.600	−10.0	−52.5
34	Bode	−0.042	+0.117	+0.992	−2.4	+6.7
35	Bode A	−0.020	+0.156	+0.988	−1.2	+9.0
36	Chladni	+0.020	+0.070	+0.997	+1.1	+4.0
37	Epigenis A	−0.003	+0.920	+0.392	+0.4	+66.9
38	Werner D	+0.051	−0.455	+0.889	+3.3	−27.1
39	Zach δ[b]	+0.055	−0.876	+0.479	+6.5	−61.2
40	Aratus	+0.072	+0.400	+0.914	+4.5	+23.5
41	Cassini C	+0.101	+0.665	+0.740	+7.7	+41.6
42	Pickering	+0.122	−0.050	+0.991	+7.0	−2.9
43	Airy A	+0.128	−0.293	+0.948	+7.7	−17.0
44	Egede A	+0.113	+0.782	+0.613	+10.4	+51.4
45	Hipparchus C	+0.142	−0.129	+0.981	+8.2	−7.4
46	Manilius ε[a]	+0.153	+0.250	+0.956	+9.0	+14.4
47	Gabulfeda F	+0.216	−0.280	+0.935	+13.0	−16.3
48	Eudoxus A	+0.239	+0.717	+0.655	+20.0	+45.8

Table B.13. (Continued)

No.	Name	ξ	η	ζ	λ	β
49	Manelaus	+0.264	+0.280	+0.923	+15.9	+16.2
50	Dionysius	+0.297	+0.049	+0.954	+17.2	+2.8
51	Nicolai A	+0.296	−0.675	+0.676	+23.7	−42.5
52	Dawes	+0.424	+0.296	+0.856	+26.3	+17.2
53	Posidonius A	+0.419	+0.525	+0.741	+29.4	+31.6
54	Polybius A	+0.432	−0.391	+0.813	+28.0	−23.0
55	Hercules G	+0.435	+0.724	+0.535	+39.0	+46.3
56	Janssen K	+0.466	+0.720	+0.514	+42.2	−46.1
57	Maury	+0.510	+0.603	+0.613	+39.7	+37.0
58	Censorinus	+0.540	−0.007	+0.842	+32.7	−0.4
59	Rosse	+0.545	−0.307	+0.780	+34.9	−17.9
60	Cepheus A	+0.547	+0.656	+0.520	+46.4	+40.9
61	Macrobius B	+0.611	+0.357	+0.707	+40.8	+20.9
62	Gutenberg A	+0.634	−0.157	+0.757	+39.9	−9.0
63	Tralles A	+0.650	+0.461	+0.604	+47.0	+27.4
64	Stevinus A	+0.667	−0.528	+0.526	+51.8	−31.9
65	Proclus	+0.702	+0.278	+0.656	+46.9	+16.1
66	Furnerius A	+0.716	−0.552	+0.427	+59.2	−33.5
67	Bellot	+0.728	−0.215	+0.651	+48.2	−12.4
68	Picard	+0.789	+0.251	+0.561	+54.5	−14.5
69	Firmicus	+0.887	+0.127	+0.444	+63.4	+7.2
70	Langrenus M	+0.903	−0.170	+0.395	+66.4	−9.8

[a] central mountain;
[b] bright spot at inner SE wall.

Table B.14. Technique and exposure data for lunar and planetary photography. Using a CCD on the planets has a fundamental advantage over conventional photography. The exposure time for each frame is 1/60 second, in photography on film 1 second or longer. With CCD the recording rate is 30 frames per second. CCD video cameras are hardly any bigger than a box of 35-mm film. The Canon Ci-20R has a detector 1/2 inch across. The sensitive rectangular area contains 380 000 pixels and offers 500 horizontal television lines of resolution. See also: Sky and Telescope Vol. 83, No. 2, February 1992 pp. 209–214.

	Moon	Planets
Exposure at Focal Point		
Focal length	from 100 cm	from 200 cm
Size of negative	from 10 mm	from 0.5 mm
Exposure	0.04-0.2s	0.2-1s
Guiding	no	no/yes
Emulsion ISO	25/15°-200/24°	25/15°-100/21°
Enlargement	10 to 25× of negative diameter	
Definition	very good	fair
Turbulence	noticeable	quite noticeable
Exposure Using Eyepiece Projection		
Focal length of telescope	from 50 cm	from 100 cm
Focal length of eyepiece	15–20 mm	20–25 mm
Size of negative	from 20 mm	from 2 mm
Exposure	0.2 to 2 s	1 to 10 s
Guiding	yes/no	yes
Emulsion ISO	50/18°-400/27°	25/15°-200/24°
Enlargement	5 to 10× of negative diameter	
Definition	very good	good
Turbulence	noticeable	quite noticeable

For **Table B.15** see pp. 266.

Table B.16. Temperature and atmospheric data for the major planets and the satellite Titan.

1. Surface temperature, gas pressure (reference value), and chemical composition of the atmospheres of the inner terrestrial planets, the Saturnian satellite Titan, and Pluto. Note that hPa = hecto Pascal (previously mb).

Planet	Surface Temperature		Atmospheric Pressure (hPa)	Chemical Composition % by Volume (without H_2O)	Mean Partial Pressure of Atmospheric H_2O
	°C	K			
Mercury	−185 ... + 425	90 ... 700	< 0.01	H, little He	
Venus	+455 ... + 525	730 ... 800	90 300	97 CO_2, 2 N_2, 0.025 He, 0.024 SO_2, trace Ar, O_2, Ne	(0, 1–0.4% by volume in clouds)
Earth	−65 ... + 60	210 ... 335	1013.3	78.084 N_2, 20.946 O_2, 0.931 Ar 0.031 CO_2, 0.0018 Ne, 0.0005 He 0.000 15 CH_4, Trace Kr, Xe, H_2, NO_2, CO, O_3	5hPa
Mars	−140 ... + 15	135 ... 290	6.1	96.13 CO_2, 1.74 N_2, 1.45 Ar, 0.11 O_2	0.0009 hPa
Titan	−180	94	1500	approx. 93 N_2, 6 N_2, 0.9 H_2 trace C_2H_6, C_3H_8, HCN	
Pluto	−235	40	0.15	CH_4 (Pluto at perihelion)	

2. Surface temperature and chemical composition of the atmospheres in the cloud decks of Jupiter and Saturn and in the reflecting layers of Uranus and Neptune.

Planet	Surface Temperature		Chemical Composition % by Volume
	°C	K	
Jupiter	−135 ... − 125	138 ... 150	89 H_2, 11 He, 0.2 CH_4, 0.02 NH_3, trace D (D = deuterium) C_2H_6, PH_3, GeH_4 (germanium)
Saturn	−190 ... − 180	138 ... 150	89 H_2, 11 He, 0.3 CH_4, 0.02 NH_3, trace C_2H_6
Uranus	−210	63	89 H_2, 11 He, CH_4, very low NH_3
Neptune	−220	55	89 H_2, 11 He, probably also CH_4

Table B.15. Planetary data.

Quantity	Mercury	Venus	Earth
Average distance from Sun (in 10^6 km)	57.909	108.209	149.598
Average distance from Sun (in AU)	0.387 10	0.723 33	1.000 00
Perihelion distance (in AU)	0.31	0.72	0.98
Aphelion distance (in AU)	0.47	0.73	1.02
Closest approach to Earth (in AU)	0.53	0.27	—
Furthest distance from Earth (in AU)	1.47	1.73	—
Circumference of orbit (in 10^6 km)	360	680	940
Average orbital speed (in km s^{-1})	47.9	35.0	29.8
Sidereal period (in Tropical years)	0.2409	0.6152	1.000 04
Inclination to the ecliptic	7°.005	3°.395	—
Orbital eccentricity	0.2056	0.0068	0.0167
Equatorial diameter (in km)	4878	12 104	12 756.280
Diameter in Earth diameters	0.382	0.949	1.000
Oblateness	0	0	1/298.258
Surface area in Earth surface areas	0.146	0.90	1.00
Volume in Earth volumes	0.056	0.85	1.00
Mass in Earth masses (without satellites)	0.05527	0.8150	1
Mass in kg	3.3022×10^{23}	4.8690×10^{24}	5.9742×10^{24}
Mean density in kg m^{-3}	5430	5240	5515
Escape speed in km s^{-1}			
(from the planetary equator)	4.2	10.4	11.2
Surface gravity in m s^{-1} (equatorial)	3.71	8.85	9.78
Rotational period (sidereal)	$58^{d}.6462$	$-243^{d}.01$	$23^{h}56^{m}04^{s}$
Day–night cycle	176^{d}	$116^{d}.75$	1^{d}
Inclination of equator to orbit	0°	177°20′	23°26″
North pole:			
Right ascension	280°.98	272°.79′	
Declination	61°.44	67°.21′	
Geometrical albedo[a]	0.106	0.65	0.367
Maximum visual apparent magnitude	−1.2	−4.28	−3.86[b]

Table B.15. (continued)

Mars	Jupiter	Saturn	Uranus	Neptune	Pluto
227.941	778.38	1424.3	2866.6	4492.3	5887.3
1.52369	5.2032	9.5210	19.162	30.029	39.354
1.38	4.95	9.00	18.27	29.71	29.7
1.67	5.45	10.04	20.06	30.34	49.1
0.38	3.95	8.00	17.29	28.71	28.7
2.67	6.45	11.04	21.07	31.31	50.1
1400	4900	9000	18 000	28 000	37 000
24.1	13.1	9.6	6.8	5.4	4.7
1.8809	11.869	29.628	84.665	165.49	251.86
$1\overset{\circ}{.}850$	$1\overset{\circ}{.}305$	$2\overset{\circ}{.}487$	$0\overset{\circ}{.}772$	$1\overset{\circ}{.}771$	$17\overset{\circ}{.}147$
0.0934	0.0482	0.0550	0.047	0.010	0.246

Mars	Jupiter	Saturn	Uranus	Neptune	Pluto
6786.8	142 796	120 000	50 800	48 600	≈ 2400
0.532	11.19	9.41	3.98	3.81	0.19
1/192.81	1/15.4300	1/9.291 88	1/33	1/38.6	?
0.28	120	82	16	14	0.034
0.15	1317	748	62	54	0.063
0.10745	317.8	95.16	14.50	17.20	0.023
6.4191×10^{23}	1.8988×10^{27}	5.6850×10^{26}	8.6625×10^{25}	1.0278×10^{26}	1.4×10^{22}
3940	1330	700	1300	1760	2000

Mars	Jupiter	Saturn	Uranus	Neptune	Pluto
5.0	59.6	35.6	21.4	23.7	1.2
3.72	24.80	10.50	9.00	11.60	0.6
$24\overset{\mathrm{d}}{.}37\overset{\mathrm{m}}{.}23^{\mathrm{s}}$	$9^{\mathrm{h}}56^{\mathrm{m}}$	$10^{\mathrm{h}}30^{\mathrm{m}}$	$-17^{\mathrm{h}}15^{\mathrm{m}}$	$18^{\mathrm{h}}25^{\mathrm{m}}$	$-153^{\mathrm{h}}17^{\mathrm{m}}$
$1\overset{\mathrm{d}}{.}027$	$0\overset{\mathrm{d}}{.}41$	$0\overset{\mathrm{d}}{.}43$			
25°11′	3°08′	26°43′	97°51′	29	117°34′
$317\overset{\circ}{.}60$	$268\overset{\circ}{.}04$	$40\overset{\circ}{.}05$	$257\overset{\circ}{.}26$	$295\overset{\circ}{.}23$	$311\overset{\circ}{.}48$
$52\overset{\circ}{.}84$	$64\overset{\circ}{.}49$	$83\overset{\circ}{.}49$	$-15\overset{\circ}{.}09$	$40\overset{\circ}{.}62$	$4\overset{\circ}{.}14$
0.150	0.52	0.47	0.51	0.41	0.52
−2.52	−2.7	−0.6	+5.3	+7.50	+13.8

The orbital data for Mercury to Mars and also the periods for the outer planets from Jupiter are averages, the orbital elements applying to the epoch 1990.0. The remaining orbital data for the outer planets are the so-called osculating elements which, because they are continuously influenced by mutual perturbations of the planets, apply strictly only for the date 1988 December 25.

[a] The geometrical albedo is the ratio of the intensity of the light, as viewed from Earth, of the planet for zero phase angle to the intensity, as viewed from the same position, of a circular, perfectly white reflecting surface with the same radius as the planet.

[b] For the Earth as seen from the Sun.

Table B.17. Planetary satellite data.

Planet	Satellite		Sidereal Period (days[a])	Semimajor Axis (10^3 km)	Orbital Eccentricity	Dimension(s) (km)	Geometrical Albedo
Earth		Moon	27.321 66	384.405	0.0549 0049	3476	0.12
Mars		Phobos	0.318 9102	9.378	0.015	19×27	0.06
		Deimos	1.262 441	23.459	0.0005	11×15	0.07
Jupiter	I	Io	1.769 137 79	422	0.004	3630	0.61
	II	Europa	3.551 181 04	671	0.009	3138	0.64
	III	Ganymede	7.154 5530	1070	0.002	5262	0.42
	IV	Callisto	16.689 018	1883	0.007	4800	0.20
	V	Amalthea	0.498 1791	181	0.003	150×270	0.05
	VI	Himalia	250.566	11 480	0.1580	186	0.03
	VII	Elara	259.653	11 737	0.2072	76	0.03
	VII	Parsiphae	735 r	23 500	0.378	50	
	XI	Sinope	758 r	23 700	0.275	36	
	X	Lysithea	259.2	11 720	0.107	36	
	XI	Carme	692 r	22 600	0.2068	40	
	XII	Ananke	631 r	21 200	0.1687	30	
	XIII	Leda	238.7	11 094	0.1476	16	
	XIX	Adrastea	0.2983	129		15×25	0.05
	XV	Thebe	0.675	222	0.015	90×110	0.05
	XVI	Metis	0.294 78	128		40	0.05
Saturn	S1	Mimas	0.942 421 81	185.6	0.020	392	0.5
	S2	Enceladus	1.370 217 86	238.1	0.0045	510	1.0
	S3	Tethys	1.887 802 16	294.7	0.0000	1060	0.9
	S4	Dione	2.736 914 74	377.5	0.002 23	1120	0.7
	S5	Rhea	4.517 500 44	527.2	0.0010	1530	0.7
	S6	Titan	15.945 4207	1221.6	0.029 19	5150	0.21
	S7	Hyperion	21.276 609	1481.0	0.104	220×410	0.3
	S8	Iapetus	79.330 183	3560.1	0.0283	1460	0.2[b]
	S9	Phoebe	550.5 r	12 954.0	0.1633	220	0.06
	S10	Janus	0.6945	151.47	0.007	160×220	0.8
	S11	Epimetheus	0.6942	151.42	0.009	100×140	0.8
	S12	1980 S6	2.737	377.5	0.005	30×34	0.7
	S13	Telesto	1.888	294.7		16×30	0.5
	S14	Calypso	1.888	294.7		22×24	0.6
	S15	Pandora (1980 S26)	0.629	141.7	0.004	66×110	0.9
	S16	Prometheus (1980 S27)	0.613	139.4	0.003	74×140	0.6
	S17	Atlas	0.602	137.7	0.000	20×40	0.9
Uranus		1986 U7	0.335 03	49.7	< 0.001	15–40	
		1986 U8	0.376 41	53.8	0.01	25–50	
		1986 U9	0.434 58	59.2	< 0.001	40–50	
		1986 U3	0.463 57	61.8	< 0.001	50–60	
		1986 U6	0.473 65	62.7	< 0.001	50–60	
		1986 U2	0.493 07	64.6	< 0.001	70–80	
		1986 U1	0.513 20	66.1	< 0.001	70–80	
		1986 U4	0.56	69.9	< 0.001	50–60	
		1986 U5	0.623 53	75.3	< 0.001	50–60	
		1986 U1	0.761 83	86.0	< 0.001	170	

Table B.17. (continued)

Planet	Satellite	Sidereal Period (days*)	Semimajor Axis (10^3 km)	Orbital Eccentricity	Dimension(s) (km)	Geometrical Albedo
Uranus	Miranda	1.413 4793	129.78	0.0027	484	0.22
	Ariel	2.520 3794	191.24	0.0034	1160	0.26
	Umbriel	4.144 177	265.97	0.0050	1190	0.11
	Titania	8.705 872	435.84	0.0022	1610	0.21
	Oberon	13.463 239	582.60	0.0008	1550	0.18
Neptune	1989 N6	0.30	48	(0)	(54)	0.06 ?
	1989 N5	0.31	50	(0)	(80)	0.06 ?
	1989 N3	0.33	52.5	(0)	(180)	0.06 ?
	1989 N4	0.43	62	(0)	(150)	(0.05)
	1989 N2	0.55	73.6	(0)	(190)	(0.06)
	1989 N1	1.12	117.6	(0)	400	0.06
	Triton	5.88 r	354.8	0.00	2760	0.7–0.9
	Nereid	360.13	5513.4	0.75	(340)	(0.14)
Pluto	Charon	6.387	19.7		≈ 1300	

[a] Given in tropical days for Saturnian satellites; r = retrograde orbit.
[b] Bright side: 0.5; dark side: 0.05.

Table B.18. The IAU Mars map 1958: Names and locations of 128 features.

Long.	Lat.	Name	Long.	Lat.	Name	Long.	Lat.	Name
30°	+45°	Acidalium M.	210°	+25°	Elysium	10°	+20°	Oxus
215.	−5	Aeolis	220.	−45	Eridania	200.	+60	Panchaia
310.	+10	Aeria	40.	−25	Erythraeum M.	340.	−25	Pandorae Fr.
230.	+40	Aetheria	220.	+22	Eunostos	155.	−50	Phaetontis
230.	+10	Aethiopis	335.	+20	Euphrates	320.	+20	Phison
140.	0	Amazonis	0.	+15	Gehorn	190.	+30	Phlegra
250.	+5	Amenthes	270.	−40	Hadriaticum M.	110.	−12	Phoenicis L.
105.	−45	Aonius S.	290.	−40	Hellas	70.	−40	Phrixi R.
330.	+20	Arabia	340.	−6	Hellespontica D.	280.	−65	Promethei S.
115.	−25	Araxes	325.	−50	Hellespontus	185.	+45	Propontis
100.	+45	Arcadia	240.	−20	Hesperia	50.	−23	Protei R.
25.	−45	Argyre	345.	+15	Hiddekel	315.	+42	Protonilus
335.	+48	Arnon	60.	+75	Hyperboreus L.	38.	−15	Pyrrhae R.
50.	−15	Auroae S.	295.	−20	Japygia	340.	−8	Sabaeus S.
250.	−40	Ausonia	130.	−40	Icaria	150.	+60	Scandia
40.	−60	Australe M.	275.	+20	Isidis R.	320.	−30	Serpentis M.
50.	+60	Baltia	330.	+40	Ismenius L.	70.	−20	Sinai
290.	+55	Boreosyrtis	40.	+10	Jamuna	155.	−30	Sirenum M.
90.	+50	Boreum M.	63.	−5	Juventae F.	245.	+45	Sithonius L.
75.	+3	Candor	200.	0	Laestrygon	90.	−28	Solis L.
260.	+40	Casius	200.	+70	Lemuria	200.	+30	Styx
210.	+50	Cebrenia	270.	0	Libya	100.	−20	Syria
320.	+60	Cecropia	65.	+15	Lunae L.	290.	+10	Syrtis Maior
95.	+20	Ceraunius	25.	−10	Margaritifer S.	70.	+50	Tanais
205.	+15	Cerberus	150.	−20	Memnonia	70.	+40	Tempe
0.	−50	Chalce	285.	+35	Meroe	85.	−35	Thaumasia
260.	−50	Chersonesus	0.	−5	Meridiani S.	255.	+30	Thoth
210.	−58	Chronium M.	350.	+20	Moab	180.	−70	Thyle I
30.	+10	Chryse	270.	+8	Moeris L.	230.	−70	Thyle II
110.	−50	Chrysokeras	72.	−28	Nector	10.	+10	Thymiamata
220.	−20	Cimmerium M.	270.	+35	Neith R.	85.	−5	Tithonius L.
110.	−35	Claritas	260.	+20	Nepenthes	80.	+30	Tractus albus
280.	+55	Copais Pa.	55.	−45	Nereidum Fr.	268.	−25	Trinacria
65.	−15	Coprates	30.	+30	Niliacus L.	198.	+20	Trivium Charontis
230.	−5	Cyclopia	55.	+30	Nilokeras			
0.	+40	Cydonia	290.	+42	Nilosyrtis	255.	−20	Tyrrhenium M.
305.	−4	Deltoton S.	130.	+20	Nix Olympica	260.	+70	Uchronia
340.	−15	Deucalionis R.	330.	−45	Noachis	290.	+50	Umbra
0.	+35	Deuteronilus	65.	−45	Ogygis R.	250.	+50	Utopia
180.	+50	Diacria	200.	+80	Olympia	15.	−35	Vulcani Pe.
320.	+50	Dioscuria	65.	−10	Ophir	50.	+10	Xanthe
345.	0	Edom	0.	+60	Ortygia	320.	−40	Yaonis R.
190.	−45	Electris	18.	+8	Oxia Pa.	195.	0	Zephyria

Abbreviations: M = mare (sea), L = lacus (lake), R = regio (region), S = sinus (bay), Pa = palus (swamp), Pe = pelagus (sea), F = fons (fountain), Fr = fretum (strait), D = depressio (depression).

Table B.19. Data on the bright minor planets. After D. Morrison and J. Meeus.

Minor Planet	m			Type	D (km)	Albedo (vis)	Rotation Period	Max. Ampl.	Elements			P
	max	avg	min						a	e	i	(d)
1 Ceres	7.3	7.6	8.1	C	1000	0.06	$9^{h}04^{m}.7$	0.04	2.77	0.08	$10^{\circ}.6$	1682
2 Pallas	7.5	8.6	10.2	C	600	0.08	$7^{h}48^{m}.4$	0.15	2.77	0.23	34.8	1683
3 Juno	8.2	9.7	11.0	S	240	0.16	$7^{h}12^{m}.8$	0.15	2.67	0.26	13.0	1594
4 Vesta	6.3	6.8	7.4	U	530	0.23	$5^{h}20^{m}.5$	0.13	2.36	0.09	7.1	1326
5 Astraea	9.8	11.0	12.0	S	115	0.15	$16^{h}48^{m}.4$	0.3	2.58	0.19	5.2	1512
6 Hebe	8.3	9.4	10.6	S	200	0.17	$7^{h}16^{m}.5$	0.2	2.43	0.20	14.8	1380
7 Iris	7.8	9.4	10.6	S	210	0.16	$7^{h}08^{m}.1$	0.3	2.39	0.23	5.5	1346
8 Flora	8.7	9.6	10.6	S	150	0.15	$13^{h}.6$	0.1	2.20	0.16	5.9	1193
9 Metis	9.1	9.9	10.6	S	150	0.14	$5^{h}03^{m}.8$	0.3	2.39	0.12	5.6	1346
11 Parthenope	9.9	10.5	11.1	S	150	0.13	$10^{h}40^{m}.0$	0.1	2.45	0.10	4.6	1403
12 Victoria	9.9	11.3	12.5	S	120	0.12	$8^{h}39^{m}.2$	0.3	2.33	0.22	8.4	1303
14 Irene	9.6	10.5	11.4	S	150	0.17	11^{h}	0.1	2.59	0.16	9.1	1522
15 Eunomia	8.6	9.5	10.6	S	270	0.15	$6^{h}04^{m}.8$	0.5	2.64	0.19	11.7	1569
16 Psyche	9.9	10.6	11.3	M	250	0.09	$4^{h}18^{m}.2$	0.1	2.92	0.14	3.1	1822
18 Melpomene	8.8	10.2	11.4	S	150	0.14	$11^{h}50^{m}.0$	0.4	2.30	0.22	10.1	1271
20 Massalia	9.2	10.1	10.9	S	130	0.17	$8^{h}05^{m}.9$	0.3	2.41	0.14	0.7	1365
44 Nysa	9.7	10.7	11.5	E	85	0.38	$6^{h}25^{m}.1$	0.5	2.42	0.15	3.7	1376
192 Nausikaa	9.3	11.0	12.3	S	95	0.17			2.40	0.25	6.8	1360
216 Kleopatra	9.9	11.6	12.8		120?				2.79	0.25	13.1	1705
324 Bamberga	9.1	11.4	13.0	C	240	0.03	8^{h}	0.1	2.69	0.34	11.2	1608
433 Eros	8.0	11.0	12.6	S	24	0.20	$5^{h}16^{m}.2$	1.5	1.46	0.22	10.8	643
1036 Ganymed	9.4	14.1	16.6		35?				2.67	0.54	26.4	1590
1620 Geographos	9.7	13.4	17.1	S	2	0.21	$5^{h}13^{m}.5$	2.0	1.24	0.33	13.3	507
1627 Ivar	9.9	15.3	17.4	S	7	0.20			1.86	0.40	8.4	930
10 Hygiea	10.2	10.7	11.3	C	450	0.04	18^{h}	0.1	3.15	0.10	3.8	2043
19 Fortuna	10.1	11.1	12.0	C	215	0.03	$7^{h}27^{m}.0$	0.6	2.44	0.16	1.6	1394
22 Kalliope	10.6	11.2	11.8	M	175	0.13	$4^{h}08^{m}.8$	0.3	2.91	0.10	13.7	1814
31 Euphrosynw	11.0	12.0	13.3	C	370	0.03			3.15	0.22	26.3	2046
65 Cybele	11.8	12.4	13.1	C	310	0.02			3.43	0.11	3.6	2324
349 Dembowska	10.6	11.0	11.5	R	145	0.26	$4^{h}42^{m}.0$	0.3	2.92	0.09	8.3	1827
354 Eleonora	10.4	11.1	11.7	S	150	0.15	$4^{h}16^{m}.6$	0.3	2.80	0.12	18.4	1708
511 Davida	10.5	11.4	12.2	C	320	0.04	10^{h}	0.3	3.19	0.17	15.8	2078
704 Interamnia	11.0	11.6	12.6	C	350	0.03	$8^{h}43^{m}.6$	0.1	3.06	0.15	17.2	1952

m = apparent magnitude (brightest, at mean opposition, and faintest).

Types: C = containing carbon, S = containing silicon, M = containing metals, E = resembling stony meteorites (enstatite achondrites), R = reddish (Fe, with or without silicates), U = unclassifiable.

D = diameter in km.

Max. Ampl. = largest observed light variation.

a = semimajor axis in AU; e = eccentricity; i = inclination to ecliptic; P = sidereal period in days.

Table B.20. Table of change of the central meridians of Mars and Jupiter.

	Mars	Jupiter			Mars	Jupiter	
		Sys. I	Sys. II			Sys. I	Sys. II
1^m	0°.2	0°.6	0°.6	1^h	14°.6	36°.6	36°.3
2	0.5	1.2	1.2	2	29.2	73.2	72.5
3	0.7	1.8	1.8	3	42.9	109.7	108.8
4	1.0	2.4	2.4	4	58.5	146.3	145.1
5	1.2	3.0	3.0	5	73.1	182.9	181.3
10	2.4	6.1	6.0	6	87.7	219.5	217.6
20	4.9	12.2	12.1	7	102.3	256.1	253.8
30	7.3	18.3	18.1	8	117.0	292.7	290.1
40	9.7	24.4	24.2	9	131.6	329.2	326.4
50	12.2	30.5	30.2	10	146.2	5.8	2.6

Table B.21. Atmospheric currents on Jupiter. Data by H. Haug and coworkers, Wilhelm-Foerster-Sternwarte, Berlin, and based on a table by P. Muller for the international program IJVTOP on the occasion of the Voyager flybys in 1979.

Code	Latitude Range			Main Flow	Latitude	Rotation System	Limits (bands, zones, sections, objects) 1	2	3	4	5
A	−90°	to	−46°.5	SPR		II	SPC	SSSTEZ	SSSTB	SSTEZ	
B	−46.5	to	−37	SSTEC	−40°	II	SSTB	S/STEZ	STEZB	SSTEZ	
C	−37	to	−28	STEC	−34[a]	II	N/STEZ	S/STB	WOS-FA	WOS-BC	WOS-DE
D	−28	to	−25	N/STB		II	STB				
E	−25	to	−22	STRZ	−34[b]	II	STRZ	STRD	RS	RSH	
F	−22	to	−15.5	SEBS	−18	II	SEBS	SEBD			
G	−15.5	to	−12	SEBZ	−13	II	SEBZ				
H	−12	to	−10	S/SEBN		II	S/SEBN				
J	−10	to	−3	SEC		I/II	N/SEBN	S/EZ			
K	−3	to	+3	CEC		I	EB	M/EZ			
L	+3	to	+10	NEC	+6	I	N/EZ	S/NEB			
M	+10	to	+14	M/NEB	+13	II	NEB				
N	+14	to	+22.5	NTRC	+17	II	N/NEB	NTRZ			
P	+22.5	to	+25	NTEC-C	+23	I/II	S/NTB				
Q	+25	to	+28	NTEC-B		II	M/NTB				
R	+28	to	+34	NTEC-A	+31	II	N/NTB	S/NTEZ			
S	+34	to	+36	MNTC-B	+35	II	S/NNTB				
T	+36	to	+42	NNTC-A		II	NNTB	NNTEZ			
U	+42	to	+47	NNNTEC		II	NNNTB				
V	+47	to	+90	NPR		II	NPB	NNTEZ			

Regions or objects may be identified by abbreviated letter–number codes (advantageous for large amounts of input data), e.g., N1 for N/NEB = north edge of NEB. Observation of an object should include an adequate determination of position. The jovigraphic latitude assigns the object to a current, and the repeated measures of longitude will then determine if the rotation rate is typical or atypical.

WOS = white oval spots (long-lived, in STB): FA, BC, DE; RS = Great Red Spot; RSH = red spot hollow; STRD = south tropical convergence zone disturbance; SEBD = disturbance in SEB.

Table B.22. Periodic comets with periods under 200 yr, observed in more than one perihelion passage (ordered by period).

Name	T	p	q	e	ω	Ω	i	m_0
Encke	1980.93	3.30	0.340	0.847	186.0	334.2	11.9	14.5
Grigg–Skjellerup	1982.37	5.09	0.989	0.666	359.3	212.6	21.1	17.5
du Toit–Hartley	1982.24	5.20	1.195	0.602	251.7	308.6	2.9	18.0
Honda–Mrkos–Pajdusakova	1980.24	5.28	0.581	0.809	184.6	232.9	13.1	17.0
Tempel 2	1983.42	5.29	1.381	0.545	190.9	119.2	12.4	15.0
Schwassmann–Wachmann 3	1979.67	5.38	0.941	0.694	198.7	69.3	11.4	12.5
Neujmin 2[a]	1927.04	5.43	1.138	0.567	193.7	328.0	10.6	11.0
Brorsen 2[a]	1879.24	5.46	0.590	0.810	14.9	102.3	29.4	9
Tempel 1	1978.03	5.50	1.497	0.520	179.1	68.3	10.5	13.5
Clark	1978.90	5.51	1.557	0.501	209.0	59.1	9.5	14.0
Tuttle–Giacobini–Kresak	1978.98	5.58	1.124	0.643	49.4	153.3	9.9	18.0
Tempel–Swift[a]	1908.76	5.68	1.153	0.638	113.5	291.1	5.4	13
Wirtanen	1974.51	5.87	1.256	0.614	351.8	83.5	12.3	15.5
West–Kohoutek–Ikemura	1981.28	6.12	1.401	0.581	358.1	84.6	30.1	15.0
Kohoutek	1981.29	6.24	1.571	0.537	169.9	273.1	5.4	13.0
Forbes	1980.73	6.27	1.479	0.565	262.6	23.0	4.7	15.5
du Toit–Neujmin–Delporte	1970.77	6.30	1.672	0.510	114.3	189.2	2.8	12
de Vico–Swift	1965.31	6.31	1.624	0.524	325.4	24.4	3.6	10.5
Pons–Winnecke	1976.91	6.36	1.254	0.635	172.4	92.7	22.3	16.0
d'Arrest	1982.70	6.38	1.291	0.625	177.0	138.9	19.4	15.5
Kopff	1977.18	6.43	1.572	0.545	162.9	120.3	4.7	13.5
Schwassmann–Wachmann 2	1981.21	6.50	2.135	0.387	357.5	125.9	3.7	14.0
Giacobini–Zinner	1979.12	6.52	0.996	0.715	172.0	195.1	31.7	9.6
Wolf–Harrington	1978.20	6.53	1.615	0.538	187.0	254.2	18.5	13.5
Churyumov–Gerasimenko	1976.27	6.59	1.299	0.631	11.3	50.4	7.1	13.0
Biela[a]	1852.73	6.62	0.861	0.756	223.2	247.3	12.6	8
Tsuchinshan 1	1978.35	6.65	1.499	0.576	22.8	96.2	10.5	16.0
Perrine–Mrkos[b]	1968.84	6.72	1.272	0.643	166.1	240.2	17.8	12.0
Reinmuth 2	1981.08	6.74	1.946	0.455	45.4	296.0	7.0	15.0
Johnson	1977.02	6.76	2.196	0.386	206.2	117.8	13.9	11.5
Borrelly	1981.14	6.77	1.319	0.631	352.8	75.1	30.2	13.0
Gunn	1982.90	6.82	2.459	0.316	197.0	67.9	10.4	10.0
Tsuchinshan 2	1978.72	6.83	1.785	0.504	203.2	287.6	6.7	14.5
Arend–Rigaux	1978.09	6.83	1.442	0.600	329.0	121.5	17.9	15.5
Harrington	1980.98	6.86	1.605	0.556	233.0	119.0	8.7	15.0
Brooks 2	1980.90	6.90	1.850	0.490	198.2	176.2	5.5	13.5
Taylor	1977.03	6.97	1.951	0.465	355.6	108.2	20.6	13.0
Finlay	1981.47	6.98	1.101	0.698	322.1	41.8	3.6	13.0
Longmore	1981.81	6.98	2.400	0.343	195.9	15.0	24.4	11.5
Holmes	1979.14	7.06	2.160	0.413	23.6	327.4	19.2	14.0
Daniel	1978.52	7.09	1.662	0.550	10.8	68.5	20.1	16.0
Shajn–Schaldach	1979.02	7.25	2.223	0.407	215.3	167.2	6.2	14.5
Faye	1977.16	7.39	1.610	0.576	203.7	199.1	9.1	14.0
Ashbrook–Jackson	1978.63	7.43	2.284	0.400	349.0	2.1	12.5	11.5
Whipple	1978.24	7.44	2.469	0.352	190.0	188.3	10.2	12 5
Harrington–Abell	1976.31	7.59	1.776	0.540	138.5	336.8	10.2	15.0
Reinmuth 1	1980.83	7.59	1.982	0.487	9.5	121.1	8.3	15.0
Kojima	1978.39	7.85	2.399	0.393	348.6	154.1	0.9	15.0

Table B.22. (Continued)

Name	T	p	q	e	ω	Ω	i	m_0
Oterma[c]	1958.44	7.85	2.399	0.393	348.6	155.1	4.0	8
Gehrels 2	1981.88	7.98	2.362	0.408	183.5	215.5	6.7	12.5
Arend	1975.39	7.98	1.847	0.538	46.9	355.7	20.0	15.0
Schaumasse	1960.30	8.18	1.196	0.705	51.9	86.2	12.0	15.5
Jackson–Neujmin	1978.98	8.37	1.425	0.654	196.3	163.1	14.1	16.5
Wolf	1976.07	8.42	2.501	0.396	161.1	203.8	27.3	13.8
Comas Solá	1978.73	8.94	1.870	0.566	42.8	62.4	13.0	12.5
Kearns–Kwee	1981.92	8.99	2.224	0.486	131.4	315.3	9.0	10.0
Denning–Fujikawa	1978.75	9.01	0.779	0.820	334.0	41.0	8.7	9
Swift–Gehrels	1981.91	9.26	1.361	0.691	84.5	314.0	9.2	14.5
Neujmin 3	1972.37	10.6	1.976	0.590	146.9	150.2	3.9	14.0
Väisälä 1	1982.58	10.9	1.800	0.633	47.9	134.5	11.6	13.5
Klemola	1976.61	10.9	1.766	0.642	148.9	181.6	10.6	14.0
Gale[b]	1938.46	11.0	1.183	0.761	209.1	67.3	11.7	11
Slaughter–Burnham	1981.88	11.6	2.544	0.504	44.2	345.9	8.2	13.5
Van Biesbroeck	1978.92	12.4	2.395	0.553	134.3	148.6	6.6	13.5
Wild 1	1973.50	13.3	1.981	0.647	167.9	358.2	19.9	14.5
Tuttle	1980.95	13.7	1.015	0.823	206.9	269.9	54.5	10.0
du Toit	1974.25	15.0	1.294	0.787	257.2	22.1	18.7	14.6
Schwassmann–Wachmann 1	1974.13	15.0	5.448	0.105	14.5	319.6	9.7	4.5
Neujmin 1	1966.94	17.9	1.543	0.775	346.8	347.2	15.0	10.2
Crommelin	1956.82	27.9	0.743	0.919	196.0	250.4	28.9	10.7
Tempel–Tuttle	1965.33	32.9	0.982	0.904	172.6	234.4	162.7	8
Stephan–Oterma	1980.93	37.7	1.574	0.860	358.2	78.5	18.0	11.0
Westphal[b]	1913.90	61.9	1.254	0.920	57.1	347.3	40.9	17.0
Olbers	1956.47	69.6	1.178	0.930	64.6	85.4	44.6	5.5
Pons–Brooks	1954.39	70.9	0.774	0.955	199.0	255.2	74.2	6
Brorsen–Metcalf	1919.79	72.0	0.484	0.972	129.6	311.2	19.2	9
Halley	1986.11	76.0	0.587	0.967	111.8	58.1	162.2	3.3
Herschel–Rigollet	1939.60	155.0	0.748	0.974	29.3	355.3	64.2	8

T = time of perihelion passage, P = period in years, q = perihelion distance from the Sun, e = eccentricity, ω = longitude of perihelion from ascending node, Ω = ascending node, i = inclination with respect to ecliptic. (For the derivation of heliocentric and geocentric positions from these elements, see Chaps. 2 and 3 in Vol. 1.)

Except for P/Halley, the elements are from the last observed perihelion prior to mid-1982. They may change owing to perturbations, especially at encounters with Jupiter. m = magnitude for $r = \Delta = 1$, and is estimated by various methods. The predicted magnitude m at any point (r, Δ) is computed by $m = m_0 + 5 \log \Delta + 2.5n \log r$, where n is generally assumed to be 4, but may range from 2 to 6 (in rare cases more). a = lost; b = probably lost; c = following a strong perturbation by Jupiter, P is now 19 yr and $q > 5$.

Table B.23. Brightness of the zodiacal light, depending on ecliptic latitude β and longitude relative to the Sun $\lambda - \lambda_\odot$, in units of S_{10} (light of a star of $m_V = 10$ per square degree); values in parentheses are extrapolated. From Levasseur-Regourd and Dumont (1980).

$\lambda-\lambda_\odot$ \ β	0	5	10	15	20	25	30	35	40	45	50	55	60	65	70
180	180	166	152	139	127	116	105	96	89	82	76	71	66	62	59
175	169	163	151	138	126	115	105	96	89	82	76	71	66	62	59
170	161	158	147	135	123	114	104	96	89	82	76	70	66	62	59
165	153	150	140	129	118	110	102	95	88	81	75	70	65	62	59
160	147	144	134	122	113	106	98	93	86	80	75	69	65	61	59
155	143	140	130	118	110	102	94	89	83	78	73	68	64	61	58
150	140	139	129	116	107	99	91	86	80	75	71	67	63	60	58
145	139	138	129	115	106	97	89	83	77	73	69	65	62	60	58
140	139	138	129	115	105	96	87	81	75	71	67	64	62	60	58
135	140	139	130	115	105	95	86	80	74	70	66	63	61	59	58
130	141	140	132	116	105	95	86	80	74	69	65	63	61	60	59
125	144	142	135	118	106	96	87	80	74	69	65	63	61	60	59
120	147	145	138	120	108	98	88	81	75	70	66	63	61	60	59
115	152	150	143	124	111	100	89	82	76	71	67	64	62	61	60
110	158	156	148	128	113	101	91	84	78	73	68	65	63	61	61
105	166	164	154	133	117	104	93	86	80	75	70	67	64	62	61
100	175	172	160	137	120	107	96	89	82	77	72	68	65	63	62
95	187	184	168	144	125	111	99	92	84	79	74	70	66	64	63
90	202	196	176	151	130	115	103	95	87	81	76	72	68	66	64
85	219	211	186	158	137	121	108	99	90	84	79	74	70	68	65
80	239	227	197	167	144	127	113	103	94	87	82	77	72	70	67
75	264	248	210	177	153	134	118	107	98	91	85	79	74	71	68
70	296	273	228	188	162	142	124	112	103	95	88	82	77	73	69
65	338	305	250	205	174	152	133	120	109	99	92	85	79	75	70
60	394	342	275	228	190	163	143	129	116	105	96	89	82	77	72
55	470	395	310	253	209	179	158	140	125	113	103	93	85	79	74
50	572	458	355	285	238	200	173	152	135	120	108	98	89	82	76
45	712	550	422	335	270	228	195	168	146	130	115	103	92	84	78
40	920	695	510	400	316	262	220	186	160	140	123	108	95	87	80
35	1275	905	630	495	384	312	250	208	177	151	132	113	99	90	82
30	1930	1210	825	615	475	370	285	230	194	162	140	119	103	93	84
25					(580)	425	320	253	209	174	150	126	107	96	86
20						(490)	355	278	226	185	157	130	111	98	88
15							(410)	310	242	196	162	136	115	100	89
10							(460)	340	260	206	167	138	117	102	90
5							(490)	355	271	212	169	139	118	103	90
0							(500)	360	275	215	170	140	118	103	90

Table B.23. (Continued)

$\lambda-\lambda_\odot$ \ β	75	80	85	90
180	58	58	60	63
175	58	58	60	63
170	58	58	60	63
165	58	58	60	63
160	58	59	61	63
155	58	59	61	63
150	58	59	61	63
145	58	59	61	63
140	59	60	61	63
135	59	60	62	63
130	60	60	62	63
125	60	61	62	63
120	60	61	62	63
115	61	61	62	63
110	61	62	62	63
105	62	62	62	63
100	62	62	62	63
95	63	62	63	63
90	64	63	63	63
85	64	63	63	63
80	65	64	63	63
75	66	64	63	63
70	67	65	64	63
65	68	65	64	63
60	69	66	64	63
55	70	66	64	63
50	71	67	65	63
45	72	67	65	63
40	74	68	65	63
35	75	69	66	63
30	76	70	66	63
25	78	72	67	63
20	79	73	67	63
15	80	73	67	63
10	80	74	68	63
5	80	74	68	63
0	80	74	68	63

Table B.24. The 88 constellations. Listed are the four-letter (older) and three-letter (current) abbreviations, translations, and areas in square degrees.

Andromeda	And	Andr	Andromeda	721
Antlia	Ant	Antl	The Air Pump	239
Apus	Aps	Apus	The Bird of Paradise	206
Aquarius	Aqr	Aqar	The Water Carrier	980
Aquila	Aql	Aqil	The Eagle	653
Ara	Ara	Arae	The Altar	238
Aries	Ari	Arie	The Ram	441
Auriga	Aur	Auri	The Charioteer	657
Boötes	Boo	Boot	The Herdsman	905
Caelum	Cae	Cael	The Sculptor's Chisel	125
Camelopardus	Cam	Caml	The Giraffe	756
Cancer	Cnc	Canc	The Crab	506
Canes Venatici	CVn	CVen	The Hunting Dogs	467
Canis Major	CMa	CMaj	The Great Dog	380
Canis Minor	CMi	CMin	The Lesser Dog	183
Capricorn	Cap	Capr	The Sea Goat	414
Carina	Car	Cari	The Keel	494
Cassiopeia	Cas	Cass	Cassiopeia	599
Centaurus	Cen	Cent	The Centaur	1060
Cepheus	Cep	Ceph	The King	588
Cetus	Cet	Ceti	The Whale	1231
Chamaeleon	Cha	Cham	The Chameleon	131
Circinus	Cir	Circ	The Compasses	93
Columba	Col	Colm	The Dove	270
Coma Berenices	Com	Coma	Berenice's Hair	386
Corona Austrina	CrA	CorA	The Southern Crown	128
Corona Borealis	CrB	CorB	The Northern Crown	179
Corvus	Crv	Corv	The Crow	184
Crater	Crt	Crat	The Cup	282
Crux	Cru	Cruc	The Southern Cross	68
Cygnus	Cyg	Cygn	The Swan	805
Delphinus	Del	Delf	The Dolphin	189
Dorado	Dor	Dora	The Swordfish	179
Draco	Dra	Drac	The Dragon	1083
Equuleus	Equ	Equl	The Little Horse	72
Eridanus	Eri	Erid	The River Eridanus	1138
Fornax	For	Forn	The Furnace	397
Gemini	Gem	Gemi	The Twins	514
Grus	Gru	Grus	The Crane	365
Hercules	Her	Herc	Hercules	1225
Horologium	Hor	Horo	The Clock	249
Hydra	Hya	Hyda	The Water Snake	1303
Hydrus	Hyi	Hydi	The Sea Serpent	243
Indus	Ind	Indi	The Indian	294
Lacerta	Lac	Lacr	The Lizard	201
Leo (major)	Leo	Leon	The Lion	947
Leo minor	LMi	LMin	The Lesser Lion	232
Lepus	Lep	Leps	The Hare	290
Libra	Lib	Libr	The Scales	538
Lupus	Lup	Lupi	The Wolf	334

Table B.24. (Continued)

Lynx	Lyn	Lync	The Lynx	545
Lyra	Lyr	Lyra	The Lyre	285
Mensa	Men	Mens	The Table Mountain	153
Microscopium	Mic	Micr	The Microscope	209
Monoceros	Mon	Mono	The Unicorn	481
Musca	Mus	Musc	The Fly	138
Norma	Nor	Norm	The Level (Square)	165
Octans	Oct	Octn	The Octant	292
Ophiuchus	Oph	Ophi	The Serpent Bearer	948
Orion	Ori	Orio	The Hunter	594
Pavo	Pav	Pavo	The Peacock	377
Pegasus	Peg	Pegs	The Winged Horse	1136
Perseus	Per	Pers	Perseus	615
Phoenix	Phe	Phoe	The Phoenix	469
Pictor	Pic	Pict	The Painter	247
Pisces	Psc	Pisc	The Fishes	890
Piscis Austrinus	PsA	PscA	The Southern Fish	245
Puppis	Pup	Pupp	The Stern	673
Pyxis	Pyx	Pyxi	The Mariner's Compass	221
Reticulum	Ret	Reti	The Rhomboidal Net	114
Sagitta	Sge	Sgte	The Arrow	80
Sagittarius	Sgr	Sgtr	The Archer	867
Scorpius	Sco	Scor	The Scorpion	497
Sculptor	Scl	Scul	The Sculptor's Workshop	475
Scutum	Sct	Scut	Sobieski's Shield	109
Serpens	Ser	Serp	The Serpent	637
Sextans	Sex	Sext	The Sextant	313
Taurus	Tau	Taur	The Bull	797
Telescopium	Tel	Tele	The Telescope	251
Triangulum Australe	TrA	TrAu	The Southern Triangle	109
Triangulum (Boreale)	Tri	Tria	The (Northern) Triangle	294
Tucana	Tuc	Tucn	The Tucan	294
Ursa Major	UMa	UMaj	The Great Bear	1279
Ursa Minor	UMi	UMin	The Lesser Bear	256
Vela	Vel	Velr	The Sail	500
Virgo	Vir	Virg	The Maiden	1294
Volans	Vol	Voln	The Flying Fish	141
Vulpecula	Vul	Vulp	The Little Fox	268

The codes LMC and SMC are also used as quasi-constellation abbreviations for objects belonging to one of the Magellanic Clouds.

Table B.25. The 170 brightest stars (to visual magnitude +3.0).

Star	α_{2000}	δ_{2000}	V	$U-V$	$B-V$	$V-R$	$V-I$	Sp/L	π	Remarks
α And	$0^{h}08^{m}23^{s}.2$	$+29°05'26''$	2.06	−0.58	−0.11	−0.02	−0.12	B8p III	$0''.024$	sd, Sp var
β Cas	0 09 10.6	+59 08 59	2.28	0.45	0.34	0.31	0.52	F2 IV	0.072	sd
γ Peg	0 13 14.1	+15 11 01	2.86	−1.08	−0.21	−0.08	−0.31	B2 IV	−0.004	
β Hyi	0 25 45.3	−77 15 16	2.80		0.62			G2 IV	0.153	
α Phe	0 26 17.0	−42 18 22	2.38		1.09	0.81	1.40	K0 III	0.035	sd
α Cas	0 40 30.4	+56 32 15	2.22	2.30	1.17	0.79	1.38	K0 II–III	0.009	irr var
β Cet	0 43 35.3	−17 59 12	2.00	1.88	1.00	0.72	1.24	K1 III	0.057	
γ Cas	0 56 42.4	+60 43 00	2.41	−1.21	−0.11	0.09	0.00	B0 IV	0.034	irr var
β And	1 09 43.9	+35 37 14	2.04	3.53	1.57	1.24	2.24	M0 III	0.043	
δ Cas	1 25 48.9	+60 14 07	2.69	0.28	0.13	0.15	0.24	A5 V	0.029	var
α Eri	1 37 42.9	−57 14 12	0.49		−0.16			B5 IV	0.023	
β Ari	1 54 38.3	+20 48 29	2.64	0.23	0.13	0.12	0.18	A5 V	0.063	sd
α Hyi	1 58 46.2	−61 34 12	2.9		0.28			F0 V	0.041	
γ And	2 03 53.9	+42 19 47	2.10	2.13	1.21	0.94	1.63	K3 II	0.011	vtr
β Tri	2 09 32.5	+34 59 14	3.00	0.30	0.16	0.15	0.22	A5 III	0.022	sd
α UMi	2 31 45.0	+89 15 51	2.04	0.97	0.59	0.50	0.81	F8 Ib	0.008	vd, sd,var
θ Eri	2 58 15.6	−40 18 17	2.90		0.13	0.15	0.22	A3 V	0.028	vd, sd
α Cet	3 02 16.7	+04 05 23	2.52	3.57	1.63	1.35	2.51	M2 III	0.003	
γ Per	3 04 47.7	+53 30 23	2.94	1.16	0.70	0.61	1.06	G8 III + A3	0.011	sd
β Per	3 08 10.1	+40 57 21	2.15	−0.44	−0.06	0.07	0.04	B8 V	0.031	str, var
α Per	3 24 19.3	+49 51 41	1.80	0.88	0.48	0.45	0.78	F5 Ib	0.029	
δ Per	3 42 55.4	+47 47 15	3.03	−0.62	−0.12	0.04	−0.07	B5 III	0.007	
η Tau	3 47 29.0	+24 06 18	2.86	−0.45	−0.10	0.04	0.00	B7 III	0.008	
ζ Per	3 54 07.9	+31 53 01	2.86	−0.45	0.10	0.16	0.25	B1 Ib	0.007	
ε Per	3 57 51.2	+40 00 37	2.89	−1.18	−0.18	−0.06	−0.24	B0.5 Ib	0.007	sd
γ Eri	3 58 01.7	−13 30 31	2.94	3.54	1.60	1.26	2.26	M1 III	0.003	
α Tau	4 35 55.2	+16 30 33	0.86	3.50	1.55	1.22	2.15	K5 III	0.048	vd
ι Aur	4 56 59.6	+33 09 58	2.67	3.29	1.54	1.06	1.88	K3 II	0.015	
β Eri	5 07 50.9	−05 05 11	2.78	0.25	0.12	0.15	0.23	A3 III	0.042	
β Ori	5 14 32.2	−08 12 06	0.15	−0.66	−0.03	0.03	0.00	B8 Ia	0.011	vd, sd
α Aur	5 16 41.3	+45 59 53	0.06	1.28	0.81	0.61	1.04	G8 III + F	0.073	sd
γ Ori	5 25 07.8	+06 20 59	1.63	−1.09	−0.21	−0.09	−0.31	B2 III	0.026	
β Tau	5 26 17.5	+28 36 27	1.66	−0.62	−0.13	−0.01	−0.09	B7 III	0.018	
β Lep	5 28 14.7	−20 45 34	2.81	1.31	0.82	0.65	1.09	G5 III	0.014	vd
δ Ori	5 32 00.3	−00 17 57	2.21	−1.27	−0.21	−0.07	−0.28	B0 II	0.006	vd, sd
α Lep	5 32 43.7	−17 49 20	2.58	0.47	0.19	0.22	0.43	F0 Ib	0.002	
ι Ori	5 35 25.9	−05 54 36	2.76	−1.30	−0.23	−0.06	−0.28	O9 III	0.021	sd
ε Ori	5 36 12.7	−01 12 07	1.70	−1.21	−0.19	−0.01	−0.19	B0 Ia	−0.007	var
ζ Tau	5 37 38.6	+21 08 33	3.03	−0.84	−0.18	0.00	−0.09	B2 IVp	0.006	
α Col	5 39 38.9	−34 04 27	2.63		−0.15	−0.01	−0.11	B8 Ve	−0.005	
ζ Ori	5 40 45.5	−01 56 34	1.74	−1.26	−0.21	−0.06	−0.27	O9.5 Ib	0.022	vd, sd
κ Ori	5 47 45.3	−09 40 11	2.06	−1.18	−0.18	−0.03	−0.21	B0.5 Ia	0.009	
α Ori	5 55 10.3	+07 24 25	0.69	3.92	1.86	1.60	3.05	M2 Iab	0.005	var
β Aur	5 59 31.7	+44 56 51	1.90	0.06	0.02	0.09	0.08	A2 V	0.037	sd
θ Aur	5 59 43.2	+37 12 45	2.63	−0.26	−0.08	−0.01	−0.04	B9.5 Vp	0.018	vd
β CMa	6 22 41.9	−17 57 22	1.98	−1.20	−0.24	−0.11	−0.35	B1 II–III	0.014	Sp var
α Car	6 23 57.2	−52 41 44	−0.72		0.15			F0 II	0.018	
γ Gem	6 37 42.7	+16 23 57	1.91	0.07	0.00	0.07	0.06	A0 IV	0.031	sd
α CMa	6 45 08.9	−16 42 58	−1.45	−0.03	−0.01	0.00	−0.03	A1 V	0.376	vd
τ Pup	6 49 56.1	−50 36 53	2.9		1.20			K0 III		sd
ε CMa	6 58 37.5	−28 58 20	1.50	−1.13	−0.21	−0.08	−0.29	B2 II		
δ CMa	7 08 23.4	−26 23 35	1.80	1.20	0.67	0.52	0.85	F8 Ia	−0.018	
π Pup	7 17 08.5	−37 05 51	2.70		1.62	1.24	2.15	K5 III	0.023	
η CMa	7 24 05.6	−29 18 11	2.41	−0.77	−0.07	0.07	0.02	B5 Ia		
β CMi	7 27 09.0	+08 17 21	2.90	−0.37	−0.09	0.03	−0.04	B7 V	0.020	
α Gem	7 34 35.9	+31 53 18	1.58	0.04	0.03	0.07	0.06	A1 V + Am	0.072	vtr, ea sd
α CMi	7 39 18.1	+05 13 30	0.35	0.47	0.43	0.41	0.66	F5 IV	0.285	vd
β Gem	7 45 18.9	+28 01 34	1.13	1.87	1.00	0.75	1.25	K0 III	0.093	

Table B.25. (Continued)

Star	α_{2000}	δ_{2000}	V	$U-V$	$B-V$	$V-R$	$V-I$	Sp/L	π	Remarks
ζ Pup	8 03 35.0	−40 00 11	2.25	−1.34	−0.29	−0.06	−0.27	O5 Ia		
ρ Pup	8 07 32.6	−24 18 15	2.82	0.63	0.44	0.37	0.58	F6 IIp	0.031	var
γ Vel	8 09 31.9	−47 20 12	1.82		−0.27	−0.02	−0.16	WC7 + O7		vd, ea sd
ε Car	8 22 30.8	−59 30 34	1.86	0.25	1.28			K0 II + B		
δ Vel	8 44 42.2	−54 42 30	1.96		0.04			A0 V	0.043	
ι UMa	8 59 12.4	+48 02 30	3.15	0.27	0.19	0.21	0.29	A7 V	0.066	vtr, sd
λVel	9 07 59.7	−43 25 57	2.20		1.62	1.24	2.19	K5 Ib	0.015	
β Car	9 13 12.1	−69 43 02	1.68		0.00			A1 IV	0.038	
ι Car	9 17 05.4	−59 16 31	2.25		0.18			F0 I	0.011	
κ Vel	9 22 06.8	−55 00 38	2.50		−0.18			B2 IV	0.007	sd
α Hya	9 27 35.2	−08 39 31	1.96	3.20	1.47	1.04	1.81	K3 III	0.017	
ε Leo	9 45 51.0	+23 46 27	2.98	1.27	0.81	0.66	1.06	G0 II	0.002	
υCar	9 47 06.1	−65 04 18	2.96		0.27			A8 Ib		vd
α Leo	10 08 22.3	+11 58 02	1.35	−0.47	−0.12	0.00	−0.09	B7 V	0.039	
γ Leo	10 19 58.3	+19 50 30	1.98	2.14	1.15	0.86	1.48	K0 IIIp	0.019	vd
θ Car	10 42 57.4	−64 23 40	2.8		−0.22			O9.5 V		
μ Vel	10 46 46.1	−49 25 12	2.68		0.88	0.69	1.18	G5 III		vd
β UMa	11 01 50.4	+56 22 56	2.38	−0.04	−0.01	0.08	0.04	A1 V	0.042	
α UMa	11 03 43.6	+61 45 03	1.80	2.01	1.07	0.81	1.39	K0 III	0.031	vd
δ Leo	11 14 06.4	+20 31 25	2.57	0.21	0.11	0.16	0.19	A4 V	0.040	
β Leo	11 49 03.5	+14 34 19	2.13	0.16	0.08	0.06	0.07	A3 V	0.076	
γ UMa	11 53 49.8	+53 41 41	2.44	0.04	0.00	0.05	0.02	A0 V	0.020	
δ Cen	12 08 21.5	−50 43 20	2.63		−0.09	0.04	−0.08	B2 Ve	0.020	var
δ Cru	12 15 08.6	−58 44 56	2.80		−0.23			B2 IV	0.003	
γ Crv	12 15 48.3	−17 32 31	2.57	−0.45	−0.10	−0.02	−0.11	B8 III		
α Cru	12 26 35.9	−63 05 56	1.05		−0.26			B1 IV		vd, sd
δ Crv	12 29 51.8	−16 30 56	2.94	−0.14	−0.05	−0.05	−0.09	B9.5 V	0.018	
γ Cru	12 31 09.9	−57 06 47	1.63					M3 III		
β Crv	12 34 23.2	−23 23 48	2.64	1.53	0.88	0.61	1.05	G5 III	0.027	
α Mus	12 37 11.0	−69 08 08	2.69		−0.20			B3 IV		
γ Cen	12 41 30.9	−48 57 34	2.16		−0.01	0.03	0.03	A0 III	0.027	vd
γ Vir	12 41 39.5	−01 26 58	2.73	0.34	0.36	0.29	0.49	F0 V	0.097	vd
β Cru	12 47 43.3	−59 41 19	1.25		−0.23			B1 III		var
ε UMa	12 54 01.7	+55 57 35	1.78	−0.02	−0.03	−0.02	−0.06	A0p	0.008	sd
α CVn	12 56 01.6	+38 19 06	2.84	−0.43	−0.10	−0.04	−0.12	A0p + F5	0.023	vd
ε Vir	13 02 10.5	+10 57 33	2.84	1.68	0.94	0.55	1.09	G9 III	0.032	
ι Cen	13 20 35.7	−36 42 44	2.73	0.05	0.02	0.06	0.05	A2 V	0.053	
ζ UMa	13 23 55.5	+54 55 31	2.27	0.01	0.02	0.06	0.03	A2 V	0.037	vd, sd
α Vir	13 25 11.5	−11 09 41	0.96	−1.18	−0.25	−0.09	−0.32	B1 IV + B2	0.021	sd
ε Cen	13 39 53.2	−53 27 59	2.99		−0.24	−0.15	−0.40	B1 III		
η UMa	13 47 32.3	+49 18 48	1.86	−0.85	−0.18	−0.07	−0.26	B3 V	0.004	
η Boo	13 54 41.0	+18 23 52	2.68	0.81	0.58	0.45	0.74	G0 IV	0.102	sd
ζ Cen	13 55 32.3	−47 17 18	2.54		−0.24	−0.13	−0.34	B2 IV		sd
β Cen	14 03 49.4	−60 22 22	0.61		−0.23			B1 III	0.016	vd, sd?
θ Cen	14 06 40.8	−36 22 12	2.05		0.96	0.76	1.29	K0 III	0.059	
α Boo	14 15 39.6	+19 10 57	−0.06	2.52	1.24	0.98	1.64	K2 IIIp	0.090	
γ Boo	14 32 04.6	+38 18 29	3.02	0.31	0.20	0.14	0.22	A7 III	0.016	
η Cen	14 35 30.3	−42 09 28	2.30		−0.21	−0.09	−0.27	B1.5 V		
α Cen	14 39 36.1	−60 50 07	−0.29		0.71			G2 V + K1V	0.750	vd
α Lup	14 41 55.7	−47 23 17	2.29		−0.21	−0.08				
ε Boo	14 44 59.1	+27 04 27	2.37	1.70	0.97	0.77	1.29	K1 III + A	0.013	vd
β UMi	14 50 42.2	+74 09 20	2.07	3.25	1.47	1.10	1.87	K4 III	0.031	
α Lib	14 50 52.6	−16 02 31	2.74	0.26	0.15	0.17	0.21	A3 IV + F4	0.056	
β Lup	14 58 31.8	−43 08 02	2.67		−0.21	−0.09	−0.26	B2 IV		
γ TrA	15 18 54.6	−68 40 46	2.89		0.00			A1 V	0.005	
β Lib	15 17 00.3	−09 22 59	2.61	−0.48	−0.11	−0.04	−0.13	B8 V	−0.012	sd
γ Lup	15 35 08.3	−41 10 00	2.77		−0.20	−0.15	−0.37	B2 V	0.010	vd
α CrB	15 34 41.2	+26 42 53	2.27	−0.06	−0.02	0.09	0.05	A0 V	0.043	sd

Table B.25. (Continued)

Star	α_{2000}	δ_{2000}	V	$U-V$	$B-V$	$V-R$	$V-I$	Sp/L	π	Remarks
α Ser	15 44 16.0	+06 25 32	2.65	2.41	1.17	0.81	1.37	K2 III	0.046	
β TrA	15 55 08.4	−63 25 50	2.85		0.29			F2 III	0.078	
π Sco	15 58 51.0	−26 06 51	2.92	−1.08	−0.19	−0.09	−0.29	B1 V	0.005	sd
δ Sco	16 00 19.9	−22 37 18	2.33	−1.00	−0.10	−0.04	−0.17	B0 V		
β Sco	16 05 26.1	−19 48 19	2.55	−0.90	−0.08	0.00	−0.10	B0.5 V + B2	0.007	vtr, sd
δ Oph	16 14 20.6	−03 41 40	2.74	3.55	1.00	1.29	2.32	M0.5 III	0.029	
η Dra	16 23 59.3	+61 30 51	2.74	1.61	0.91	0.61	1.07	G8 III	0.043	vd
α Sco	16 29 24.4	−26 25 55	0.89	3.17	1.83	1.56	2.79	M2 Ib + B5	0.019	vd
β Her	16 30 13.1	+21 29 22	2.77	1.63	0.94	0.64	1.11	G8 III	0.017	sd
τ Sco	16 35 52.9	−28 12 58	2.82	−1.27	−0.25	0.11	−0.36	B0 V	0.014	
ζ Oph	16 37 09.4	−10 34 02	2.57	−0.82	0.02	0.10	0.06	O9.5 V	0.003	
ζ Her	16 41 17.1	+31 36 10	2.81	0.86	0.66	0.54	0.85	G0 IV	0.102	vd
α TrA	16 48 39.9	−69 01 40	1.92		1.44			K2 III	0.024	
ε Sco	16 50 09.7	−34 17 36	2.28		1.17	0.86	1.46	K2 III	0.022	
η Oph	17 10 22.6	−15 43 29	2.42	0.13	0.05	0.06	0.05	A2.5 V	0.047	vd
β Ara	17 25 17.9	−55 31 47	2.85		1.46			K3 Ib	0.026	
υSco	17 30 45.7	−37 17 45	2.70		−0.22	−0.16	−0.40	B3 Ib		
α Ara	17 31 50.4	−49 52 34	2.94		−0.18	−0.10	−0.34	B2.5 V	0.001	
β Dra	17 30 25.8	+52 18 05	2.78	1.63	1.00	0.68	1.16	G2 II	0.009	
λSco	17 33 36.4	−37 06 13	1.62		−0.18	−0.17	−0.45	B1 V		sd
α Oph	17 34 56.0	+12 33 36	2.07	0.25	0.14	0.14	0.22	A5 IV	0.069	vd
θ Sco	17 37 19.0	−42 59 52	1.86		0.41	0.35	0.55	F0 II	0.020	
κ Sco	17 42 29.1	−39 01 48	2.41	−1.08	−0.20	−0.08	−0.30	B2 III		
β Oph	17 43 28.3	+04 34 02	2.77	2.40	1.16	0.81	1.38	K2 III	0.023	
ι^1Sco	17 47 35.0	−40 07 37	3.03		0.51			F2 Ia	0.013	
γ Dra	17 56 36.3	+51 29 20	2.23	3.39	1.51	1.16	2.02	K5 III	0.017	
γ Sgr	18 05 48.4	−30 25 27	2.98		1.02	0.73	1.24	K0 III	0.018	
δ Sgr	18 20 59.6	−29 49 41	2.69	2.89	1.38	1.00	1.68	K2 III	0.039	
ε Sgr	18 24 10.3	−34 23 05	1.84		−0.02	0.00	−0.01	A0 V	0.015	
λSgr	18 27 58.1	−25 25 18	2.82	1.94	1.04	0.76	1.32	K2 III	0.046	
α Lyr	18 36 56.2	+38 47 01	0.00	0.03	0.00	−0.04	−0.07	A0 V	0.123	
σ Sgr	18 55 15.8	−26 17 48	2.07	−0.95	−0.22	−0.11	−0.31	B2 V		
ζ Sgr	19 02 36.6	−29 52 49	2.60	0.17	0.08	0.04	0.05	A2 III	0.020	vd
ζ Aql	19 05 24.5	+13 51 48	2.98	0.03	0.03	0.04	0.04	A0 V	0.036	
π Sgr	19 09 45.7	−21 01 25	2.87	0.57	0.33	0.34	0.59	F2 II–III	0.016	vtr
δ Cyg	19 44 58.4	+45 07 51	2.87	−0.12	−0.02	0.00	−0.02	B9.5 IV	0.021	vd
γ Aql	19 46 15.5	+10 36 48	2.71	3.21	1.52	1.07	1.82	K3 II	0.006	
α Aql	19 50 46.9	+08 52 06	0.74	0.31	0.23	0.14	0.27	A7 V	0.197	
γ Cyg	20 22 13.6	+40 15 24	2.23	1.21	0.67	0.50	0.84	F8 Ib	−0.006	
α Pav	20 25 38.8	−56 44 07	1.94		−0.20			B3 IV		sd
α Cyg	20 41 25.8	+45 16 49	1.25	−0.12	0.09	0.12	0.22	A2 Ia	−0.013	
ε Cyg	20 46 12.6	+33 58 13	2.46	1.91	1.03	0.72	1.29	K0 III	0.044	
α Cep	21 18 34.7	+62 35 08	2.47	0.33	0.21	0.21	0.32	A7 V	0.063	
β Aqr	21 33 33.4	−05 34 16	2.85	1.42	0.84	0.61	1.02	G0 Ib	0.000	
ε Peg	21 44 11.1	+09 52 30	2.38	3.21	1.53	1.05	1.81	K2 Ib	0.005	
δ Cap	21 47 02.3	−16 07 38	2.81	0.40	0.29	0.24	0.40	A6p	0.065	sd
α Gru	22 08 13.9	−46 57 40	1.73		−0.17	−0.08	−0.14	B7 IV	0.051	
α Tuc	22 18 13.9	−60 15 35	2.86		1.39			K3 III	0.019	sd
β Gru	22 42 40.0	−46 53 05	2.10		1.62	1.91	3.68	M5 III	0.003	
η Peg	22 43 00.1	+30 13 17	2.95	1.46	0.87	0.64	1.11	G8 II + F0	−0.002	sd
α PsA	22 57 39.0	−29 37 20	1.16	0.14	0.08	0.11	0.13	A3 V	0.144	
β Peg	23 03 46.4	+28 04 58	2.42	3.62	1.67	1.51	2.83	M2 II-III	0.015	
α Peg	23 04 45.6	+15 12 19	2.47	−0.06	−0.03	0.01	−0.01	B9.5 III	0.030	

The table contains the names and coordinates; magnitudes and colors in the 5-color Arizona–Tonantzintla photometry system (H.L. Johnson: *Sky & Telescope* **30**, 24 (1956)), or V and $B-V$ from other sources; spectral types and luminosity classes (see Chap. 24); and parallax π. (Note that most parallax measures are old, with sd of $\pm0''.01$ or more, and also so small that systematic errors up to about $0''.005$ become significant; this also explains occasional negative measurements.) vd, sd = visual or spectroscopic duplicity; tr = triple; var = variable.

Table B.26. The nearest stars (within 6 pc = 19.5 light years).

α_{2000}	δ_{2000}	Name	π	μ	V_R	m_V	Sp	Companion
$14^h29^m\!.7$	−62°41′	Proxima Cen	0″.772	3″.85	−16	11.2	M5 Ve	
14 39.6	−60 51	α Cen	0.750	3.67	−22	−0.1	G2 V	1.3 K0 V*
17 57.8	+04 42	+4°3561	0.546	10.32	−108	9.5	M5 V	
10 56.5	+07 01	Wolf 359	0.419	4.67	+13	13.5	M6 Ve	
11 03.3	+35 58	+36°2147	0.394	4.80	−86	7.5	M2 V	
06 45.1	−16 43	α CMa (Sirius)	0.376	1.32	−8	−1.5	A1 V	8.4 DA *
01 39.0	−17 57	LDS 838	0.376	3.31	+30	12.5	M6 Ve	12.9 *
18 49.8	−23 50	Ross 154	0.342	0.74	−4	10.4	M4 Ve	
23 41.9	+44 11	Ross 248	0.319	1.60	−81	12.2	M5 Ve	
03 32.9	−09 28	ε Eri	0.307	0.98	+16	3.7	K2 V	
11 47.7	+00 48	Ross 128	0.302	1.40	−13	11.1	M4 V	
22 38.6	−15 17	L 789-6	0.294	3.27	−60	12.8	M5 Ve	13.3 *
00 18.4	+45 01	+43°44	0.291	2.90	+14	8.1	M3 V	11.0 M6 Ve
22 03.4	−56 47	ε Ind	0.290	4.69	−40	4.7	K5 V	
18 42.8	+59 38	+59°1915	0.290	2.27	+1	8.9	M4 V	9.7 M4 V*
21 06.9	+38 45	61 Cyg	0.289	5.21	−64	5.2	K5 V	6.0 K7 V*
01 43.1	−15 56	τ Cet	0.287	1.92	−16	3.5	G8 V	
23 05.9	−35 51	−36°15693	0.285	6.90	+10	7.3	M2 V	
07 39.3	+05 13	α CMi (Procyon)	0.285	1.25	−3	0.4	F5 IV	11 DF *
08 29.8	+26 47	G 51-15	0.276	1.26		14.8	M7 V	
07 27.4	+05 13	+5°1668	0.266	3.76	+26	9.8	M4 V	
01 12.6	−17 00	L 725-32	0.265	1.36	+28	11.8	M6 Ve	
05 11.7	−44 01	−44°1905	0.260	8.82	+245	8.8	M1 VI	
21 17.2	−38 52	−39°14192	0.260	3.47	+21	6.7	M0 Ve	
22 28.0	+57 42	+56°2783	0.253	0.90	−26	9.9	M3 V	11.5 M4 Ve*
06 29.4	−02 49	Ross 614	0.242	0.97	+24	11.2	M4 Ve	14.7 *
16 30.3	−12 40	−12°4523	0.241	1.17	−13	10.1	M5 V	
12 33.3	+09 01	Wolf 424	0.233	1.76	−5	13.2	M6 V	13.2 *
00 49.2	+05 23	Van Maanen 2	0.231	2.98	+54	12.4	DG	
02 00.2	+13 03	L 1159-16	0.223	2.08		12.2	M5 Ve	
00 05.4	−37 21	−37°15492	0.223	6.11	+23	8.6	M3 Ve	
10 44.5	−61 12	L 143-23	0.223	1.66		13.9	dM	
17 28.7	−46 54	−46°11540	0.221	1.14		9.4	M3	
10 48.2	−11 20	LP 731-58	0.220	1.64		15.6	M7 V	
11 45.7	−64 50	L 145-141	0.219	2.69		11.4	DC	
17 36.4	+68 20	+68°946	0.216	1.31	−22	9.1	M3 V	**
21 33.6	−49 01	−49°13515	0.216	0.78	+8	8.7	M2 V	
10 11.3	+49 27	+50°1725	0.214	1.45	−26	6.6	K2 V	
00 06.7	−07 32	G 158-27	0.214	2.06		13.7	M5 V	
22 53.3	−14 18	−15°6290	0.213	1.12	+9	10.2	M4 V	
17 37.1	−44 19	−44°11909	0.212	1.14		11.0	M5 V	
19 53.9	+44 25	G 208-44/45	0.212	0.69		13.4	M6 Ve	15.5 dM*
								14.3 dM
04 15.3	−07 39	40 Eri	0.210	4.04	−42	4.4	K0 V	9.5 DA
						11.2	M4 Ve	*
10 19.6	+19 52	+20°2465	0.205	0.49	+11	9.4	M4 V	
18 05.5	+02 30	70 Oph	0.201	1.12	−7	4.2	K0 V	6.0 K5 V*
11 05.5	+43 32	+44°2051	0.200	4.51	+65	8.7	M2 V	14.4 M5 Ve
19 50.8	+08 52	α Aql (Altair)	0.197	0.66	−26	0.8	A7 V	

Table B.26. (Continued)

α_{2000}	δ_{2000}	Name	π	μ	V_R	m_V	Sp	Companion
22 46.8	+44 20	+43°4305	0.197	0.83	−2	10.0	M5 Ve	
11 47.6	+78 41	AC+79°3888	0.192	0.87	−119	10.8	M4 V	
08 58.3	+19 46	G 9-38	0.192	0.87		14.1	M7 Ve	14.9
13 45.7	+14 53	+15°2620	0.186	2.30	+15	8.5	M2 Ve	
06 00.1	+02 42	G 99-49	0.185	0.30		11.6	dM	
17 15.4	−26 36	36 Oph	0.185	1.23	−1	5.1	K0 V	5.1 K1 V
								6.3 K5 V
19 32.4	+69 40	σ Dra	0.180	1.83	+27	4.7	K0 V	
04 31.2	+58 59	G 175-34	0.178	2.40		11.1	M4 V	12.5 DC
20 11.2	−36 06	−36°13940	0.178	1.63	−130	5.3	K3 V	11.5 M2 V
19 20.7	−45 33	L 347-14	0.177	2.94		12.2	M5 V	
06 10.5	−21 52	−21°1377	0.177	0.73	+4	8.1	M1 V	
14 57.5	−21 25	−20°4123/25	0.177	1.99	+26	5.7	K4 V	7.9 M2 V
20 08.7	−66 11	δ Pav	0.176	1.65	−22	3.5	G7 V	
23 49.2	+02 24	+1°4774	0.176	1.38	−65	9.0	M2 Ve	
06 54.8	+33 16	Wolf 294	0.174	0.85	+36	9.9	M4 V	
05 31.5	−03 41	−3°1123	0.173	2.23	+11	8.0	M1 Ve	
00 15.5	−16 08	L 722-22	0.172	0.86		11.5	dM	12.0 *
19 16.9	+05 10	+4°4048	0.172	1.48	+33	9.1	M3 V	17.5 M6
08 12.7	−21 33	L 674-15	0.171	0.73		12.1	M4	
00 49.1	+57 49	η Cas	0.170	1.22	+10	3.4	G0 V	7.5 M0 *
05 42.2	+12 42	Ross 47	0.167	2.54	+103	11.6	M4 V	
17 46.5	−57 20	L 205-128	0.167	1.73		10.7	M4 V	

π = parallax (distance in pc = $1/\pi$).
μ = annual proper motion (tangential velocity = $4.74\mu/\pi$ km s^{-1}).
V_R = radial velocity in km s^{-1}.
m_V = apparent visual magnitude. (The true luminosity L in visual light, in units of the solar luminosity, is computed from $\log L = -0.4 m_V - 2\log\pi - 0.08$. For red stars, particularly of type M, the total luminosity is 3 to 6 times higher than the visual one, since most of the radiation lies in the infrared.)
* binary orbit known.
** presumed faint companion of still uncertain magnitude.

Including the Sun, there are—according to present data—93 stars in 67 systems: seven of solar type or brighter, six degenerates, and eighty red dwarfs. There are no indications of invisible companions for any of these stars. The list includes 16 of the 25 stars or systems with annual proper motions exceeding 3″.

Table B.27. Absolute magnitudes of stars in the MK system. From Schmidt-Kaler: Landolt-Börnstein, New Series, *Astronomy and Astrophysics*, Vol. 2b, Springer, Berlin 1982, p. 18.

Sp	M_v							
	V	IV	III	II	Ib	Iab	Ia	Ia–0
O 3	$-6^{\mathrm{M}}.0$						$-6^{\mathrm{M}}.8$	
4	−5.9	$-6^{\mathrm{M}}.1$	$-6^{\mathrm{M}}.5$				−6.8	
5	−5.7	−6.0	−6.3				−6.8	
6	−5.5	−5.8	−6.1			$-6^{\mathrm{M}}.5$	−6.8	
7	−5.2	−5.5	−5.9	$-6^{\mathrm{M}}.0$	$-6^{\mathrm{M}}.3$	−6.5	−6.8	
8	−4.9	−5.4	−5.8	−6.0	−6.2	−6.5	−6.8	
9	−4.5	−5.2	−5.6	−5.9	−6.2	−6.5	−6.8	
B 0	−4.0	−4.7	−5.1	−5.7	−6.1	−6.4	−6.9	$-8^{\mathrm{M}}.2$
1	−3.2	−3.8	−4.4	−5.4	−5.8	−6.4	−6.9	−8.3
2	−2.45	−3.1	−3.9	−4.8	−5.7	−6.4	−6.9	−8.3
3	−1.6	−2.4	−3.0	−4.5	−5.5	−6.3	−7.0	−8.3
5	−1.2	−1.7	−2.2	−4.0	−5.4	−6.2	−7.0	−8.4
7	−0.6	−1.1	−1.5	−3.5	−5.3	−6.2	−7.1	−8.4
8	−0.25	−0.7	−1.2	−3.1	−5.2	−6.2	−7.1	−8.5
9	+0.2	−0.2	−0.6		−5.2	−6.2	−7.1	−8.5
A 0	+0.65	+0.3	+0.0	−3.0	−5.2	−6.3	−7.1	−8.5
1	+1.0	+0.7	+0.2	−3.0	−5.2	−6.4	−7.2	−8.5
2	+1.3	+1.0	+0.3	−2.9	−5.2	−6.5	−7.2	−8.6
3	+1.5	+1.2	+0.5	−2.8	−5.2	−6.5	−7.2	−8.7
5	+1.95	+1.3	+0.7	−2.8	−5.1	−6.6	−7.4	−8.8
7	+2.2	+1.7	+1.1	−2.7	−5.1		−7.7	−8.9
8	+2.4	+2.0	+1.2	−2.6	−5.1		−7.8	−8.9
F 0	+2.7	+2.2	+1.5	−2.5	−5.1	−6.6	−8.0	−9.0
2	+3.6	+2.4	+1.7	−2.4	−5.1	−6.6	−8.0	−9.0
5	+3.5	+2.5	+1.6	−2.3	−5.1	−6.6	−8.0	−9.0
8	+4.0	+2.8		−2.3	−5.1	−6.5	−8.0	−9.0
G 0	+4.4	+3.0	+1.0	−2.3	−5.0	−6.4	−8.0	−8.9
2	+4.7	+3.0	+0.9	−2.3	−5.0	−6.3	−8.0	−8.8
5	+5.1	+3.1	+0.9	−2.3	−4.6	−6.2	−7.9	−8.6
8	+5.5	+3.1	+0.8	−2.3	−4.4	−6.1	−7.8	−8.5
K 0	+5.9	+3.1	+0.7	−2.3	−4.3	−6.0	−7.7	−8.5
1	+6.15	+3.1	+0.6	−2.3	−4.3	−6.0	−7.6	
2	+6.4		+0.5	−2.3	−4.3	−5.9	−7.6	
3	+6.65		+0.3	−2.3	−4.3	−5.9	−7.5	
4	+7.0		0.0	−2.3	−4.3	−5.8	−7.5	
5	+7.35		−0.2	−2.3	−4.4	−5.8	−7.5	
7	+8.1		−0.3	−2.3	−4.4	−5.7	−7.4	
M 0	+8.8		−0.4	−2.5	−4.5	−5.6	−7.0	−8.0
1	+9.3		−0.5	−2.5	−4.6	−5.6	−7.0	−8.0
2	+9.9		−0.6	−2.6	−4.7	−5.6	−6.9	−8.0
3	+10.4		−0.6	−2.6	−4.8	−5.6	−6.9	−8.0
4	+11.3		−0.5	−2.6	−4.8	−5.6	−6.8	−8.0
5	+12.3		−0.3		−4.8	−5.6	−6.8	

Table B.28. Unreddened colors $(B-V)_0$ of stars in the MK system. From Schmidt-Kaler: Landolt-Börnstein, New Series, *Astronomy and Astrophysics*, Vol. 2b, Springer, Berlin 1982, p. 15.

Sp	$(B-V)_0$					
	V	III	II	Ib	Iab	Ia
O 5	$-0\overset{\mathrm{m}}{.}33$	$-0\overset{\mathrm{m}}{.}32$	$-0\overset{\mathrm{m}}{.}32$	$-0\overset{\mathrm{m}}{.}32$	$-0\overset{\mathrm{m}}{.}31$	$-0\overset{\mathrm{m}}{.}31$
6	− 0.33	− 0.32	− 0.32	− 0.32	− 0.31	− 0.31
7	− 0.32	− 0.32	− 0.32	− 0.31	− 0.31	− 0.31
8	− 0.32	− 0.31	− 0.31	− 0.29	− 0.29	− 0.29
9	− 0.31	− 0.31	− 0.31	− 0.28	− 0.27	− 0.27
B 0	− 0.30	-0.29_5	− 0.29	− 0.24	− 0.23	− 0.23
1	-0.26_5	-0.26_5	− 0.26	− 0.20	− 0.19	− 0.19
2	− 0.24	− 0.24	− 0.23	− 0.18	− 0.17	− 0.16
3	-0.20_5	-0.20_5	− 0.20	− 0.14	− 0.13	− 0.12
5	− 0.17	− 0.17	− 0.16	− 0.10	− 0.10	− 0.08
6	− 0.15	− 0.15	− 0.14	− 0.08	− 0.08	− 0.06
7	-0.13_5	− 0.13	− 0.12	− 0.06	− 0.05	− 0.04
8	− 0.11	− 0.11	− 0.10	− 0.04	− 0.03	− 0.02
9	-0.07_5	-0.07_5	− 0.07	− 0.03	− 0.02	0.00
A 0	− 0.02	− 0.03	− 0.03	− 0.01	− 0.01	+ 0.02
1	+ 0.01	+ 0.01	+ 0.01	+ 0.03	+ 0.02	+ 0.02
2	+ 0.05	+ 0.05	+ 0.03	+ 0.04	+ 0.03	+ 0.03
3	+ 0.08	+ 0.08	+ 0.07	+ 0.06	+ 0.06	+ 0.05
5	+ 0.15	+ 0.15	+ 0.11	+ 0.09	+ 0.09	+ 0.09
7	+ 0.20	+ 0.22	+ 0.16	+ 0.13	+ 0.12	+ 0.12
8	+ 0.25	+ 0.25	+ 0.18	+ 0.14	+ 0.14	+ 0.14
F 0	+ 0.30	+ 0.30	+ 0.25	+ 0.19	+ 0.17	+ 0.17
2	+ 0.35	+ 0.35	+ 0.30	+ 0.23	+ 0.23	+ 0.22
5	+ 0.44	+ 0.43	+ 0.38	+ 0.33	+ 0.32	+ 0.31
8	+ 0.52	+ 0.54	+ 0.58	+ 0.56	+ 0.56	+ 0.56
G 0	+ 0.58	+ 0.65	+ 0.71	+ 0.76	+ 0.76	+ 0.75
2	+ 0.63	+ 0.77	+ 0.81	+ 0.86	+ 0.87	+ 0.87
5	+ 0.68	+ 0.86	+ 0.89	+ 1.00	+ 1.02	+ 1.03
8	+ 0.74	+ 0.94	+ 0.99	+ 1.14	+ 1.14	+ 1.17
K 0	+ 0.81	+ 1.00	+ 1.08	+ 1.20	+ 1.25	+ 1.25
1	+ 0.86	+ 1.07	+ 1.14	+ 1.24	+ 1.32	+ 1.32
2	+ 0.91	+ 1.16	+ 1.29	+ 1.33	+ 1.36	+ 1.36
3	+ 0.96	+ 1.27	+ 1.40	+ 1.46	+ 1.46	+ 1.46
5	+ 1.15	+ 1.50	+ 1.49	+ 1.59	+ 1.60	+ 1.60
7	+ 1.33	+ 1.53	+ 1.57	+ 1.61	+ 1.63	+ 1.63
M 0	+ 1.40	+ 1.56	+ 1.58	+ 1.64	+ 1.67	+ 1.67
1	+ 1.46	+ 1.58	+ 1.59	+ 1.68	+ 1.69	+ 1.69
2	+ 1.49	+ 1.60	+ 1.59	+ 1.68	+ 1.71	+ 1.71
3	+ 1.51	+ 1.61	+ 1.60	+ 1.69	+ 1.69	+ 1.69
4	+ 1.54	+ 1.62		+ 1.75	+ 1.76	+ 1.76
5	+ 1.64	+ 1.63			+ 1.80	
6	+ 1.73	+ 1.52				
7	+ 1.80:	+ 1.50				
8	+ 1.93:	+ 1.50				

Table B.29. Abbreviations of some chemical elements.

H	Hydrogen	Si	Silicon	Zr	Zirconium
He	Helium	Cl	Chlorine	Tc	Technetium
Li	Lithium	Ar	Argon	Ru	Ruthenium
C	Carbon	Ca	Calcium	In	Indium
N	Nitrogen	Ti	Titanium	Xe	Xenon
O	Oxygen	V	Vanadium	Ba	Barium
Ne	Neon	Cr	Chromium	La	Lanthanum
Na	Sodium	Mn	Manganese	Eu	Europium
Mg	Magnesium	Fe	Iron	Gd	Gadolinium
Al	Aluminum	Co	Cobalt	Au	Gold
P	Phosphorus	Ni	Nickel	Hg	Mercury
S	Sulfur	Sr	Strontium	Pb	Lead

Table B.30. Wavelengths of selected spectral lines in visible light.

Element	λ (nm)	Element	λ (nm)	Element	λ (nm)
Hα	656.282	He I	447.148	Na I	589.594 D_1
Hβ	486.133		412.081		588.997 D_2
Hγ	434.047		402.619		
Hδ	410.174		396.473	Mg I	457.110
Hε	397.007	He II	656.010	Mg II	448.133
H_8	397.007		541.152		448.113
H_9	383.539		468.568		
H_{10}	379.790		544.159	Ca I	657.278
H_{11}	377.063		419.983		422.673
H_{12}	375.015		402.560	Ca II	396.847 H
H_{13}	373.437		392.348		393.366 K
H_{14}	372.194				
H_{15}	371.197	Hg I	579.006	Cr I	425.435
			576.960		
He I	587.565		546.074	Mn I	403.076
	587.562		435.834		
	501.568		404.656	Ba II	493.409
	471.314		365.014		455.403

Table B.31. Radial velocities for selected stars.

a Proposed standard stars

Name	α(2000)	δ(2000)	V	Type	V_R (km s^{-1})
α Cas	$00^h40^m.5$	56°32′	$2^m.2$	K0 II–III	− 3.9
19 Tau[1]	03 45.2	24 28	4.3	B6 V	3.1
20 Tau[1]	03 45.8	24 22	3.9	B7 III	7.6
β Gem	07 45.2	28 01	1.2	K0 III	3.3
ε Leo	09 45.8	23 46	3.0	G0 II	4.8
β Vir	11 50.7	01 46	3.6	F8 V	5.0
α Boo	14 15.7	19 11	0.1	K2 III	− 5.3
β Oph	17 43.5	04 34	2.8	K2 III	−12.0
γ Aql	19 46.2	10 37	2.6	K3 II	− 2.1
ι Psc	23 39.9	05 38	4.1	F7 V	5.3

[1] IAU standard star.

b Spectroscopic double stars

Name	α(2000)	δ(2000)	V	Type	Period	V_R (km s^{-1})	RV Ampl.[2] (km s^{-1})
β Tri	$02^h09^m.5$	34°59′	$3^m.0$	A5 III	$31^d.39$	15.2	67/138
o Per	03 44.3	32 17	3.8	B1 III	4.42	19.8	219/319
λ Tau	04 00.7	12 29	3.4	B3 V + A4 IV	3.42	15.2	111
π Ori	04 54.2	02 27	3.7	B2 III	3.70	24.2	116
η Ori	05 24.5	−02 23	3.4	B1 V	7.98	35.9	290
δ Ori	05 32.0	−00 18	2.2	O9.5 II	5.73	20.1	202
ι Ori	05 35.4	−05 55	2.8	O9 III + O9 III	29.14	26.6	230/392
β Aur	05 59.5	44 57	1.9	A2 IV + A2 IV	3.96	−17.1	215/223
o Leo	09 41.1	09 54	3.5	A2 + F6 III	14.50	27.1	108/126
ζ[2] UMa	13 23.9	54 56	2.3	A2 V	20.54	− 5.6	138/135
α Vir	13 25.2	−11 09	1.0	B1 V + B3 V	4.01	0.0/2.0[3]	240/378
α Dra	14 04.4	64 22	3.7	A0 III	51.42	−13.0	94
α CrB	15 34.7	26 43	2.2	A0 V	17.36	1.5	72
π Sco	15 58.9	−26 07	2.9	B1 V + B1 V	1.57	− 4.0	262/394
β[1] Sco	16 05.4	−19 48	2.6	B0.5 V	6.83	− 1.0/39.2[3]	258/430
ε Her	17 00.3	30 55	3.9	A0 V	4.02	−24.2	141/224
θ Aql	20 11.3	−00 49	3.2	B9 III + B9 III	17.12	−27.9	102/127
δ Cap	21 47.0	−16 08	2.8	Am	1.02	− 0.2	142
ι Peg	22 07.0	25 20	3.8	F5 V	10.21	− 4.6	98

[1] System velocity (γ).
[2] Full amplitude (2K); if known, given for both components.
[3] Radial velocity for both primary and secondary components given instead of γ velocity.

Table B.32. Double stars.

The selection of 320 visually double and multiple systems contains bright stars (primarily those with Flamsteed numbers and with companions no fainter than magnitude 8.5) over the separation range 1–30″, and north of declination −15°. Some closer and fainter pairs have been included for test purposes.

Successive columns give: (1) ADS number, (2) discoverer code, (3) coordinates for 2000, (4) magnitudes, (5) spectral class (of combined light or of primary star), (6) position angle θ, (7) separation ρ (in cases of substantial orbital motion, these are the data for 1990, and the increase or decrease is indicated by signs of + or −), and (8) remarks, which provide star names and indicate additional components, if detected. For instance, C 8.5 7″ indicates that a third star of magnitude 8.5 is at 7″ of separation, or A-BC (0″.2) identifies the companion as a close pair BC with 0″.2 of separation. Components fainter than magnitude 10 and spectroscopic duplicities are not mentioned.

N 1–30 are double stars in the north polar region (numbered in Fig. 26.1, Chap. 26) which may serve for comparison purposes.
P is the period of revolution in years, if known (or ≈ indicates approximately known). Stars without any *P* value usually show very little orbital motion.

ADS	Name		α_{2000}	δ_{2000}	m	Sp	θ	ρ	Remarks
32	Σ	3056	$00^{h}\ 04^{m}.6$	+34° 16′	7.9–7.9	K	145	0.7	
61	Σ	3062	06.2	+58 26	6.5–7.3	G	308+	1.5	$P = 107$
102	Σ	2	09.1	+79 43	6.8–7.1	A	20−	0.7	$P \approx 450$ N29
191	Σ	12	14.9	+08 49	6.1–7.7	F	148	11.5	35 Psc
207	Σ	13	16.2	+76 57	6.8–7.3	B	54	0.9	$P \approx 1000$
558	Σ	46	39.9	+21 26	5.6–8.5	K	194	6.5	55 Psc
671	Σ	60	49.1	+57 49	3.5–7.5	G	312	12	η Cas P480
683	Σ	61	49.8	+27 43	6.3–6.3	F	296	4.4	65 Psc
710	Σ	65	52.8	+68 51	7.9–7.9	A	39	3.1	
824	Σ	79	01 00.0	+44 43	6.0–6.8	B	192	7.8	
875	Σ	84	03.8	+01 22	6.2–8.7	F	254	16	26 Cet
899	Σ	88	05.6	+21 28	5.6–5.8	A	159	30	74 Psc
903	Σ	90	05.8	+04 55	6.7–7.6	F	83	33	77 Psc
923	Σ	91	07.2	−01 44	7.4–8.2	F	317	4.2	
996	Σ	100	13.7	+07 33	5.6–6.5	A	63	23	ζ Psc A-BC (C 11 1.4)
1030	OΣ	28	19.1	+80 52	8.0–8.0	F	296	0.9	N26
1081	Σ	113	19.8	−00 29	6.4–7.1	F	12	1.6	42 Cet A-BC (0″.1 P21)
1339	Σ	147	41.8	−11 19	6.1–7.4	F	90	2.2	χ^1 Cet
1438	Σ	162	49.2	+47 54	6.5–7.0	A	207	2.0	A-BC (0″.2)
1411	OΣ	34	49.8	+80 53	7.8–8.1	A	278+	0.5	$P \approx 400$ N30
1457	Σ	174	50.1	+22 17	6.2–7.4	F	166	2.8	1 Ari
1507	Σ	180	53.5	+19 18	4.7–7.8	A	0	7.8	γ Ari
1504	Σ	170	55.4	+76 14	7.4–8.2	A	246	3.3	
1538	Σ	186	55.9	+01 51	6.8–6.8	G	57	1.3	P170
1563	H V	12	58.1	+23 25	4.9–7.6	A	46	38	λ Ari
1615	Σ	202	02 02.0	+02 46	4.6–5.6	A	278	1.9	α Psc $P \approx 1000$
1630	Σ	205	03.9	+42 20	2.3–5.1	K	63	10	γ And A-BC (0″.6 P61)
1659	OΣ	37	10.4	+81 30	6.9–9.1	A	210	1.2	N23
1683	Σ	222	10.8	+39 03	6.0–6.7	A	36	17	59 And
1697	Σ	227	12.4	+30 18	5.5–6.9	G	72	3.9	ι Tri
1703	Σ	231	12.7	−02 24	5.9–7.7	G	232	16	66 Cet
—		—	12.7	+79 42	6.5–7.1	A	275	55	N 2
1860	Σ	262	29.0	+67 24	4.7–7.6	A	232	2.4	ι Cas AB $P \approx 1000$ (C 8.5 7″)
1878	Σ	268	29.4	+55 32	6.9–8.2	A	129	2.8	

Table B.32. (Continued)

ADS	Name		α_{2000}	δ_{2000}	m	Sp	θ	ρ	Remarks
1477	Σ	93	31.5	+89 16	2.1–9.1	F	220	18.4	α UMi N8
2046	Σ	295	41.2	−00 42	5.8–9.0	F	309	4.1	84 Cet
2080	Σ	299	43.3	+03 14	3.6–7.4	A	296	2.8	γ Cet
2122	Σ	305	47.5	+19 22	7.3–8.2	G	308	3.7	$P \approx 800$
2151	Σ	311	49.3	+17 28	5.3–8.7	B	120	3.2	π Ari
2157	Σ	307	50.6	+55 54	3.9–8.5	K	300	28	η Per
2185	Σ	314	52.9	+53 00	7.1–7.3	B	309	1.6	AB-C (0″.2)
2257	Σ	333	59.2	+21 20	5.2–5.5	A	207	1.5	ε Ari
2336	Σ	346	03 05.4	+25 15	6.2–6.2	B	249	0.3	52 Ari $P230$ (C 10 5″)
2294	Σ	320	06.2	+79 25	5.8–9.0	M	229	4.6	47 Cep N14
2348	Σ	327	11.7	+81 28	6.0–10	A	283	24	N5
2390	Σ	360	12.1	+37 12	8.1–8.3	G	127	2.6	
2459	AC	2	18.3	−00 56	5.6–7.5	K	252+	1.3	95 Cet $P \approx 300$
2563	Σ	389	30.1	+59 22	6.4–7.4	A	69	2.6	
2616	Σ	412	34.4	+24 28	6.5–6.6	A	3	0.7	7 Tau $P \approx 600$ (C 10 22″)
2628	β	533	35.6	+31 41	7.6–7.6	F	43	1.1	
2668	Σ	425	40.0	+34 08	7.6–7.6	F	76	1.8	
2850	Σ	470	54.3	−02 57	4.9–6.3	A	347	6.8	32 Eri
2867	OΣ	67	57.2	+61 06	5.3–8.5	K	47	1.9	
2888	Σ	471	57.8	+40 01	3.0–8.2	B	10	8.8	ε Per
2926	Σ	479	04 00.9	+23 13	6.9–7.8	B	128	7.3	
2999	Σ	495	07.6	+15 09	6.0–8.8	A	222	3.8	
2963	Σ	460	10.0	+80 42	5.6–6.3	F	125	0.8	$P420$ N27
3082	OΣ	77	15.9	+31 42	8.1–8.1	F	280+	0.7	$P200$
3188	OΣ	81	24.6	+33 58	5.9–8.7	F	27	4.4	56 Per
3264	Σ	554	30.1	+15 38	5.7–8.0	F	15	1.7	80 Tau $P170$
3274	Σ	550	32.0	+53 55	5.9–7.0	B	308	10	1 Cam
3297	Σ	559	34.6	+18 01	7.0–7.1	B	277	3.1	
3353	Σ	572	38.5	+26 56	7.2–7.2	F	194	4.0	
3358	Σ	566	40.0	+53 24	5.6–7.4	A	214−	0.8	2 Cam $P420$ (AB 0″.1 $P27$)
3409	Σ	590	43.6	−08 48	6.7–6.8	F	317	9.2	55 Eri
3572	Σ	616	59.3	+37 54	5.1–8.0	A	0	5.1	ω Aur
3597	Σ	627	05 00.5	+03 37	6.6–6.9	A	260	21	
3711	OΣ	98	07.9	+08 30	5.8–6.5	A	351−	0.7	14 Ori $P200$
3734	Σ	644	10.4	+37 18	6.9–7.0	B	221	1.6	
3764	Σ	652	11.7	+01 02	6.3–7.8	F	182	1.7	
3797	Σ	654	13.3	+02 52	4.7–8.5	K	64	7.0	ρ Ori
3800	Σ	661	13.3	−12 57	4.5–7.4	B	358	2.5	κ Lep
3823	Σ	668	14.5	−08 12	0.3–7.0	B	202	9.4	β Ori A-BC (0″.2)
3824	Σ	653	15.5	+32 41	5.3–7.5	A	226	15	14 Aur
3962	Σ	696	22.8	+03 32	5.0–7.1	B	28	32	23 Ori
3991	Wnc	2	23.9	−00 53	6.7–7.0	F	162	2.5	A-BC (0″.2)
4002	Da	5	24.4	−02 24	3.8–4.8	A	63	23	ζ Ori AB-C (0″.1)
4068	Σ	716	29.2	+25 09	5.9–6.7	A	206	4.8	118 Tau
4123	Σ	729	31.2	+03 17	5.8–7.1	B	27	2.0	33 Ori
4134	Hei	42	32.0	−00 18	3.2–3.3	B	142	0.2	δ Ori AB (C 6 53″)
4179	Σ	738	35.1	+09 56	3.7–5.6	O	43	4.4	λ Ori
4186	Σ	748	35.3	−05 23	6.8–7.9	O	32	9/13	
					5.4–6.8	O	61	/13	θ^1 Ori AB/AC/CD
4188	Σ	I 16	35.4	−05 25	5.2–6.5	B	92	52	θ^2 Ori
4193	Σ	752	35.4	−05 55	2.9–7.1	O	141	11	ι Ori
4208	Σ	749	37.2	+26 56	6.4–6.5	B	327	1.1	
4241	Σ	762	38.8	−02 36	3.8–7.5	O	84	13	σ Ori AB-D (E 6.5 42″)
					4.2–5.1	O	137−	0.2	σ Ori AB $P170$
4263	Σ	774	40.8	−01 56	2.0–4.2	B	163	2.5	ζ Ori

Table B.32. (Continued)

ADS	Name		α_{2000}	δ_{2000}	m	Sp	θ	ρ	Remarks
4390	Σ	795	48.0	+06 27	6.0–6.0	A	215	1.3	52 Ori
4376	Σ	3115	49.1	+62 48	6.5–7.6	A	354	0.9	
4566	OΣ	545	59.7	+37 14	2.7–7.2	A	314	3.4	θ Aur
—	Kpr	23	06 04.1	+23 16	4.7–5.3	G	193+	0.2	1 Gem P 13
4773	Σ	845	11.7	+48 43	6.1–6.8	A	356	7.8	41 Aur
4841	β	1008	14.9	+22 31	3v –8.8	A	258	1.6	η Gem
4991	Σ	899	22.8	+17 34	7.1–8.1	A	19	2.3	
5012	Σ	900	23.8	+04 36	4.5–6.5	A	27	13	ε Mon
5107	Σ	919	28.9	−07 03	4.7–5.2	B	132	7.3	β Mon AB (BC 5.6 3″)
5166	Σ	924	32.4	+17 47	7.1–8.0	F	210	20	20 Gem
5322	Σ	950	41.0	+09 53	4.8–7.6	O	213	2.9	15 Mon
—	Mlr	318	42.5	+66 12	7.1–8.6	F	308	1.7	
5400	Σ	948	46.3	+59 27	5.3–6.2	A	75	1.7	12 Lyn AB (C 7.4 9″)
5436	Σ	958	48.3	+55 43	6.3–6.3	F	257	4.7	
—	Cou	1877	53.2	+38 27	6.5–8.0	F	141	0.6	60 Aur
5559	Σ	982	54.6	+13 10	4.8–7.7	F	147	7.1	38 Gem
5605	Σ	997	56.1	−14 03	5.2–8.5	G	340	3.0	μ CMa
5586	OΣ	159	57.3	+58 25	4.7–5.6	G	66	0.4	15 Lyn
5746	Σ	1009	07 05.6	+52 45	6.9–7.0	A	149	4.1	
5983	Σ	1066	20.2	+21 59	3.5–8.5	F	223	6.0	δ Gem $P \approx 1200$
6004	Σ	1065	22.2	+50 10	7.3–7.4	F	254	15	20 Lyn
6012	Σ	1062	22.8	+55 17	5.6–6.5	B	315	15	19 Lyn
6126	Σ	1104	29.4	−14 59	6.2–7.8	F	20	2.1	
6175	Σ	1110	34.6	+31 53	1.9–2.9	A	77−	3.0	α Gem P467 (C 9.1 71″)
6321	OΣ	179	44.5	+24 24	3.7–8.4	G	239	7.0	κ Gem
6348	Σ	1138	45.5	−14 41	6.1–6.8	A	339	17	2 Pup
6381	Σ	1146	48.0	−12 12	5.7–7.7	F	4	2.4	
6420	β	101	51.8	−13 54	5.6–6.3	G	291+	0.6	9 Pup P23
6454	Σ	1157	54.5	−02 48	7.9–7.9	F	207	0.8	
6569	Σ	1177	08 05.6	+27 32	6.5–7.4	B	351	3.5	
6623	Σ	1187	09.5	+32 14	7.1–8.0	F	26	2.7	
6650	Σ	1196	12.2	+17 39	5.6–6.0	G	180−	0.7	ζ Cnc AB P60 (C 6.3 5.″5)
6811	Σ	1224	26.7	+24 33	7.1–7.6	A	48	5.8	24 Cnc A-BC (0.″2 P22)
6815	Σ	1223	26.8	+26 56	6.3–6.3	A	218	5.0	23 Cnc
6886	Σ	1245	35.8	+06 37	6.0–7.1	F	25	10	
6977	Σ	1270	45.4	−02 36	6.5–7.5	F	263	4.7	
6988	Σ	1268	46.7	+28 46	4.2–6.6	G	307	30	ι Cnc
6993	Σ	1273	46.8	+06 25	3.5–6.9	F	289	3.1	ε Hya A-BC (0.″2 P15) $P \approx 900$
7034	Σ	1282	50.9	+35 04	7.5–7.5	F	278	3.6	
7071	Σ	1291	54.2	+30 34	6.2–6.5	K	313	1.6	57 Cnc
7093	Σ	1295	55.5	−07 58	6.7–6.9	A	0	4.3	17 Hya
7137	Σ	1298	09 01.4	+32 15	6.0–8.1	A	137	4.5	66 Cnc
7187	Σ	1311	07.5	+22 59	6.9–7.3	F	200	7.6	
7203	Σ	1306	10.4	+67 08	4.9–8.2	F	0	3.3	13 UMa $P \approx 1200$
7251	Σ	1321	14.4	+52 41	8.1–8.1	K	90	17.5	$P \approx 1000$
7286	Σ	1333	18.5	+35 22	6.4–6.7	A	48	1.8	
7292	Σ	1334	18.9	+36 49	3.9–6.5	A	229	2.8	38 Lyn
—	Cou	10	20.1	+88 35	7.1–10	A	65	1.7	4 UMi N21
7307	Σ	1338	21.0	+38 12	6.5–6.7	F	270+	1.0	$P \approx 350$
7324	Σ	1340	22.5	+49 32	7.0–8.7	B	319	6.2	39 Lyn
7348	OΣ	200	24.9	+51 34	6.6–8.3	G	335	1.5	
7380	Σ	1355	27.2	+06 14	7.5–7.5	F	347	2.2	
7446	Σ	1362	38.1	+73 06	7.2–7.2	F	128	4.9	
7545	OΣ	208	52.1	+54 04	5.3–5.4	A	187+	0.2	ϕ UMa P105
7555	AC	5	52.5	−08 06	5.6–6.0	A	72	0.6	γ Sex P78

Table B.32. (Continued)

ADS	Name	α_{2000}	δ_{2000}	m	Sp	θ	ρ	Remarks
7704	ΟΣ 215	10 16.3	+17 44	7.2–7.4	F	182	1.4	*P*500
7705	Σ 1415	17.9	+71 04	6.7–7.3	A	167	17	
7724	Σ 1424	20.0	+19 50	2.6–3.8	K	124	4.3	γ Leo $P \approx 620$
7837	Σ 1450	35.1	+08 39	5.8–8.5	A	157	2.2	49 Leo
7902	Σ 1466	43.4	+04 45	6.3–7.4	K	240	6.8	35 Sex
7936	Σ 1476	49.3	−04 01	6.9–7.7	A	10	2.3	40 Sex
7979	Σ 1487	55.6	+24 45	4.5–6.3	A	110	6.5	54 Leo
8119	Σ 1523	11 18.2	+31 32	4.4–4.9	G	58−	1.3	ξ UMa *P*60
8131	Σ 1529	19.4	−01 39	6.9–7.9	F	254	9.4	
8148	Σ 1536	23.9	+10 32	4.0–6.7	F	125−	1.5	ι Leo *P*183
8175	Σ 1543	29.1	+39 20	5.3–8.3	A	358	5.5	57 UMa
8197	ΟΣ 235	32.4	+61 05	5.8–7.1	F	287+	0.6	*P*73
8220	Σ 1552	34.7	+16 48	6.0–7.3	B	209	3.4	90 Leo
8250	Σ 1561	38.9	+45 07	6.5–8.5	G	252	10	
8347	Σ 1579	55.1	+46 29	6.8–8.0	A	40	3.8	65 UMa AB-C(0″.3) D 6.8 63″
8406	Σ 1596	12 04.3	+21 28	6.0–7.5	F	237	3.7	2 Com
—	Sh 136	11.0	+81 43	6.5–8.5	K	76	67	N1
8489	Σ 1622	16.1	+40 40	5.9–8.2	K	260	12	2 CVn
8494	Σ 1625	16.2	+80 08	7.3–7.8	F	219	14.4	N9
8505	Σ 1627	18.1	−03 56	6.7–7.0	F	196	20	
8519	Σ 1633	20.6	+27 04	7.0–7.1	F	245	9.0	
8539	Σ 1639	24.4	+25 35	6.7–7.8	A	325	1.6	$P \approx 750$
8561	Σ 1645	28.1	+44 48	7.4–8.0	F	158	10	
8600	Σ 1657	35.1	+18 23	5.2–6.7	K	271	20	24 Com
8627	Σ 1669	41.3	−13 01	6.0–6.1	F	309	5.4	
8630	Σ 1670	41.7	−01 27	3.7–3.7	F	286−	2.8	γ Vir *P*169
8682	Σ 1694	49.3	+83 24	5.3–5.8	A	326	22	32 Cam N6
8695	Σ 1687	53.3	+21 15	5.1–7.3	G	173+	1.1	35 Com *P*510 (C 9.0 29″)
8375	ΟΣ 258	54.1	+82 31	7.0–11	K	70	10.5	N12
8706	Σ 1692	56.1	+38 18	2.9–5.4	A	228	20	α CVn
8710	Σ 1695	56.3	+54 05	6.0–7.9	A	283	3.5	
8708	ΟΣ 256	56.4	−00 57	7.2–7.5	F	276	0.9	
8739	β 1082	13 00.7	+56 22	5.0–7.2	F	65+	1.5	78 UMa *P*107
8801	Σ 1724	10.0	−05 32	4.4–9.0	A	343	7.1	θ Vir AB-C (0″.1)
—	Kpr 61	12.4	+80 28	6.3–10	G	178	1.0	N25
8891	Σ 1744	23.9	+54 56	2.4–4.0	A	151	14	ζ UMa
8949	Σ 1757	34.3	−00 19	7.8–8.7	K	119	2.2	*P*460
8974	Σ 1768	37.5	+36 17	5.1–7.0	F	100−	1.8	*P*230
8972	Σ 1763	37.6	−07 53	7.9–7.9	K	241	2.8	81 Vir
8991	Σ 1772	40.7	+19 58	5.7–8.8	A	136	4.7	1 Boo
9000	Σ 1777	43.1	+03 32	5.7–8.0	K	229	3.0	84 Vir
9031	Σ 1785	49.1	+26 59	7.6–8.0	K	166+	3.4	*P*156
9053	Σ 1788	55.0	−08 04	6.5–7.7	F	95	3.4	
9173	Σ 1821	14 13.5	+51 47	4.6–6.6	A	236	13	κ Boo
9174	Σ 1816	13.9	+29 06	7.5–7.6	F	90	0.8	
9182	Σ 1819	15.3	+03 08	7.9–8.0	F	221−	0.9	*P*220
9192	Σ 1825	16.6	+20 07	6.6–8.3	F	162	4.4	
9247	Σ 1835	23.4	+08 27	5.1–6.6	A	192	6.4	A-BC (0″.2 *P*40)
9273	Σ 1846	28.2	−02 14	5.0–9.5	K	110	4.8	ϕ Vir
9277	Σ 1850	28.6	+28 17	7.0–7.4	A	262	25	
9358	Σ 1915	32.8	+85 56	7.0–10	K	321	2.4	N18
9338	Σ 1864	40.7	+16 26	4.9–5.8	A	108	5.7	π Boo
9343	Σ 1865	41.1	+13 44	4.5–4.6	A	305	1.0	ζ Boo *P*125
9392	Σ 1883	48.9	+05 57	7.5–7.5	F	287	0.7	*P*228
9396	β 106	49.3	−14 09	5.8–6.7	A	355	1.8	μ Lib

Table B.32. (Continued)

ADS	Name	α_{2000}	δ_{2000}	m	Sp	θ	ρ	Remarks
9406	Σ 1890	49.6	+48 43	6.1–6.8	F	45	3.0	39 Boo
9413	Σ 1888	51.4	+19 06	4.8–6.9	G	326–	7.0	ξ Boo *P*152
9445	Hu 908	53.0	+78 11	6.5–10	K	254	1.2	N24
9425	OΣ 288	53.4	+15 43	6.9–7.5	F	169	1.3	*P*260
9494	Σ 1909	15 03.8	+47 39	5.2–6v	G	47+	1.7	44 Boo *P*225
9507	Σ 1910	07.5	+09 14	7.4–7.4	G	211	4.3	
9626	Σ 1938	24.5	+37 21	7.0–7.6	G	11–	2.1	μ Boo BC *P*246 (A 4.5 108″)
9696	Σ 1972	29.1	+80 37	6.9–7.7	G	80	31	π^1 UMi N3
—	Cou 610	32.9	+31 21	4.2–6v	B	203	0.6	θ CrB
9701	Σ 1954	34.8	+10 32	4.2–5.2	F	177	3.8	δ Ser
9737	Σ 1965	39.4	+36 38	5.0–6.0	B	304	6.2	ζ CrB
9769	Σ 1989	39.6	+79 59	7.4–8.2	F	30	0.7	π^2 UMi *P*170 N28
9778	Σ 1970	46.2	+15 25	3.7–9.5	A	265	30	β Ser
9853	Σ 2034	48.5	+83 37	7.6–8.1	A	115	1.4	N22
9909	Σ 1998	16 04.4	−11 26	4.8–5.1	F	43+	0.8	ξ Sco *P*46 (C 7.2 8″)
9933	Σ 2010	08.1	+17 03	5.3–6.5	G	12	30	κ Her
9969	Σ 2021	13.3	+13 33	7.5–7.7	K	347	4.2	49 Ser
9979	Σ 2032	14.7	+33 51	5.7–6.7	G	235	7.0	σ CrB $P \approx 1000$
10052	Σ 2054	23.8	+61 42	5.9–7.1	G	355	1.1	
10058	OΣ 312	24.0	+61 31	2.9–8.3	F	141	5.2	η Dra
10129	Σ 2078	36.3	+52 55	5.6–6.6	A	108	3.3	17 Dra
10149	Σ I 31	40.6	+04 13	5.7–6.9	A	230	70	37/36 Her
10157	Σ 2084	41.3	+31 36	2.9–5.9	G	83–	1.2	ζ Her *P*34
10214	Hu 917	43.0	+77 30	6.1–9.4	F	188	2.9	N17
10312	Σ 2114	17 01.9	+08 27	6.5–7.7	A	187	1.3	
10345	Σ 2130	05.3	+54 28	5.8–5.8	F	31–	2.1	μ Dra $P \approx 670$
10418	Σ 2140	14.6	+14 23	3.5–5.4	M	107	4.5	α Her
10526	Σ 2161	23.7	+37 08	4.5–5.5	A	316	4.0	ρ Her
10597	Σ 2180	29.0	+50 53	7.6–7.8	F	261	3.1	
10598	Σ 2173	30.4	−01 04	6.0–6.1	G	335–	1.2	*P*46
10628	Σ I 35	32.2	+55 11	5.0–5.0	A	312	62	ν Dra
10728	Σ 2218	40.3	+63 40	7.1–8.3	F	320	1.5	
10759	Σ 2241	41.9	+72 09	4.9–6.1	F	15	30	ψ Dra
10750	Σ 2202	44.6	+02 35	6.2–6.6	A	93	20	61 Oph
10786	Σ 2220	46.5	+27 43	3.5–9.8	A	247	34	μ Her A-BC (1″.5 *P*43)
10850	OΣ 338	52.0	+15 20	7.2–7.3	K	357	0.8	
10875	β 130	53.3	+40 00	5.2–8.5	K	115	1.7	90 Her
10905	Σ 2245	56.4	+18 20	7.3–7.3	A	293	2.6	
11061	Σ 2308	18 00.1	+80 00	5.8–6.2	F	232	19.2	40/41 Dra N7
10993	Σ 2264	01.5	+21 36	5.1–5.2	A	258	6.2	95 Her
11005	Σ 2262	03.1	−08 11	5.3–6.0	F	280	1.8	τ Oph *P*280
11046	Σ 2272	05.5	+02 30	4.2–6.0	K	229–	1.5	70 Oph *P*88
11077	AC 15	07.0	+30 34	5.1–8.5	F	23+	1.0	99 Her *P*56
11089	Σ 2280	07.8	+26 06	5.9–6.0	A	183	14	100 Her
11123	Σ 2289	10.1	+16 29	6.5–7.5	F	224	1.2	
11336	Σ 2323	23.9	+58 48	4.9–7.9	A	352	3.7	39 Dra (C 7.4 90″)
11324	AC 11	24.9	−01 35	6.8–7.1	G	355	0.8	*P*240
11353	Σ 2316	27.2	+00 12	5.4–7.7	G	318	3.8	59 Ser
11483	OΣ 358	35.9	+16 58	6.7–7.1	G	160	1.7	$P \approx 300$
11558	Σ 2368	38.9	+52 21	7.5–7.7	A	322	1.9	
11635	Σ 2382	44.3	+39 40	5.1–6.0	A	353	2.7	ε Lyr AB (AC 208″)
11635	Σ 2383	44.3	+39 37	5.1–5.4	A	84	2.3	CD $P \approx 600$
11639	Σ I 38	44.8	+37 36	4.3–5.9	A	150	44	ζ Lyr
11640	Σ 2375	45.5	+05 30	6.3–6.7	A	119	2.4	AB-CD (each 0″.1)
11667	Σ 2379	46.5	−00 58	5.8–7.5	A	121	13	5 Aql

Table B.32. (Continued)

ADS	Name	α_{2000}	δ_{2000}	m	Sp	θ	ρ	Remarks
11745	Σ I 39	50.1	+33 21	3v –6.7	B	149	46	β Lyr
11853	Σ 2417	56.2	+04 12	4.5–5.4	A	104	22	θ Ser
—	Hei 568	19 07.0	+11 04	5.8–7.1	B	308	0.4	18 Aql
12061	Σ 2461	07.4	+32 30	5.1–9.2	F	299	3.5	17 Lyr
12169	Σ 2486	12.1	+49 51	6.6–6.8	G	213	8.6	
12469	β 142	28.2	−12 09	8.1–8.2	G	209+	0.5	*P*161
12608	Σ 2571	29.5	+78 16	7.6–8.3	F	19	11.3	N11
12540	Σ I 43	30.7	+27 58	3.2–5.4	K	54	35	β Cyg AB-C (0.″4)
12594	Σ 2540	33.3	+20 25	7.2–8.7	A	147	5.1	
12815	Σ I 46	41.8	+50 31	6.3–6.4	G	133	39	16 Cyg
12880	Σ 2579	45.0	+45 08	2.9–6.3	A	227	2.4	δ Cyg $P \approx 800$
12913	Σ 2580	46.4	+33 44	5.0–9.0	F	70	26	17 Cyg
13007	Σ 2603	48.2	+70 16	4.0–7.6	K	13	3.1	ε Dra
12962	Σ 2583	48.7	+11 49	6.1–6.9	F	109	1.3	π Aql
13087	Σ 2594	54.6	−08 14	5.8–6.5	B	170	36	57 Aql
13148	Σ 2605	55.6	+52 26	4.9–7.4	A	178	3.2	ψ Cyg
13256	Σ 2613	20 01.4	+10 45	7.5–7.7	F	352	4.0	
13277	OΣ 395	02.0	+24 56	5.9–6.3	F	120	0.9	16 Vul
13312	Σ 2624	03.5	+36 01	7.2–8.0	O	174	2.0	C 9.1 42″
13524	Σ 2675	08.9	+77 43	4.4–8.4	B	122	7.4	κ Cep N13
13442	Σ 2637	09.9	+20 55	6.4–8.7	F	325	12	θ Sge
13506	Σ 2644	12.6	+00 52	6.9–7.2	A	208	3.0	
13708	Σ 2694	14.5	+80 32	6.8–11	A	344	4.0	N15
13672	Σ 2666	18.1	+40 44	6.0–8.2	B	245	2.8	
14073	β 151	37.5	+14 36	4.0–4.9	F	181+	0.3	β Del *P*27
14158	Σ 2716	41.0	+32 18	5.9–8.0	K	47	2.7	49 Cyg
14259	Σ 2726	45.7	+30 43	4.3–9.2	K	67	6.2	52 Cyg
14279	Σ 2727	46.7	+16 07	4.5–5.5	G	268	10	γ Del
14296	OΣ 413	47.4	+36 29	4.9–6.1	B	10	10	λ Cyg AB-C (0.″1) $P \approx 450$
14336	Σ 2732	48.7	+51 55	6.5–8.5	B	74	4.1	
14421	OΣ 418	54.8	+32 42	8.1–8.2	G	286	1.1	
14504	Σ 2741	58.5	+50 28	5.8–7.1	B	27	1.9	
14499	Σ 2737	59.1	+04 18	5.8–6.1	F	285	1.0	ε Eql *P*101 (C 7.4 10″)
14556	Σ 2742	21 02.2	+07 11	7.4–7.4	F	216	2.8	2 Eql
14575	Σ 2751	02.2	+56 40	6.1–7.1	B	354	1.6	
14573	Σ 2744	03.0	+01 32	6.7–7.2	F	122	1.5	
14592	Σ 2745	04.1	−05 49	5.9–7.3	F	192	2.3	12 Aqr
14636	Σ 2758	06.9	+38 45	5.4–6.2	K	148	30	61 Cyg $P \approx 720$
14682	Σ 2762	08.6	+30 12	5.7–7.7	A	306	3.5	
14749	Σ 2780	11.8	+59 59	6.0–7.0	B	217	1.0	
14787	AGC 13	14.8	+38 03	3.8–6.4	F	11−	0.5	τ Cyg *P*49
14845	Σ 2796	15.6	+78 36	7.2–10	A	42	26	N4
14916	Σ 2801	18.6	+80 21	7.8–8.5	F	270	2.1	N20
15032	Σ 2806	28.7	+70 34	3v –8.0	B	250	13	β Cep
15007	Σ 2799	28.9	+11 05	7.4–7.4	F	89	1.6	
15184	Σ 2816	39.0	+57 29	5.6–8.0	O	121	12	C 8.0 20″
15176	β 1212	39.5	−00 03	7.2–7.6	F	256+	0.5	24 Aqr *P*49
15270	Σ 2822	44.1	+28 44	4.7–6.2	F	303	2.0	μ Cyg $P \approx 710$
15407	Σ 2843	51.6	+65 45	7.1–7.3	A	143	1.7	
15571	Σ 2873	58.4	+82 52	7.0–7.5	F	69	13.7	N10
15600	Σ 2863	22 03.8	+64 38	4.8–5.3	A	275	8.1	ξ Cep AB-C (0.″1)
15767	Σ 2878	14.5	+07 59	6.8–8.3	A	122	1.5	
15971	Σ 2909	28.8	−00 01	4.3–4.5	F	204−	1.9	ζ Aqr $P \approx 760$
15987	Σ I 58	29.1	+58 24	4v –7.5	G	191	41	δ Cep
15988	Σ 2912	30.0	+04 26	5.8–7.1	F	116	0.6	37 Peg *P*130

Table B.32. (Continued)

ADS	Name	α_{2000}	δ_{2000}	m	Sp	θ	ρ	Remarks
16095	Σ 2922	35.9	+39 38	5.8–6.5	B	186	22	87 Lac
16243	OΣ 481	43.8	+78 31	7.5–9.3	A	272	2.3	N19
16228	Σ 2942	44.1	+39 28	6.3–8.5	K	278	2.9	
16294	OΣ 482	47.5	+83 09	5.0–9.7	K	33	3.5	N16
16317	Σ 2950	51.4	+61 42	6.1–7.4	G	295	1.8	
16428	OΣ 483	59.2	+11 44	6.0–7.3	A	314+	0.6	52 Peg P280
16538	OΣ 489	23 07.9	+75 23	4.6–6.6	G	345	1.0	π Cep P150
16666	Σ 3001	18.6	+68 07	4.9–7.2	G	216	3.1	o Cep $P \approx 800$
16775	Σ 3017	27.7	+74 07	7.5–8.6	F	25	1.7	
16836	β 720	33.9	+31 20	5.6–5.8	K	90	0.5	72 Peg P240
17020	OΣ 507	48.6	+64 53	7.0–7.5	A	310	0.7	
17022	OΣ 508	48.8	+62 13	5.7–8.2	A	195	1.8	6 Cas
17140	Σ 3049	59.0	+55 45	5.1–7.2	B	326	3.0	σ Cas
17149	Σ 3050	59.5	+33 43	6.5–6.7	G	322+	1.7	$P \approx 360$

Table B.33. The Messier list of nebulae and clusters (1784).

Consecutive columns below give (1) the Messier number, (2) the NGC number (I=IC number), (3) the coordinates for 2000.0, (4) the object type, (5) the integrated visual apparent magnitude, and (6) remarks. The coded types are as follows: O open star cluster, N galactic (diffuse) nebula, P planetary nebula, G globular star cluster, E extragalactic system. The object M40 is a double star mistaken for a nebula but retained in Messier's list. M47, M48, and M91 were not specified in the list but were retrieved by later authors from Messier's observing records. The entries M101 and M102 referred originally to the same system. M103 through M109 are not in the published catalog but were added from handwritten additions in his working copy.

M	NGC	α	δ	Type	m_V	Remarks
1	1952	$5^h\ 34^m$	$+22\overset{\circ}{.}0$	N	8^m	Crab Nebula
2	7089	21 34	−0.9	G	6	
3	5272	13 42	+28.4	G	6	
4	6121	16 24	−26.5	G	6	
5	5904	15 19	+2.1	G	6	
6	6405	17 40	−32.2	O	5	
7	6475	17 54	−34.8	O	4	
8	6523	18 03	−24.4	N	6	Lagoon nebula
9	6333	17 20	−18.5	G	7	
10	6254	16 58	−4.1	G	7	
11	6705	18 51	−6.2	O	6	
12	6218	16 48	−2.0	G	7	
13	6205	16 43	+36.5	G	6	Cluster in Hercules
14	6402	17 38	−3.2	G	8	
15	7078	21 30	+12.1	G	6	
16	6611	18 20	−13.8	O	6	
17	6618	18 21	−16.2	N	7	Omega nebula
18	6613	18 20	−17.2	O	8	
19	6273	17 03	−26.3	G	7	
20	6514	18 02	−23.0	N	7	Trifid nebula
21	6531	18 05	−22.5	O	7	
22	6656	18 37	−23.8	G	6	
23	6494	17 57	−19.0	O	7	
24	6603	18 20	−18.4	O	5	
25	4725I	18 32	−19.3	O	7	
26	6694	18 46	−9.4	O	9	
27	6853	20 01	+22.7	P	8	Dumbbell nebula
28	6626	18 25	−24.9	G	7	
29	6913	20 24	+38.6	O	7	
30	7099	21 42	−23.2	G	8	
31	224	00 43	+41.3	E	4	Andromeda galaxy
32	221	00 43	+40.9	E	9	Andromeda companion
33	598	01 34	+30.6	E	6	Triangulum galaxy
34	1039	02 42	+42.8	O	6	
35	2168	06 09	+24.3	O	5	
36	1960	05 36	+34.1	O	6	
37	2099	05 52	+32.5	O	6	
38	1912	05 28	+35.8	O	7	
39	7092	21 33	+48.4	O	5	
40	—	12 35	+58.2	–	–	Double star
41	2287	06 47	−20.8	O	5	

Table B.33. (Continued)

M	NGC	α	δ	Type	m_V	Remarks
42	1976	05 36	−5.4	N	3	Orion nebula
43	1982	05 36	−5.3	N	9	
44	2632	08 40	+19.8	O	4	Praesepe
45	—	03 47	+24.2	O	2	Pleiades
46	2437	07 42	−14.8	O	6	
47	2422	07 37	−14.6	O	5	
48	2548	08 14	−5.8	O	6	
49	4472	12 30	+8.0	E	9	
50	2323	07 03	−8.4	O	6	
51	5194	13 30	+47.2	E	8	
52	7654	23 25	+61.6	O	7	
53	5024	13 13	+18.1	G	8	
54	6715	18 55	−30.5	G	7	
55	6809	19 40	−31.0	G	8	
56	6779	19 18	+30.2	G	8	
57	6720	18 54	+33.1	P	9	Ring nebula
58	4579	12 38	+11.8	E	8	
59	4621	12 42	+11.6	E	9	
60	4649	12 44	+11.5	E	9	
61	4303	12 22	+4.5	E	10	
62	6266	17 02	−30.2	G	9	
63	5055	13 16	+42.0	E	10	
64	4826	12 57	+21.6	E	7	
65	3623	11 19	+13.1	E	10	
66	3627	11 21	+13.0	E	9	
67	2682	08 51	+11.8	O	6	
68	4590	12 40	−26.8	G	9	
69	6637	18 31	−32.4	G	9	
70	6681	18 44	−32.3	G	10	
71	6838	19 54	+18.7	G	9	
72	6981	20 54	−12.5	G	10	
73	6994	20 59	−12.6	O		Only 4 stars
74	628	01 37	+15.8	E	10	
75	6864	20 06	−22.0	G	8	
76	650	01 42	+51.6	P	12	
77	1068	02 43	0.0	E	9	
78	2068	05 47	+0.1	N	8	
79	1904	05 24	−24.6	G	8	
80	6093	16 18	−23.0	G	8	
81	3031	09 56	+69.1	E	8	
82	3034	09 56	+69.7	E	9	
83	5236	13 37	−29.8	E	10	
84	4374	12 26	+12.9	E	9	
85	4382	12 26	+18.2	E	10	
86	4406	12 27	+12.9	E	10	
87	4486	12 31	+12.4	E	9	
88	4501	12 32	+14.4	E	10	
89	4552	12 36	+12.5	E	10	
90	4569	12 37	+13.1	E	10	
91	4567	12 37	+11.2	E	10	
92	6341	17 18	+43.2	G	6	

Table B.33. (Continued)

M	NGC	α	δ	Type	m_V	Remarks
93	2447	07 45	−23.9	O	8	
94	4736	12 51	+41.1	E	8	
95	3351	10 44	+11.7	E	10	
96	3368	10 47	+11.8	E	9	
97	3587	11 15	+55.0	P	12	Owl nebula
98	4192	12 14	+14.9	E	11	
99	4254	12 19	+14.4	E	10	
100	4321	12 23	+15.8	E	11	
101	5457	14 03	+54.4	E	10	
102	5866	15 06	+65.8	E	11	
103	581	01 33	+60.8	O	7	
104	4594	12 40	−11.7	E	9	Sombrero nebula
105	3379	10 48	+12.6	E	10	
106	4258	12 20	+47.3	E	9	
107	6171	16 33	−13.1	G	9	
108	3556	11 12	+65.4	E	11	
109	3992	11 58	+53.4	E	11	

Table B.34. Open clusters.

Name	Equatorial Coord.		Galactic Coord.		Diameter	Integrated Magnitude[a]
	α(2000)	δ(2000)	l(2000)	b(2000)	D	m_V
NGC 188	$0^h44^m.7$	$+85°20'$	$122°.8$	$+22°.5$	14′	11.5
Stock 2	2 14.8	+ 59 16	133.4	− 1.9	60	8.0
NGC 869 = h Per	2 19.0	+ 57 09	134.6	− 3.7	30	6.5
NGC 884 = χ Per	2 22.5	+ 57 06	135.1	− 3.6	30	8.5
Tr 2	2 37.4	+ 55 58	137.4	− 3.9	20	8.0
NGC 1039 = M 34	2 42.0	+ 42 47	143.6	− 15.6	35	8.0
Mel 20 = α Persei	3 22.0	+ 48 37	147.0	− 7.1	185	3.0
Pleiades = M 45	3 46.9	+ 24 07	166.6	− 23.5	110	3.0
Hyades	4 26.7	+ 15 51	180.1	− 22.4	330	3.5
NGC 1647	4 46.0	+ 19 04	180.4	− 16.8	45	8.5
NGC 1778	5 08.1	+ 37 03	168.9	− 2.0	7	9.0
NGC 1912 = M 38	5 28.7	+ 35 50	172.3	+ 0.9	21	9.0
NGC 1960 = M 36	5 36.1	+ 34 08	174.5	+ 1.0	12	9.0
NGC 2099 = M 37	5 52.3	+ 32 33	177.7	+ 3.1	24	10.5
NGC 2158	6 07.4	+ 24 05	186.6	+ 1.8	5	11.0
NGC 2168 = M 35	6 08.8	+ 24 20	186.6	+ 2.2	28	8.0
NGC 2169	6 08.5	+ 13 57	195.6	− 2.3	7	7.0
NGC 2244	6 32.3	+ 4 52	206.4	− 2.0	24	6.5
NGC 2264	6 41.0	+ 9 53	202.9	+ 2.2	20	7.5
NGC 2287 = M 41	6 47.0	− 20 44	231.1	− 10.2	38	6.0
Cr 121	6 54.1	− 24 38	235.4	− 10.4	50	6.0
NGC 2323 = M 50	7 03.0	− 8 21	221.7	− 1.3	16	8.0
NGC 2360	7 17.8	− 13 38	229.8	− 1.4	13	9.0
NGC 2362	7 18.8	− 24 37	238.2	− 5.5	8	7.0
Cr 140	7 23.8	− 32 12	245.2	− 7.9	42	5.5
NGC 2420	7 38.4	+ 21 34	198.1	+ 19.6	10	11.0
NGC 2451	7 45.4	− 37 59	252.4	− 6.7	45	4.0
NGC 2477	7 52.3	− 38 32	253.6	− 5.8	27	10.0
NGC 2533	8 07.0	− 29 54	247.8	+ 1.3	4	9.0
NGC 2547	8 10.6	− 49 16	264.6	− 8.6	20	6.5
NGC 2546	8 11.9	− 37 38	254.9	− 2.0	41	6.5
NGC 2632 = Praesepe	8 40.1	+ 19 59	205.5	+ 32.5	95	6.5
NGC 2682 = M 67	8 50.5	+ 11 42	215.6	+ 31.7	30	10.0
NGC 3114	10 02.7	− 60 07	283.3	− 3.8	35	6.0
IC 2581	10 27.4	− 57 39	284.6	+ 0.0	8	7.0
IC 2602	10 43.0	− 64 23	289.6	− 4.9	50	3.0
NGC 3532	11 06.5	− 58 40	289.6	+ 1.5	55	7.5
NGC 3766	11 36.2	− 61 36	294.1	+ 0.1	12	8.0
IC 2944	11 36.6	− 63 31	294.6	− 1.4	15	6.5
NGC 4349	12 24.5	− 61 53	299.8	+ 0.8	16	8.5
Mel 111 = Coma	12 25.0	+ 26 07	221.1	+ 84.1	275	5.0
NGC 4755	12 53.7	− 60 21	303.2	+ 2.5	10	7.0
Cr 285 = UMa Group	14 37.3	+ 69 34	110.3	+ 44.9		2.0
NGC 6067	16 13.2	− 54 13	329.8	− 2.2	13	8.5
NGC 6087	16 18.9	− 57 54	327.8	− 5.4	12	7.5
NGC 6231	16 54.0	− 41 48	343.5	+ 1.2	15	6.5
NGC 6405 = M 6	17 40.0	− 32 12	356.6	− 0.7	15	7.0
IC 4665	17 46.3	+ 5 43	30.6	+ 17.1	41	7.0

Table B.34. (Continued)

Name	Equatorial Coord.		Galactic Coord.		Diameter	Integrated Magnitude [a]
	$\alpha(2000)$	$\delta(2000)$	$l(2000)$	$b(2000)$	D	m_V
NGC 6475 = M 7	$17^h54^m.2$	$-34°48'$	355°.9	$-$ 4°.5	80′	6.0
NGC 6530	18 04.7	$-$ 24 23	6.1	$-$ 1.1	15	6.0
IC 4725 = M 25	18 31.7	$-$ 19 15	13.6	$-$ 4.5	32	8.0
NGC 6705 = M 11	18 51.1	$-$ 6 16	27.3	$-$ 2.8	14	10.5
NGC 6791	19 20.8	+ 37 50	70.0	+ 10.9	16	13.5
NGC 6939	20 31.4	+ 60 38	95.9	+ 12.3	8	10.0
NGC 7063	21 24.5	+ 36 30	83.1	$-$ 9.9	8	9.0
NGC 7092 = M 39	21 32.2	+ 48 27	92.5	$-$ 2.3	32	6.5
TR 37 = IC 1396	21 39.0	+ 57 29	99.3	+ 3.7	50	7.0
NGC 7160	21 53.8	+ 62 36	104.0	+ 6.5	7	8.0
NGC 7209	22 05.2	+ 46 29	95.5	$-$ 7.3	25	10.0
NGC 7380	22 47.0	+ 58 06	107.1	$-$ 0.9	12	8.5
NGC 7789	23 57.0	+ 56 43	116.6	+ 1.0	16	10.0

[a] Magnitude of the brightest star in the cluster.

Table B.35. Globular clusters.

Name	Equatorial Coord.		Galactic Coord.		V[a]	K[b]	Diameter
	$\alpha(2000)$	$\delta(2000)$	$l(2000)$	$b(2000)$			D
NGC 104 = 47 Tuc	$0^h24^m.1$	− 72°05′	305°.9	− 44°.9	4.0	3	30′
NGC 362	1 03.2	− 70 51	301.5	− 46.3	6.6	3	13
NGC 3201 = Dun 445	10 17.6	− 46 25	277.2	+ 8.6	6.8	10	18
NGC 4833 = LacI–4	12 59.6	− 70 53	303.6	− 8.0	7.4	8	14
NGC 5024 = M 53	13 12.9	+ 18 10	333.0	+ 79.8	7.7	5	13
NGC 5139 = ω Cen	13 26.8	− 47 29	309.1	+ 15.0	3.6	8	36
NGC 5272 = M 3	13 42.2	+ 28 23	42.2	+ 78.7	6.4	6	16
NGC 5286 = Dun 388	13 46.4	− 51 22	311.6	+ 10.6	7.6	5	9
NGC 5904 = M 5	15 18.6	+ 2 05	3.9	+ 46.8	5.8	5	17
NGC 5986 = Dun 552	15 46.1	− 37 47	337.0	+ 13.3	7.1	7	10
NGC 6093 = M 80	16 17.0	− 22 59	352.7	+ 19.5	7.2	2	9
NGC 6121 = M 4	16 23.6	− 26 32	351.0	+ 16.0	5.9	9	26
NGC 6205 = M 13	16 41.7	+ 36 28	59.0	+ 40.9	5.9	5	17
NGC 6218 = M 12	16 47.2	− 1 57	15.7	+ 26.3	6.6	9	14
NGC 6254 = M 10	16 57.1	− 4 06	15.1	+ 23.1	6.6	7	15
NGC 6266 = M 62	17 01.2	− 30 07	353.6	+ 7.3	6.6	4	14
NGC 6273 = M 19	17 02.6	− 26 16	356.9	+ 9.4	7.2	8	14
NGC 6341 = M 92	17 17.1	+ 43 08	68.4	+ 34.9	6.5	4	11
NGC 6388	17 36.3	− 44 44	345.5	− 6,7	6.8	3	9
NGC 6397	17 40.7	− 53 40	338.2	− 12.0	5.6	9	26
NGC 6402 = M 14	17 37.6	− 3 15	21.3	+ 14.8	7.6		12
NGC 6441	17 50.2	− 37 03	353.5	− 5.0	7.4	3	8
NGC 6626 = M 28	18 24.5	− 24 52	7.8	− 5.6	7.0	4	11
NGC 6637 = M 69	18 31.4	− 32 21	1.7	− 10.3	7.7	5	7
NGC 6656 = M 22	18 36.4	− 23 54	9.9	− 7.6	5.1	7	24
NGC 6715 = M 54	18 55.1	− 30 29	5.6	− 14.1	7.7	3	9
NGC 6723 = Dun 573	18 59.6	− 36 38	0.1	− 17.3	7.3	7	11
NGC 6752 = Dun 295	19 10.9	− 59 59	336.5	− 25.6	5.4	6	20
NGC 6809 = M 55	19 40.0	− 30 58	8.8	− 23.3	6.9	11	19
NGC 7078 = M 15	21 30.0	+ 12 10	65.0	− 27.3	6.4	4	12
NGC 7089 = M 2	21 33.5	− 0 49	53.4	− 35.8	6.5	2	13
NGC 7099 = M 30	21 40.4	− 23 11	27.2	− 46.8	7.5	5	11

[a] V = integrated apparent visual magnitude.
[b] K = concentration class after Shapley.

Table B.36. Planetary nebulae.

Perek and Kohoutek Identification No.	NGC (or IC)	Equatorial Coord.		Diameter D	Integrated Magnitude of Nebula m_{pg}	Popular Name
		α(2000)	δ(2000)			
120 + 9.1	40	0 13.0	+ 72 32	37″	10.7	
118 − 74.1	246	0 47.0	− 11 53	225	8.0	
165 − 15.1	1514	4 09.2	+ 30 47	114	10	
206 − 40.1	1535	4 14.2	− 12 44	44	9.6	
215 − 24.1	IC418	5 27.5	− 12 42	12	10.7	
205 + 14.1		7 29.0	+ 13 15			Medusa Nebula
197 + 17.1	2392	7 29.2	+ 20 55	44	9.9	Eskimo Nebula
231 + 4.2	2438	7 41.8	− 14 44	66	10.1	
234 + 2.1	2440	7 41.9	− 18 13	32	10.8	
278 − 5.1	2867	9 21.4	− 58 19	11	9.7	
272 + 12.1	3132	10 07.7	− 40 26	47	8.2	
261 + 32.1	3242	10 24.8	− 18 38	1250	8.6	
294 + 4.1	3918	11 50.3	− 57 11	12	8.4	
294 + 43.1	4361	12 24.5	− 18 48	110	10.3	
307 − 3.1	5189	13 33.5	− 65 59	153	10.3	
319 + 15.1	IC4406	14 22.4	− 44 09	28	10.6	
327 + 10.1	5882	15 16.8	− 45 39	7	10.5	
25 + 40.1	IC 4593	16 12.2	+ 12 04	120	10.9	
43 + 37.1	6210	16 44.5	+ 23 49	14	9.3	
0 + 12.1	IC4634	17 01.6	− 21 50	9	10.7	
9 + 14.1	6309	17 14.1	− 12 55	66	10.8	
96 + 29.1	6543	17 58.6	+ 66 38	350	8.8	
34 + 11.1	6572	18 12.1	+ 6 51	8	9.0	
63 + 13.1	6720	18 53.6	+ 33 02	150	9.7	Ring Nebula in Lyra
3 − 14.1		18 55.6	− 32 16	4	10.9	
33 − 2.1	6741	19 02.6	− 0 27	6	10.8	
37 − 6.1	6790	19 23.2	+ 1 31	7	10.2	
64 + 5.1		19 34.8	+ 30 31	8	9.6	
25 − 17.1	6818	19 44.0	− 14 09	17	9.9	
83 + 12.1	6826	19 44.8	+ 50 31	140	9.8	
60 − 3.1	6853	19 59.6	+ 22 43	910	7.6	Dumbbell Nebula
37 − 34.1	7009	21 04.2	− 11 22	100	8.3	Saturn Nebula
84 − 3.1	7027	21 07.1	+ 42 14	15	10.4	
106 − 17.1	7662	23 25.9	+ 42 33	130	9.2	

Table B.37. Diffuse galactic nebulae.

Reference in AGN [a]	Name [b]	Equatorial Coordinates for 2000.0		Galactic Coordinates		Diameter	B [c]	Nebula Type [d]	Notes [e]
		α	δ	l	b	D			
GN 00.02.1.01	S 171	$0^h02^m.1$	$+66°53'$	$118°.4$	$+4°.7$	180′	3	HII	NGC 7822
GN 00.08.0	VDB 1	0 10.6	+ 58 45	117.6	− 3.7	9	3	RN	LBN 578
GN 00.49.9	S 184	0 52.8	+ 56 36	123.1	− 6.3	40	3	HII	NGC 281
GN 00.56.9	S 185	1 00.0	+ 60 59	124.0	− 1.9	120	2	HII + RN	γ Cone Nebula
GN 01.27.4	S 188	1 30.6	+ 58 22	128.1	− 4.1	9	2	HII?	LBN 633
GN 02.29.3	S 190	2 33.1	+ 61 28	134.8	+ 1.0	150	3	HII	IC 1795 + 1805
GN 02.50.6	S 199	2 54.5	+ 60 24	137.6	+ 1.1	120	3	HII	IC 1848
GN 02.59.2	S 201	3 03.1	+ 60 29	138.5	+ 1.6	5	3	HII	LBN 675
GN 03.03.4	LBN 679	3 07.2	+ 56 39	140.8	− 1.4	60	1	HII	
GN 03.06.6	S 200	3 10.8	+ 62 49	138.1	+ 4.1	6	1	HII	LBN 674
GN 03.14.8	S 202	3 18.8	+ 59 38	140.6	+ 1.9	170	1	HII	
GN 03.25.0	VDB 14	3 29.1	+ 59 56	141.5	+ 2.9	50	1	RN	LBN 681
GN 03.27.8	GK Per	3 31.2	+ 43 54	151.0	− 10.1	1	2	Nova	Nova shell
GN 03.26.1	VDB 17	3 29.2	+ 31 23	158.3	− 20.4	8	3	RN	in HH 7,8, . .
GN 03.41.5	VDB 19	3 44.6	+ 32 09	160.5	− 17.8	8	3	RN	IC 348 + 1985
GN 03.41.8	Pleiades	3 46.9	+ 24 07	166.6	− 23.5	110	3	RN	M 45
GN 03.50.0	LBN 774	3 53.0	+ 25 17	166.8	− 21.7	130	1	RN	
GN 03.57.0	S 220	4 00.7	+ 36 37	160.1	− 12.3	320	3	HII	California Nebula
GN 03.59.4	S 206	4 03.2	+ 51 20	150.6	− 0.9	50	3	HII	NGC 1491
GN 04.26.9	S 222	4 30.2	+ 35 16	165.3	− 9.0	6	3	RN	NGC 1579
GN 04.28.4	S 239	4 31.3	+ 18 07	178.9	− 20.1	5	1	HH	HH 102
GN 04.36.8	S 212	4 40.7	+ 50 28	155.4	+ 12.6	5	3	HII	NGC 1624
GN 04.45.0	LBN 917	4 47.5	− 5 55	203.5	− 30.2	50	1	RN	
GN 04.52.5.01	S 260	4 55.2	+ 5 40	193.4	− 22.7	22	1	HII	LBN 860
GN 04.52.5.02	VDB 31	4 55.8	+ 30 33	172.5	− 8.0	9	1	RN	
GN 05.04.0	IC 2118	5 06.4	− 7 16	207.3	− 26.6	155	2	RN	DG 52
GN 05.04.3	NGC 1788	5 06.8	− 3 20	203.5	− 24.7	4	3	RN	VDB 33 = DG 51
GN 05.16.4	S 227	5 19.8	+ 38 57	168.7	+ 1.0	20	1	HII	LBN 781
GN 05.13.0	S 229	5 16.3	+ 34 28	172.0	− 2.2	65	3	HII	IC 405
GN 05.19.3	S 236	5 22.6	+ 33 22	173.6	− 1.7	55	3	HII	IC 410
GN 05.34.0.02		5 36.5	− 2 58	206.9	− 18.0	720	1	HII	Barnard's Loop
GN 05.23.0	S 224	5 26.6	+ 43 02	166.0	+ 4.3	30	1	SNR	
GN 05.24.8	S 234	5 28.1	+ 34 26	173.4	− 0.2	12	3	HII	IC 417
GN 05.32.5.01	S 264	5 35.2	+ 9 56	195.1	− 12.0	390	1	HII	
GN 05.32.5.02	M 42	5 35.0	− 5 28	209.0	− 19.5	60	3	HII	Orion Nebula
GN 05.30.8	DG 58	5 33.4	− 0 38	204.3	− 17.6	40	1	RN	IC 423
GN 05.31.5	M 1	5 34.5	+ 22 01	184.6	− 5.8	7	3	SNR	Crab Nebula
GN 05.38.4.02	VDB 51	5 40.9	− 1 30	206.1	− 16.3	8	3	RN	
GN 05.36.6	VDB 49	5 39.2	+ 4 07	200.7	− 14.0	20	2	RN	LBN 894
GN 05.37.7.02	S 235	5 41.1	+ 35 52	173.6	+ 2.8	10	3	HII	LBN 808
GN 05.36.8	Shajn 147	5 39.9	+ 27 46	180.3	− 1.7	200	1	SNR	S 240
GN 05.34.1	NGC 1999	5 36.5	− 6 43	210.4	− 19.7	1	3	RN	VDB 46
GN 05.33.9.01	HH 1	5 36.4	− 6 45	210.4	− 19.8	0.1	2	HH	in NGC 1999
GN 05.38.2	S 277	5 40.7	− 2 27	206.9	− 16.8	120	3	HII	IC 434
GN 05.39.1.02	NGC 2023	5 41.6	− 2 16	206.9	− 16.5	4	3	RN	VDB 52
GN 05.44.2	NGC 2068	5 46.7	+ 0 05	205.3	− 14.3	7	3	RN	VDB59
GN 05.48.7	S 242	5 51.8	+ 27 01	182.4	+ 0.2	7	2	HII	
GN 05.51.0.02	VDB 62	5 53.6	+ 1 45	204.6	− 12.0	3	2	RN	in LDN 1622

Table B.37. (Continued)

Reference in AGN[a]	Name[b]	Equatorial Coordinates for 2000.0		Galactic Coordinates		Diameter	B[c]	Nebula Type[d]	Notes[e]
		α	δ	l	b	D			
GN 06.01.1	VDB 66	6 03.5	− 9 44	216.3	− 15.0	3′	1	RN	NGC 2149
GN 06.04.9	LkHα 208	6 07.8	+ 18 40	191.4	− 0.8	2	3	BN	DG 87
GN 06.06.0.02	VDB 70	6 08.4	− 5 20	212.8	− 12.0	6	1	HII + RN	
GN 06.06.1	S 261	6 09.0	+ 15 48	194.1	− 1.9	45	2	HII	
GN 06.06.7.02	S 252	6 09.7	+ 20 30	190.0	+ 0.5	40	3	HII	NGC 2174
GN 06.09.9	S 257	6 12.8	+ 17 58	192.6	− 0.1	3	3	HII	
GN 06.11.8.02	S 269	6 14.6	+ 13 50	196.4	− 1.7	4	3	HII	LBN 876
GN 06.14.1	IC443	6 17.1	+ 22 36	189.0	+ 3.0	50	3	SNR	S 248
GN 06.16.4	VDB 75	6 19.4	+ 23 16	188.7	+ 3.8	3	2	RN	LBN 840
GN 06.17.9	S 249	6 20.9	+ 23 06	189.0	+ 4.0	80	2	HII	IC444
GN 06.28.5.02	VDB 80	6 30.8	− 9 39	219.3	− 8.9	6	1	RN	
GN 06.30.2	VDB 81	6 32.9	+ 7 20	204.3	− 0.8	18	1	RN	
GN 06.29.1	S 275	6 31.8	+ 4 56	206.3	− 2.1	100	3	HII	Rosette Nebula
GN 06.31.7	S 280	6 34.3	+ 2 33	208.7	− 2.6	40	2	HII	
GN 06.35.0	VMT 10	6 37.7	+ 6 27	205.6	− 0.1	200	1	SNR	Monoceros SNR
GN 06.36.4.02	NGC 2261	6 39.1	+ 8 45	203.7	+ 1.3	4	3	kom	Hubble's Nebula
GN 06.38.1.01	NGC 2264	6 40.8	+ 9 54	202.9	+ 2.2	250	3	HII	Cone Nebula
GN 06.42.5	S 284	6 45.1	+ 0 14	212.0	− 1.3	80	2	HII	LBN 984
GN 06.51.9	S 303	6 54.0	− 22 26	233.4	− 9.5	90	2	HII	
GN 06.52.1.02	S 308	6 54.2	− 23 57	234.8	− 10.1	35	2	HII	LBN 1051
GN 06.58.1	VDB 87	7 00.5	− 8 51	221.8	− 2.0	5	2	HII + RN	
GN 07.03.5	S 296	7 05.8	− 11 13	224.5	− 1.9	200	3	HII	NGC 2327
GN 07.02.0	S 292	7 04.4	− 10 28	223.7	− 1.9	15	3	HII	IC2177
GN 07.02.9	S 297	7 05.2	− 12 20	225.5	− 2.6	7	3	HII	
GN 07.07.6	S 301	7 09.8	− 18 29	231.5	− 4.4	9	3	HII	
GN 07.11.0	S 310	7 13.1	− 24 35	237.3	− 6.5	410	1	HII	NGC 2362
GN 07.16.3	S 298	7 18.6	− 13 12	227.7	− 0.1	22	3	HII	NGC 2359
GN 07.29.4	S 302	7 31.6	− 16 58	232.6	+ 0.9	21	2	HII	NGC 2409
GN 07.50.3	S 311	7 52.4	− 26 27	243.2	+ 0.4	22	3	HII	NGC 2467
GN 07.34.5	VDB 98	7 36.5	− 25 20	240.4	− 2.2	10	2	RN	LBN 1062
GN 08.13.5	RCW 19	8 15.4	− 35 51	253.8	− 0.5	48	3	HII	
GN 08.32.0		8 33.7	− 45 10	263.4	− 3.0	270	1	SNR	Vela SNR
GN 09.01.0	RCW 40	9 02.7	− 48 39	269.2	− 1.4	8	3	HII	
GN 09.07.0	LBN 1083	9 09.1	− 28 22	254.8	+ 13.1	300	1	HII	
GN 09.22.9	RCW 42	9 24.6	− 51 59	274.0	− 1.1	9	3	HII	
GN 10.15.0	RCW 48	10 16.8	− 57 57	283.6	− 1.0	15	3	HII	
GN 10.22.0	RCW 49	10 23.8	− 57 42	284.2	− 0.3	90	3	HII	
GN 10.22.1		10 24.5	− 18 43	261.1	+ 32.0	15	1	HII	
GN 10.41.9	NGC 3372	10 43.8	− 59 52	287.5	− 0.9	18	3	HII	η Carinae Nebula
GN 11.26.5	RCW 60	11 28.8	− 62 47	293.6	− 1.4	50	3	HII	
GN 11.28.1	RCW 61	11 30.4	− 63 49	294.1	− 2.3	15	3	HII	
GN 11.35.0	RCW 62	11 37.3	− 63 11	294.7	− 1.5	80	3	HII	IC2944
GN 13.16.5	RCW 75	13 19.8	− 62 31	306.2	+ 0.2	18	2	HII	
GN 15.51.7	RCW 98	15 55.6	− 54 39	327.5	− 0.8	6	3	HII	
GN 15.55.8	DG 130	15 58.8	− 26 07	347.2	+ 20.2	50	1	RN	S 1
GN 16.06.3	RCW 105	16 10.0	− 49 08	332.9	+ 1.9	45	3	HII	
GN 16.09.1	IC4592	16 12.0	− 19 28	354.6	+ 22.7	180	1	RN	VDB 100
GN 16.14.0	RCW 102	16 18.5	− 51 55	331.9	− 1.1	12	3	HII	

Table B.37. (Continued)

Reference in AGN[a]	Name[b]	Equatorial Coordinates for 2000.0		Galactic Coordinates		Dia-meter	B[c]	Nebula Type[d]	Notes[e]
		α	δ	l	b	D			
GN 16.18.1.02	S 9	16 21.2	− 25 35	351.3	+ 17.0	80′	2	HII	VDB 104
GN 16.22.4	VDB 105	16 25.4	− 24 28	352.9	+ 17.0	15	2	RN	IC4603
GN 16.22.6	VDB 106	16 25.6	− 23 27	353.7	+ 17.7	65	2	RN	IC4604
GN 16.27.2	VDB 108	16 30.2	− 25 07	353.1	+ 15.8	15	2	RN	IC4605
GN 16.26.3	VDB 107	16 29.4	− 26 26	351.9	+ 15.1	120	2	RN	Star: Antares
GN 16.28.1	LBN 30	16 30.8	− 7 49	7.7	+ 26.5	300	2	HII + RN	in LBN 32, 35, 39
GN 17.31.4	S 12	17 34.7	− 32 35	355.7	+ 0.1	120	2	HII	
GN 16.36.0	RCW 108	16 39.8	− 48 57	336.4	− 1.5	40	3	HII	
GN 16.50.5	RCW 110	16 54.1	− 45 10	340.9	− 0.9	7	3	HII	
GN 17.08.9	S 3	17 12.3	− 38 29	348.2	+ 0.5	12	3	HII	
GN 17.14.2	S 10	17 17.5	− 34 09	352.4	+ 2.1	30	2	HII	
GN 17.17.0	S 8	17 20.4	− 35 51	351.3	+ 0.7	120	3	HII	NGC 6334
GN 17.17.3	S 5	17 20.7	− 38 01	349.6	− 0.6	60	3	HII	NGC 6337
GN 17.22.0	S 11	17 26.0	− 34 16	353.3	+ 0.6	35	3	HII	NGC 6357
GN 17.43.4	S 16	17 46.6	− 29 18	359.8	− 0.4	20	2	HII	
GN 18.03.5		18 06.6	− 23 57	6.7	− 1.5	5	3	RN	
GN 18.03.3.02		18 06.3	− 20 52	9.3	+ 0.0	4	1	RN	
GN 17.59.5	S 30	18 02.5	− 23 00	7.0	− 0.3	20	3	HII	M20 = Trifid Nebula
GN 18.00.8	S 25	18 03.9	− 24 20	6.0	− 1.2	95	3	HII	M8 = Lagoon Nebula
GN 18.06.3	S 29	18 09.3	− 24 00	6.9	− 2.1	40	2	HII	
GN 18.06.7	S 32	18 09.8	− 23 39	7.3	− 2.0	8	2	HII	
GN 18.07.1	S 31	18 10.2	− 23 48	7.2	− 2.2	8	2	HII	
GN 18.18.9		18 21.5	− 2 03	27.7	+ 5.7	1	1	RN	
GN 18.13.8	S 39	18 16.7	− 18 39	12.5	− 1.1	3	2	HII	
GN 18.12.9.01	S 41	18 15.8	− 18 14	12.7	− 0.7	90	2	HII	
GN 18.13.6	S 44	18 16.5	− 16 44	14.1	− 0.1	60	3	HII	
GN 18.14.0	VDB 118	18 16.9	− 19 47	11.5	− 1.6	5	3	RN	
GN 18.14.1	VDB 119	18 17.1	− 19 52	11.4	− 1.7	6	3	RN	
GN 18.14.8	S 37	18 17.8	− 19 40	11.7	− 1.8	20	2	HII	
GN 18.15.1	S 54	18 17.9	− 11 44	18.7	+ 2.0	140	3	HII	
GN 18.15.8	S 49	18 18.6	− 13 58	16.8	+ 0.8	80	3	HII	M16 = NGC 6611
GN 18.17.9.01	S 45	18 20.8	− 16 11	15.1	− 0.7	60	3	HII	M17 = NGC 6618
GN 18.27.4		18 29.9	+ 1 14	31.6	+ 5.3	2	1	RN	Serpens Object
GN 18.28.7.02	VDB 124	18 31.4	− 10 48	21.1	− 0.5	18	2	RN	IC1287
GN 18.41.9	S 69	18 44.5	− 0 17	31.9	+ 1.4	20	1	HII	
GN 19.01.3	S 72	19 03.8	+ 2 19	36.4	− 1.7	25	1	HII	
GN 19.24.1	VDB 126	19 26.2	+ 22 45	57.0	+ 3.0	3	2	RN	
GN 19.28.1	S 82	19 30.3	+ 18 16	53.6	+ 0.0	9	2	HII	
GN 19.41.0	S 86	19 43.8	+ 23 17	59.4	− 0.1	40	3	HII	NGC 6820
GN 19.58.1	S 101	20 00.0	+ 35 17	71.6	+ 2.8	20	2	HII	
GN 20.02.6	VDB 128	20 04.6	+ 32 13	69.5	+ 0.4	8	2	RN	
GN 20.07.1.02	S 109	20 09.0	+ 35 56	73.1	+ 1.6	380	2	HII	
GN 20.10.2.01	S 105	20 12.0	+ 38 21	75.5	+ 2.4	18	3	HII	NGC 6888

Table B.37. (Continued)

Reference in AGN[a]	Name[b]	Equatorial Coordinates for 2000.0		Galactic Coordinates		Diameter	B[c]	Nebula Type[d]	Notes[e]
		α	δ	l	b	D			
GN 20.14.8		20 16.5	+ 41 28	78.5	+ 3.4	120′	3	HII	Part of Cygnus X
GN 20.17.3		20 19.1	+ 39 16	77.0	+ 1.8	60	3	HII	Part of Cygnus X
GN 20.18.3		20 20.2	+ 37 10	75.4	+ 0.4	0.5	3	RN	cometary
GN 20.20.0		20 01.6	+ 44 56	80.0	+ 7.6	330	3	HII	Part of Cygnus X
GN 20.20.8	S 108	20.22.6	+ 40 16	78.2	+ 1.8	180	3	HII	γ Cygni Nebula
GN 20.22.9		20 24.7	+ 42 29	80.2	+ 2.8	5	3	RN	
GN 20.25.6	S 106	20 27.5	+ 37 24	76.4	− 0.6	3	3	HII	Bipolar
GN 20.32.2	S 112	20 33.9	+ 45 39	83.8	+ 3.3	15	3	HII	in Cygnus X
GN 20.49.5		20 51.6	+ 30 56	74.3	− 8.5	210	3	SNR	Cirrus Nebula
GN 20.53.0	LBN 550	20 51.7	+ 78 11	112.0	+ 20.8	165	1	RN	very faint
GN 20.57.0	S 117	20 59.5	+ 44 20	85.5	− 1.0	240	3	HII	North America Nebula
GN 21.01.1	VDB 139	21 01.6	+ 68 10	104.1	+ 14.2	15	3	RN	NGC 7023
GN 21.05.1	LBN 289	21 07.1	+ 38 04	81.8	− 6.3	130	1	HII	in LBN 290, 321
GN 21.10.5	S 129	21 11.8	+ 59 57	98.5	+ 8.0	140	1	HII	
GN 21.37.5	S 131	21 39.0	+ 57 30	99.3	+ 3.7	17	2	HII	in VDB 142
GN 21.41.8.01	VDB 146	21 42.9	+ 66 06	105.4	+ 9.9	5	3	RN	NGC 7129
GN 21.51.6	S 125	21 53.5	+ 47 16	94.4	− 5.5	9	3	HII	Cocoon Nebula
GN 22.03.0	LBN 420	22 05.1	+ 42 45	93.2	− 10.3	190	1	RN	very faint
GN 22.08.8	VDB 150	22 09.7	+ 73 23	111.9	+ 14.1	12	2	RN	LBN 546
GN 22.12.2	VDB 152	22 13.4	+ 70 15	110.3	+ 11.4	3	3	RN	LBN 531
GN 22.23.6		22 25.0	+ 69 39	110.8	+ 10.3	5	1	RN	
GN 22.17.5	S 140	22 19.1	+ 63 17	106.8	+ 5.3	30	3	HII	
GN 22.31.2	S 126	22 33.4	+ 38 35	95.4	− 16.8	160	1	HII	very faint
GN 22.45.6	S 142	22 47.6	+ 58 04	107.1	− 1.0	30	3	HII	NGC 7380
GN 22.54.8	S 155	22 56.8	+ 62 37	110.2	+ 2.6	60	3	HII	
GN 23.18.5	S 162	23 20.7	+ 61 11	112.2	+ 0.2	40	3	HII	NGC 7635
GN 23.25.0	LBN 471	23 27.5	+ 32 02	102.9	− 27.6	250	1	RN	very faint
GN 23.59.1	S 170	0 01.7	+ 64 38	117.6	+ 2.3	20	2	HII	

[a] AGN = Atlas Galaktischer Nebel
[b] Naming codes are as follows: DG = Dorschner, J., and Gürtler, J.: Verzeichnis von Reflexionsnebeln, 1964; HH = Herbig, G. H.: Draft Catalog of Herbig-Haro Objects, 1975; IC = Dreyer, J. L. E.: Index Catalogue, 1895; LBN = Lynds, B.: Catalogue of Bright Nebulae, 1965; LkHα = Lick Hα Survey; NGC = Dreyer, J. L. E.: New General Catalogue, 1888; RCW = Rodgers, A. W., Campbell, C. T., Whiteoak, J. B.: Hα Emission Regions in the Southern Milky Way, 1960; S = Sharpless, S.: A Catalogue of HII Regions; VDB = Van den Bergh, S.: A Study of Reflection Nebulae, 1966
[c] H = brightness (1 = weak, 2 = medium, 3 = bright)
[d] HII = HII region, RN = reflection nebula, SNR = supernova remnant, HH = Herbig-Haro object
[e] Description or alternate identification

Table B.38. The strongest emission lines in HII regions, with intensities given for the Orion nebula, normalized to $I(H\beta) = 100$.

Wavelength (Å)	Element	Relative Intensity	Wavelength (Å)	Element	Relative Intensity
3187.74	He I	6.3	5577.3	O I	15
3726.05	[O II]	127	5754.6	[N II]	15
3828.80	[O II]	127	6300.3	[O I]	10
3770.63	H	5.4	6312.1	[S III]	15
3797.90	H	7.8	6562.8	Hα	350
3835.39	H	10.9	6583.4	[N II]	55
3868.76	[Ne III]	19.7	6678.1	He I	15
3889.05	H	18.1	6716.4	[S II]	10
3888.65	He I	18.1	6730.8	[S II]	10
3967.47	[Ne III]	34.4	7065.3	He I	10
3967.4	O II	24.4	7281.3	He I	10
3970.07	Hε	24.4	9069.0	[S III]	72
4101.74	Hδ	25	9229.0	H	5.8
4340.47	Hγ	41	9531.8	[S III]	181
4861.33	Hβ	100	9546.0	H	8
4958.92	[O III]	113	10049.4	H	10
5006.85	[O III]	342	10830.3	He I	70
5517.7	[Cl III]	10	10938.1	H	20
5537.6	[Cl III]	10			

Table B.39. Selected abbreviations of organisations and periodicals.

AA	= Acta Astronomica, Poland
A&A	= Astronomy and Astrophysics, Germany/France (S = Supplements)
AAA	= Astronomy and Astrophysics Abstracts, Germany
AAS	= American Astronomical Society, USA
AAVSO	= American Association of Variable Star Observers, USA
AG	= Astronomische Gesellschaft, Germany
AJ	= Astronomical Journal, New Haven, USA
AJB	= Astronomischer Jahresbericht, Germany
ALPO	= Association of Lunar and Planetary Observers, USA
AN	= Astronomische Nachrichten, Germany
ApJ	= Astrophysical Journal (S = Supplements), USA
ASP	= Astronomical Society of the Pacific, USA
AZh	= Astronomicheskij Zhurneal, Russia
BA	= Bulletin Astronomique, France
BAA	= British Astronomical Association, England
BAN	= Bulletin of the Astronomical Institutes of the Netherlands, Netherlands
BAV	= Berliner Arbeitsgemeinschaft für veränderliche Sterne, Germany
BJS	= British Interplanetary Society, England
BSAF	= L'Astronomie, Bulletin de la Société Astronomique de France, France
CdT	= Connaissance des Temps, France
DOB	= Documentation des Observateurs, France
ESA	= European Space Agency, France
ESO	= European Southern Observatory, Germany
Gaz.astr.	= Gazette astronomique, Antwerpen, Belgique
HMSO	= Her Majesty's Stationary Office, England
IAU	= International Astronomical Union, France
JBAA	= Journal of the British Astronomical Association, England
JO	= Journal des Observateurs, France
JPL	= Jet Propulsion Laboratory, USA
MfP	= Mitteilungen für Planetenbeobachter, Germany
MN	= Monthly Notice of the Royal Astronomical Society, England
MPI	= Max-Planck-Institut, Germany
NASA	= National Aeronautics and Space Administration, USA
Obs	= The Observatory, England
QJRAS	= Quarterly Journal of the Royal Astronomical Society, England
PASJ	= Publications of the Astronomical Society of Japan, Japan
PASP	= Publication of the Astronomical Society of the Pacific, USA
PAT	= Populär Astronomisk Tidsskrift, Sweden
RAS	= Royal Astronomical Society, England
SAG	= Schweizerische Astronomische Gesellschaft, Switzerland
S&T	= Sky and Telescope, USA
SuW	= Sterne und Weltraum, Germany
VdS	= Vereinigung der Sternfreunde, Germany
ZfA	= Zeitschrift für Astrophysik, Germany
ZfI	= Zeitschrift für Instrumentenkunde, Germany

Table B.40. Abbreviations of reference catalogues.

AC	= Astrographic Catalogue (also: Carte du Ciel): Photographic star positions, mostly in plate coordinates; epoch about 1900. Over 150 vols
BD	= Bonner Durchmusterung: Survey of approximate positions (equator 1855) of 450,000 stars north of $\delta = -23°$. Catalogs and maps reprinted by G. Dümmler Verlag, Bonn. The southern extensions are CoD = Cordoba and CPD = Cape photographic Durchmusterung (equator 1875; not recently reprinted)
BS	= Bright Star Catalogue, 4th edition (1982) and Supplement (1983), Yale Univ. Obs: General data on 11,600 stars brighter than 7^m
FK5	= Fifth Fundamental Catalogue, G. Braun Verlag 1988–91: 4600 star positions and proper motions, serving as coordinate-system reference points
GCVS	= General Catalogue of Variable stars (Obshchtij katalog peremennykh zvozd), 4th edition, Nauka, Moscou 1985–90
HD	= Henry Draper Catalogue: Magnitudes and Harvard spectral types of 250,000 stars. Harvard Annals vols 91–99. (HD numbers are still widely used.)
NGC	= New General Catalogue of Nebulae and Clusters (and two supplements IC = Index Catalogues), Mem.RAS vols 49, 51, 59. Revised edition: Univ. of Arizona Press 1973
PPM	= Positions and proper motions, about 400,000 stars. Spektrum Akademischer Verlag 1991–93. (Supersedes the AGK, SAOC, and NZC catalogues.)

Updated versions of some other catalogues are available commercially on CD-ROM but not in print.

Supplemental Reading List for Vol. 3

The following is a list of suggested readings to supplement the references given at the end of each chapter in Vol. 3. The list is composed mostly of recent (up to early 1992) British and American bookprints, some of which may have already been referred to in the individual chapters. Older books are included as far as there is a fair chance of obtaining them by interlibrary loan. Few non-English books are included, since they are seldom obtainable; even university libraries carry few foreign books in the sciences because of the language barrier. Names of frequently-referred-to publishers appear in abbreviated form; the coding is given in Sect. A.7 in Vol. 1.

Chapter 24

- Cohen, M.: *In Darkness Born—The Story of Star Formation*, CaUP 1988.
- Clark, D.: *Superstars*, MGrH 1984.
- Cooper, A., Walker, N.: *Getting the Measure of the Stars*, AdHg 1989.
- Duley, W.W., Williams, D.A.: *Interstellar Chemistry*, AcdP 1984.
- Greenstein, G.: *Frozen Star*, Macdonald, London 1984.
- Grindlay, J.E., Philip, A.G.D. (eds.): *The Harlow Shapley Symposium on Globular Cluster Systems in Galaxies (IAU Symposium No. 126)*, ReiP 1988.
- Kippenhahn, R.: *100 Billion Suns*, Unwin, London 1985.
- Kippenhahn, R., Weigert, A.: *Stellar Structure and Evolution*, SpVg 1989.
- Mihalas, D.: *Stellar Atmospheres*, FreC 1971.
- Payne-Gaposchkin, C.: *Stars and Clusters*, HaUP 1979.
- Rowan-Robinson, M.: *The Cosmological Distance Scale—Distance and Time in the Universe*, FrmC 1985.
- Zeilik, M., Gibson, D.M., (eds.): *Cool Stars, Stellar Systems, and the Sun*, Lecture Notes in Physics Vol. 254, SpVg 1986.
- Zim, H., Baker, R.: *Stars*, Golden, New York 1956. Available from SkyP.

Chapter 25

- Bode, M. (ed.): *Classical Novae*, JoWS 1989.
- Campbell, L, Jacchia, L.: *The Story of Variable Stars* (The Harvard Books on Astronomy), Blakiston, Philadelphia 1945.
- Chapman, C.R., Morrison, D.: *Cosmic Catastrophies*, PlnP 1989.
- Davidson, K., Moffat, A.F.J., Lamers, H.J.G. (eds.): *Physics of Luminous Blue Variables*, KlwP 1989.
- Drechsel, H., Kondo, Y., Rahe, J. (eds.): *Cataclysmic Variables: Recent Multi-frequency Observations and Theoretical Developments*, ReiP 1987.
- Duerbeck, H.W., Seitter, W.C.: Variable Stars. In: Schaifers, K., Voigt, H.H. (eds.): Landolt-Börnstein, New Series, Group VI, *Astronomy and Astrophysics*, Vol. 2b: *Stars and Star Clusters*, SpVg 1982.

- Genet, R.M., Hayes, D.S., Hall, D.S., Genet, D.R.: *Supernova 1987A: Astronomy's Explosive Enigma*, FrbP 1987.
- Goldsmith, D.: *Supernova! The Exploding Star of 1987*, St. Martin's Press, New York 1989.
- Hoffmeister, C., Richter, G., Wentzel, D.: *Variable Stars*, SpVg 1985.
- Imshennik, V.S., Nadezhin, D.K.: *Supernova 1987A in the Large Magellanic Cloud*, Harwood Academic, Philadelphia 1989.
- Isles, J.E.: *Webb Society Deep-Sky Observer's Handbook*, Volume 8: *Variable Stars*, EnsP 1990.
- Kholopov, P.N. (ed.): *General Catalogue of Variable Stars* (4th edn.), Nauka, Moscow 1985.
- Kenyon, S.J.: *The Symbiotic Stars*, CaUP 1986.
- Levy, D.H.: *Observing Variable Stars*, CaUP 1989.
- Lyne, A.G., Graham-Smith, F.: *Pulsar Astronomy*, CaUP 1989.
- Manchester, R.N., Taylor, J.H.: *Pulsars*, FrmC 1977.
- Marschall, L.A.: *The Supernova Story*, PlnP 1988.
- Murdin, P.: *End in Fire: The Supernova in the Large Magellanic Cloud*, CaUP 1990.
- Murdin, P., Murdin, L.: *Supernovae*, CaUP 1985.
- Percy, J.R.: *The Study of Variable Stars Using Small Telescopes*, CaUP 1986.
- Petschek, A.G. (ed.): *Supernovae*, SpVg 1990.
- Scovil, C.E.: *AAVSO Variable Star Atlas*, SkyP 1980.
- Shklovsky, I.S.: *Supernovae*, JoWS 1968.
- Strohmeier, W.: *Variable Stars*, PrmP 1972.
- Thompson, G.D.: *The Supernova Search Charts and Handbook*, CaUP 1990.
- Tsesevittch, V.P., Kazanasmas, M.S.: *Atlas of Finding Charts of Va riable Stars*, Nauka, Moscow 1972.
- Tsesevittch, V.P.: *Eclipsing Variable Stars*, JoWS 1973.

Chapter 26

- Batten, A.H. (ed.): *Algols*, KlwP 1989.
- Couteau, P.: *Ces Astronomes Fous du Ciel ou l'histoire de l'observation des étoiles doubles*, Edisud, Aix-en-Provence 1988.
- Heintz, W.D.: *Double Stars*, KlwA 1978.
- Jones, K.G. (ed.): *Webb Society Deep-Sky Observer's Handbook*, Volume 1: *Double Stars*, EnsP 1986.
- Meeus, J.: *Some Bright Visual Binary Stars*. Booklet available from SkyP.
- Pringle, J.E., Wade, R.A.: *Interacting Binary Stars*, CaUP 1985.
- Worley, C.E.: *Visual Observing of Double Stars*. Booklet available from SkyP.

Chapter 27

- Blitz, L., Kockman, F.J. (eds.): *The Outer Galaxy*, SpVg 1988.
- Bohren, C.F., Huffmann, D.R.: *Absorption and Scattering of Light by Small Particles*, JoWS 1983.
- Bok, B., Bok, P.: *The Milky Way*, HaUP 1981.
- Davis, J.: *Journey to the Center of Our Galaxy*, Comtemporary Books, 1991.
- Gilmore, G., Carswell, B. (eds.): *NATO ASI Series C: The Galaxy*, KlwA 1987.
- Gilmore, G., King, I.R., van der Kruit, P.C.: *The Milky Way as a Galaxy*, UScB 1990.
- Henning, R., Stecklum, B. (eds.): *The Role of Dust in Dense Regions of Interstellar Matter*, ReiP 1986.
- Hollenbach, D.J., Thronson, H.A. Jr. (eds.): *Interstellar Processes*, ReiP 1987.
- Mihalas D., Binney, J.: *Galactic Astronomy* (2nd edn.), FrmC 1981.
- Moran, J.M., Ho, P.P. (eds.): *Intersteller Matter*, GdBr 1989.

- Norman, C.A., Renzini, A., Tosi, M.: *Stellar Populations*, Space Telescope Science Institute, CaUP 1986.
- Philip, A.G.D., Lu, P.K. (eds.): *The Gravitational Force Perpendicular to the Galactic Plane*, L. Davis, Schenectady 1989.
- Saslaw, W.C.: *Gravitational Physics of Stellar and Galactic Systems*, CaUP 1985.
- Scheffler, H., Elsässer, H.: *Physics of the Galaxy and Interstellar Matter*, SpVg 1987.
- Smoluchowski, R., Bahcall, J.N., Matthews, M.S. (eds.): *The Galaxy and the Solar System*, UAzP 1987.
- Spitzer, L.: *Physical Properties in the Interstellar Medium*, JoWS 1978.
- Tayler, R.J.: *Galaxies: Structure and Evolution*, Wykeham, London 1978.
- Tayler, R.J.: *The Hidden Universe*, Ellis Horwood, 1991.
- Verschuur, G.L.: *Interstellar Matters*, SpVg 1989.
- Woerden, H. van, Allen, R.J. (eds.): *The Milky Way Galaxy*, Symposium No. 6 of the International Astronomical Union, ReiP 1985.
- Wynn-Williams, G.: *The Fullness of Space: Nebulae, Stardust, and the Interstellar Medium*, CaUP 1992.

Chapter 28

- Arp, H.: *Quasars, Redshifts, and Controversies*, CaUP 1987.
- Barrow, J.D., Tipler, F.J.: *The Anthropic Cosmological Principle*, OxUP 1986.
- Barrow, J.D., Peebles, P.J.E., Sciama, D.W.: *The Material Content of the Universe*, CaUP 1988.
- Bartusiak, M.: *Thursday's Universe*, Tempus Books, Microsoft Press, Redmond, Washington 1988.
- Bergia, S., Bertotti, B., Balbinot, R., Messina, A.: *Modern Cosmology in Retrospect*, CaUP 1990.
- Berry, M.: *Principles of Cosmology and Gravitation*, AdHg 1989.
- Bernstein, J.: *Kinetic Theory in the Expanding Universe*, CaUP 1988.
- Bernstein, J.: *The Tenth Dimension*, MGrH 1989.
- Bernstein, J., Feinberg, G. (eds.): *Cosmological Constants*, CoUP 1989.
- Bertotti, B., Balbinot, R., Bergia, S., Messina, A. (eds.): *Modern Cosmology in Retrospect*, CaUP 1990.
- Binney, J., Tremaine, S.: *Galactic Dynamics*, PrUP 1988.
- Börner, G.: *The Early Universe: Facts and Fiction*, SpVg 1988.
- Cadogan, P.: *From Quark to Quasar*, CaUP 1985.
- Chaisson, E.: *Relatively Speaking: Relativity, Black Holes, and the Fate of the Universe*, NorC 1988.
- Cohen, N.: *Gravity's Lens: Views of the New Cosmology*, JoWS 1989.
- Collins, P.D.B., Martin, A.D., Squires, E.J.: *Particle Physics & Cosmology*, JoWS 1989.
- Cornell, J. (ed.): *Bubbles, Voids, and Bumps in Time: The New Cosmology*, CaUP 1989.
- Davies, P.: *The Cosmic Blueprint*, SiSh 1988.
- Davies, P.C.W., Brown, J.: *Superstrings: A Theory of Everything*, CaUP 1988.
- Disney, M.: *The Hidden Universe*, Dent, London 1984.
- Ellis, G.F.R., Williams, R.M.: *Flat and Curved Space-Times*, OxUP 1988.
- Ferris, T.: *Coming of Age in the Milky Way*, William Morrow, New York 1988.
- Ferris, T.: *Galaxies*, Stewart, Tabori, & Chang, 1980.
- Gould, S.J.: *Time's Arrow, Time's Cycle*, PngB 1988.
- Gribbin, J.: *The Omega Point. The Search for the Missing Mass and the Ultimate Fate of the Universe*, Heinemann, London 1987.
- Gribbin, J.: *In Search of the Big Bang*, Heinemann, London 1986.
- Harrison, E.R.: *Cosmology*, CaUP 1981.
- Harrison, E.R.: *Darkness at Night: A Riddle of the Universe*, HaUP 1988.

- Hawking, S.W.: *A Brief History of Time: From the Big Bang to Black Holes*, Bamtam, New York 1988.
- Hawking, S.W., Ellis, G.F.R.: *The Large Scale Structure of Space-Time*, CaUP 1973.
- Heidmann, J.: *Cosmic Odyssey*, CaUP 1989.
- Henbest, N.: *Universe: A Computer-Generated Voyage Through Time and Space*, McMi 1992.
- Hewitt, A., Burbidge, G., Fang, L.Z. (eds.): *Observational Cosmology*, IAU Symposium No. 124, KlwA 1987.
- Hodge, P.W.: *Galaxies*, HaUP 1986.
- Hodge, P.W. (ed.): *The Universe of Galaxies: Readings from Scientific American*, FrmC 1984.
- Hoyle, F.: *The Quasar Controversy Resolved*, University College Cardiff Press, Cardiff 1981.
- Iyer, B.R., Kembhavi, A., Narlikar, J.V., Vishveshwara, C.V. (eds.): *Highlights in Gravitation and Cosmology*, CaUP 1989.
- Kaufmann III, W.J.: *Galaxies and Quasars*, FrmC 1976.
- Kenyon, I.R.: *General Relativity*, OxUP 1990.
- Kerszberg, P.: *The Invented Universe: The Einstein–De Sitter Controversy (1916–17) and the Rise of Relativistic Cosmology*, OxUP 1989.
- Kippenhahn, R.: *Light from the Depths of Time*, SpVg 1987.
- Kolb, E.W., Turner, M.S.: *The Early Universe*, AdWs 1989.
- Krauss, L.M.: *The Fifth Essence: The Search for Dark Matter in the Universe*, Basic, 1989.
- Kron, R.G., Renzini, A. (eds.): *Towards Understanding Galaxies at Large Redshift*, KlwA 1988.
- Landolt-Börnstein: New Series, Group VI, *Astronomy and Astrophysics*, Vol. 2c, *Interstellar Matter, Galaxy, Universe*, SpVg 1982.
- Layzer, D.: *Constructing the Universe*, Scientific American Library, FrmC 1984.
- Layzer, D.: *Cosmogenesis: The Growth of Order in the Universe*, OxUP 1989.
- Lederman, L.M., Schramm, D.N.: *From Quarks to the Cosmos*, Scientific American Library, FrmC 1989.
- Lightman, A.: *Ancient Light: Our Changing View of the Universe*, HaUP 1991.
- Maffei, P.: *The Universe in Time*, MITP 1989.
- Mellier, Y., Fort, B., Soucail, G. (eds.): *Gravitational Lensing*, SpVg 1990.
- Miller, H.R., Wiita, P.J. (eds.): *Active Galactic Nuclei*, Lecture Notes in Physics, No. 307, SpVg 1988.
- Mitton, S.: *Exploring the Galaxies*, ScrS 1976.
- Naber, G.L.: *Spacetime and Singularities. An Introduction*, CaUP 1989.
- Narlikar, J.V.: *Introduction to Cosmology*, Jones and Bartlett, Boston 1983.
- Narlikar, J.V.: *The Primeval Universe*, OPUS Series, OxUP 1988.
- Narlikar, J.V., Padmanabhan, T.: *Gravity, Gauge Theories, and Quantum Cosmology*, ReiP 1986.
- Novikov, I.: *Black Holes and the Universe*, CaUP 1990.
- Oegerle, W., Fitchett, M., Danly, L. (eds.): *Clusters of Galaxies*, CaUP 1990.
- Ogilvie, M.B.: *Women in Science, Antiquity Through the Nineteenth Century: A Bibliographic Dictionary with Annotated Biography*, MITP 1990.
- Parker, B.: *Creation: The Story of the Origin and Evolution of the Universe*, PlnP 1988.
- Poundstone, W.: *The Recursive Universe: Cosmic Complexity and the Limits of Scientific Knowledge*, OxUP 1987.
- Preston, R.: *First Light: The Search for the Edge of the Universe*, Macdonald, London 1991.
- Prokhovnik, S.J.: *Light in Einstein's Universe*, ReiP 1985.
- Reeves, H.: *Atoms of Silence: An Exploration of Cosmic Evolution*, MITP 1984.
- Riordin, M., Schramm, D.: *The Shadows of Creation*, FrmC 1991.
- Robson, J.M. (ed.): *Origin and Evolution of the Universe. Evidence for Design?*, McGill-Queens University Press, 1987.
- Ronan, C.A.: *The Natural History of the Universe*, McMi 1991.
- Rowan-Robinson, M.: *The Cosmological Distance Ladder*, FrmC 1985.

- Rozental, I.L.: *Big Bang, Big Bounce, How Particles and Fields Drive Cosmic Evolution*, SpVg 1988.
- Rubin, V.C., Coyne, G.V. (eds.): *Large-Scale Motions in the Universe*, PrUP 1989.
- Saslaw, W.C.: *Gravitational Physics of Stellar and Galactic Systems*, CaUP 1985.
- Schatzman, E.: *Our Expanding Universe*, MGrH 1992.
- Seitter, W.C., Duerbeck, H.W., Tacke, M. (eds.): *Large-scale Structures in the Universe: Observational and Analytical Methods*, SpVg 1988.
- Sersic, J.L.: *Extragalactic Astronomy*, ReiP 1982.
- Shipman, H.L.: *Black Holes, Quasars, and the Universe*, HoMf 1980.
- Silk, J.: *The Big Bang*, FrmC 1988.
- Stoeger, W.R. (ed.): *Theory and Observational Limits in Cosmology*, Specola Vaticana, Rome 1988.
- Swarup, G., Kapahi, V.K. (eds.): *Quasars*, ReiP 1986.
- Trefil, J.: *The Dark Side of the Universe: Searching for the Outer Limits of the Cosmos*, ScrS 1988.
- Tucker, W., Tucker, K.: *The Dark Matter: Contemporary Science's Quest for the Mass Hidden in Our Universe*, Morrow, New York 1991.
- van den Bergh, S., Pritchet, C.J.: *The Extragalactic Distance Scale: Proceedings of the ASP 100th Anniversary Symposium*, Brigham Young University Press, Provo 1989.
- Vorontsov-Vel'yaminov, B.A.: *Extragalactic Astronomy*, Harwood Academic, New York 1988.
- Wald, R.M.: *Space, Time, and Gravity* (2nd edn.), UChP 1992.
- Weedman, D.W.: *Quasar Astronomy*, CaUP 1986.
- Wheeler, J.A.: *A Journey into Gravity and Spacetime*, FrmC 1990.
- Wright, A., Wright, H.: *At the Edge of the Universe*, Ellis Horwood, Chichester 1989.
- Zee, A.: *An Old Man's Toy: Gravity at Work and Play in Einstein's Universe*, McMi 1989.

Index to Volume 3

Springer-Verlag and the Environment

We at Springer-Verlag firmly believe that an international science publisher has a special obligation to the environment, and our corporate policies consistently reflect this conviction.

We also expect our business partners – paper mills, printers, packaging manufacturers, etc. – to commit themselves to using environmentally friendly materials and production processes.

The paper in this book is made from low- or no-chlorine pulp and is acid free, in conformance with international standards for paper permanency.